SETTING OUT
FOR CONSTRUCTION

A Practical Guide
to Site Surveying

Saffron Grant

CH
PUBLISHING

Published in 2020 by Costello House Publishing

ISBN Paperback: 978-1-9160684-0-7
Ebook: 978-1-9160684-1-4

A CIP catalogue copy of this book can be found in the British Library.

Published with the help of Indie Authors World
www.indieauthorsworld.com

For my amazing family

WHAT OTHER PEOPLE HAVE SAID ABOUT THIS BOOK

"An excellent reference for new and experienced engineers alike. Anyone who thinks they know everything or have seen it all, will one day find assumption is the mother of all mistakes!"

PHIL LANGRISHE, MANGING DIRECTOR, APPS IN CADD.

"Much of the source literature available as reference books focuses on aspects of surveying and site engineering that are too complex and beyond the scope of the reader for the most part. Setting Out For Construction is commendable because it lays out in simple terms how to do certain site engineering tasks, why to do them and what the potential benefits and drawbacks might be. In other words, a simple reference text that does not wander off in to pages of mostly inexplicable algebra and complex maths calculations!"

DAVID CLINCH, SENIOR CONSTRUCTION SUPERVISOR, ORSTED.

"Very impressive and comprehensive! It's the kind of thing I wish I had around when I was a trainee."

DAN BARKER, SENIOR ENGINEER, WINVIC CONSTRUCTION.

"I found the book to be well written and easy to understand. It covers the entire spectrum of setting out and would be a welcome addition to any project. Each section has been well thought out and every process is thoroughly explained. Anybody who has a good read will find it incredibly useful".

CHARLIE CALDICOTT, SENIOR ENGINEER.

Acknowledgements

THANKS GO TO THE FOLLOWING:

Alan Penrose, Jason Smith, Optical

Alasdair Cruickshank, Senior Technical Manager Survey, Morgan Sindall Group

Alisoun MacKenzie, Give to Profit

Brian Core, Key Account Manager, Sitetech

Chris Grant, Digital Construction Skills

David Clinch, Senior Construction Supervisor

Gail McEwen, Setting Out For Construction

Janice Gilbert, WordPerfect

Kenny Urquhart, West Dunbartonshire Council

Kenny Cook, Engineering Surveyor

Kim and Sinclair Macleod, Indie Authors World

Mike Skicko, Gary Kelly and Mark Francis, Leica Geosystems

Nicola MacGuire and Craig Stoddart, ESS Safeforce

Peter Roberts, Technical Support Manager and Petya Spencer, Topcon

Phil Langrishe, Applications in CADD Ltd

Richard Selby, Dominic Gregory and Simon Harley, KOREC Group

Sian Harris, One Red Sock Design

Simon Hunt, Charlie Caldicott, Tim Reeve, Dan Barker, Winvic Construction Ltd

CONTENTS

CHAPTER 1 - INTRODUCTION

CHAPTER 2 - WHEN YOU START ON A NEW SITE

CHAPTER 3 - QUALITY MANAGEMENT AND DRAWINGS

CHAPTER 4 - DRAWINGS

CHAPTER 5 - ACCURACY, ERRORS AND TOLERANCES

CHAPTER 6 - TOOLS, MARKERS AND EQUIPMENT

CHAPTER 7 - SAFETY

CHAPTER 8 - LEVELLING

CHAPTER 9 - ERRORS IN LEVELLING

CHAPTER 10 - INSTRUMENT CHECKS ON THE LEVEL

CHAPTER 11 - TOTAL STATION

CHAPTER 12 - INSTRUMENT CHECKS ON THE TOTAL STATION

CHAPTER 13 - TOTAL STATION FUNCTIONS

CHAPTER 14 - GNSS EQUIPMENT

CHAPTER 15 - TAPE MEASURES AND STRING LINES

CHAPTER 16 - PIPE LASERS AND MACHINE CONTROL

CHAPTER 17 - SURVEY SOFTWARE - COMPUTER AIDED DESIGN (CAD)

CHAPTER 18 - STAGES OF SETTING OUT

CHAPTER 19 - HORIZONTAL CONTROL POINTS

CHAPTER 20 - VERTICAL CONTROL POINTS

CHAPTER 21 - COORDINATES, BASELINES AND OFFSET LINES

CHAPTER 22 - PROFILE BOARDS

CHAPTER 23 - BATTER RAILS

CHAPTER 24 - CONSTRUCTION ACTIVITIES

CHAPTER 25 - CONCRETE DEFECTS

APPENDICES

INTRODUCTION FROM THE AUTHOR

WHY I DECIDED TO WRITE THIS BOOK

The purpose of this book is to get essential information about setting out for construction into the hands of those who need it. Advances in technology, the decline of the traditional mentoring system and stretched resources in the further and higher education sectors mean that there are a growing number of people performing setting out activities without having a full understanding of the fundamental principles.

The idea behind this book is to raise awareness about what constitutes good practice in setting out and to facilitate its widespread application by offering clear and specific instructions along with the relevant supporting theory and knowledge. It is intended as a support material for Engineering Surveying modules of further and higher education courses, the practical training courses I run and the online courses I have developed, as well as being a standalone manual for those who have not been provided with the relevant training yet are expected to carry out setting out duties.

In an ideal world, every engineer would be given the training they need, either as part of their academic studies or once they are working in the industry. If an employer puts an individual into a position of responsibility, then they are duty bound to ensure that they are fully trained to carry out the tasks expected of them. However, training in setting out is considered by many employers as a luxury rather than a necessity.

I have been building my personal collection of surveying and setting out books for many years and now have a shelf full of them in my office. I have even gathered books from around the world to help me understand how setting out terminology and practices differ from country to country. There are many excellent books available about Engineering Surveying, however I have not yet found a book which covers everything that a Site Engineer would need to know about setting out in one volume. The books in common use by site engineers tend to be text books aimed at Civil Engineering students or specialist land surveyors. Although the quality of the information in these books is exceptional, they tend to present the information more in the context of land surveying and topics such as the science of the inner workings of the equipment, derivations of principles and formulae and wider issues such as mapping techniques. Although they do include sections on setting out, most do not go into enough detail to be of practical use to the reader. I wanted to write a book to fill this gap.

There are many excellent engineers and experts in the industry, but I don't believe that currently, the critical skills and knowledge are sufficiently widespread. Site engineers and the construction industry are both losing out as a result. I believe that if a large enough number of people have access to the right information, a tipping point will be achieved, and good practice will become normal practice. This will benefit the construction industry by improving productivity and profits, and will benefit site engineers by making their job more enjoyable and allowing them to achieve their potential.

THE AIM OF THIS BOOK

The aim of this book is to enable site engineers to follow good practice, to record all their work in a standardised way with all the correct checks in place and to keep evidence of them. Good setting out is not just about achieving the required accuracy, but getting the work done quickly, efficiently and professionally and with all the paperwork in place.

In this book I share the things I wish I had known as a young engineer. I understand the pressure and responsibility that comes with setting out, the stress and anxiety that it can cause and the impact those can have on your personal wellbeing and your enjoyment of your job. I want you to be able to leave work in the evenings and not have to worry about whether the work you did during the day was correct.

I want you to feel empowered and fully armed with the information you need, enabling you to take full responsibility for the work you do. I want you to feel you have got a reliable reference book in which you will be able to find the information you need quickly and easily, allowing you to work independently and confidently. I want this book to help you flourish and develop and reach your full potential and not be held back in your career by misinformation or lack of knowledge.

For those of you who attended college or university, this book can help you put the theory you learnt into the context of your practical activities and responsibilities on site. For those of you who have practical experience of the construction industry but no formal training in setting out, this book provides the underpinning theory.

Many site engineers are embarrassed to ask questions that they think they should know the answer to. They may spend a disproportionate amount of time trying to work out the answer, coming up with an alternative and possibly sub-standard solution or trying to find a way to avoid the task altogether. This book is designed to help you work out solutions for yourself, or at the very least, to give you enough information that when you do ask for help, you will be able to ask the questions in the right way. This will show that you have thought about the problem and will minimise disruption to your colleagues and mentors.

This book is designed to give you choices and flexibility in how you approach a task. Although it is nearly always possible to find a way around having to do calculations or understand the theory, this strategy can weaken your position as a credible engineer. This book covers the basics so that you are not reliant on designers, equipment or electronic drawings to provide you with solutions. If you take the time to understand the first principles, it will repay you dividends over the course of your career. You will be more confident and more credible and most importantly, the better your understanding of the maths and theory, the less likely you are to make errors.

The content of this book is based on my own industry and teaching experience as well as in consultation industry experts and people who will use this book. As well as aiming to provide a solid foundation in the theory, it aims to address the common questions and misconceptions which I have encountered as a trainer.

WHO THIS BOOK IS FOR

Setting out duties are no longer limited to those with a formal qualification in Civil Engineering, as was traditionally the case. Nowadays, setting out is frequently undertaken by other site personnel such as Site Managers or tradespeople. This book is for anyone who is involved in setting out, including those who are responsible for checking other people's work. It will also be useful for those who supervise setting out activities or recruit setting out or Site Engineers. This book is aimed at:

- → Complete beginners
- → Students at college or university who are studying related modules
- → Trainee Engineers
- → Graduate engineers and Construction Managers
- → Site Engineers
- → Site Agents
- → Overseas engineers who would like to learn the terminology in English

→ Engineers' assistants

→ Site Managers

→ Anyone involved in setting out in any capacity

→ Managers or supervisors who are ultimately responsible when things go wrong

→ Those who have some setting out experience but have not had formal training

→ Those who have some setting out experience but find that they are prone to errors

→ Those who have construction industry experience who would like to get involved in setting out.

THE FORMAT OF THIS BOOK

This book is written in plain English in a question and answer format. The information has been presented in a logical sequence with each section building on the previous one. It has been designed so that you can either read it in order from start to finish, or quickly find the answer to a specific question. I have avoided large paragraphs of text as much as possible, and broken the information down into short stand-alone sections. The following icons are used to highlight key skills, total station instructions, troubleshooting tips and useful checklists:

- At the beginning of each section, I have listed the key skills required to demonstrate full competency in that area.

– The main body of the book has been kept generic in terms of the different instrument types where possible. This icon indicates that full step-by-step instructions for the specific task can be found in the appendices for the following total stations:

→ Leica (TS06, TS07, TS09, TS15, TS16, ICON)

→ Trimble (Any instrument which runs 'Access' software)

→ Topcon (Any instrument which runs 'Magnet Field' software)

– The first aid icon indicates troubleshooting tips. If something isn't going right, these tips should help you figure out what's going wrong and put it right.

✔ – The tick icon draws your attention to checklists which will help you make sure you have everything in place.

You will also find a wealth of:

→ Worked examples

→ Booking examples

→ Step-by-step instructions

→ Diagrams

→ Tables

→ Practical tips

→ Questions and answers

Revised editions will be published regularly so if you can't find what you are looking for, please do get in touch to let me know what you would like to see included in future editions. See 'How to contact the author'.

WHAT THIS BOOK COVERS

This book focuses on the most common tasks that setting out engineers do for the majority of the time. It is highly focused on employers' needs and is concerned with the setting out aspects of construction projects rather than the mapping and surveying aspects of the land surveying profession.

This book covers everything from using basic tools such as tape measure, string line, chalk line and spirit level to more complex equipment such as the optical level, rotating laser level and total station.

The techniques covered in this book can be applied to setting out any type of construction works including earthworks, highways, bridges, buildings, drainage, water projects, reinforced concrete, structural steel work, foundations, rail.

WHAT THIS BOOK DOES NOT COVER

This is book is not intended as an academic tome. There are already several good books which address this need. Instead, this book focuses on the typical setting out tasks required of a Site Engineer on a daily basis.

Topics not covered in this book include:

→ Specific types of survey software, but gives an overview of the functionality you can expect from a good survey software CAD package.

→ Setting out of high-rise buildings. This is a specialist subject which would warrant a whole book in itself. This book gives an overview of the specific challenges faced.

→ Details about the inner workings of the instruments, but gives enough information to allow you to ensure that your equipment is functioning correctly and gives information about what do if they are not.

→ Cable Avoidance Tools. Although these may be thought of as surveying equipment, they are safety critical and not within the scope of this book.

→ Surveying in the context of mass information gathering, specialist monitoring, installation of control over sites larger than 1km, laser scanners, hydrographic surveys, drone technology.

ABOUT THE TOTAL STATION INSTRUCTIONS IN THE APPENDICES

The appendices of this book contain instructions for the total stations which are in common use in the UK and Ireland at the time of publishing. From time to time, manufacturers may release new firmware, which may lead to minor differences between the instructions given here and the actual workflow in the instruments, however it is unlikely that there will be any major differences. New editions of this book will be released periodically in order to keep up to date with the latest equipment on the market.

Instructions are given for:

→ Leica TS06

→ Leica TS07

→ Leica TS09

→ Leica TS15

→ Leica TS16

→ Leica iCON

→ Topcon (Magnet software which is on all Topcon mechanical and robotic total stations)

→ Trimble (Access software which is on a large proportion of Trimble mechanical and robotic total stations)

Step-by-step instructions are given for the following:

→ Set the position and orientation of the total station by setting up over a known point and backsighting to a known point

→ Set the position and orientation of the total station by setting up at an unknown point and backsighting to known points (resection)

→ Measure the distance from the total station to another point

→ Measure the distance and level difference between two points which are not occupied by the total station

→ Take a survey

→ Set out (stake out) points of known coordinates relative to control points

→ Set out points in relation to a line defined by coordinates

→ Set out points in relation to a defined line which is created by sighting to the beginning and end points of the line

→ Measure an area

→ Measure a volume

ABOUT THE AUTHOR

Over the last 20 years Saffron Grant has experienced the construction industry from a range of perspectives working for designers, Clients, Contractors and as a business owner and trainer.

After graduating from Coventry University in 1999 with a BEng (Hons) Civil Engineering with Industrial Training Placement, Saffron started her career with Glasgow City Council Land Services before moving into a design role in the private sector. Whilst working for the bridges team with Bullen Consultants, (now part of Aecom), she gained experience in bridge inspections and assessment, structural steel and reinforced concrete design. She went on to work as a Site Engineer for Costain on the construction of a new water treatment works for the Manchester ring main. She then worked for JMP Consulting, (now part of Systra Group), in a Design, Project Management and Site Supervision role within the highways team. She worked as a freelance Site Engineer before setting up her own construction training business specialising in setting out courses.

From the start of her career she was committed to promoting Civil Engineering as a career to young people as well as being actively involved with professional bodies. In 2001, she was part of a team of three graduate engineers to win the Henry Palmer Award, a nationwide competition run by the ICE to design a scheme to attract young people into a career in Civil Engineering. She also spent some time as the publicity secretary for the ICE Glasgow and West of Scotland Graduates and Students Committee.

Since setting up her own training business 10 years ago, Saffron has been focused on raising the standards of site and setting out engineers. She is an Associate Lecturer at

Glasgow Caledonian University and an Approved Trainer for the CITB. She has delivered Engineering Surveying modules to FE colleges and has trained colleges and university lecturers to enable them to deliver the modules themselves. She has worked with colleges and universities to help them review the content of their surveying modules to make them relevant and employer-focused. She is currently working on behalf of the Scottish Civils Training Group to carry out research into the employability and skills of Scottish construction industry graduates, and is representing a group of Scottish Contractors in the review of the SQA Engineering Surveying modules at HNC and HND level.

In her endeavours to reach as many people as possible with essential information about setting out, she developed an online course containing step-by-step videos on different setting out activities, which now has over 600 students enrolled.

Saffron is an INLPTA certified NLP Master Practitioner and uses these skills in every aspect of her technical training and business, including executive coaching and the personal and professional development courses that she runs.

HOW TO CONTACT THE AUTHOR

You can connect with Saffron Grant via:

> → Website: settingoutforconstruction.com

> → Email: saffron@settingoutforconstruction.com

> → LinkedIn: linkedin.com/in/saffrongrant

> → Facebook: facebook.com/settingoutforconstruction

HOW TO WORK WITH THE AUTHOR

If you would like to work with Saffron Grant, or her business Setting Out for Construction, you can find out more at www.settingoutforconstruction.com.

Here are some examples of services available:

> → Scheduled courses (CITB Assured Products) ranging from 1 to 5 days (covering levelling, total station, CAD, robotic total station, GNSS and machine control)

> → One-to-one training based on the content of any standard course, or bespoke to your needs

> → Module and curriculum design for colleges and universities

> → Teaching delivery and assessment and field trip support for colleges and universities

- → Short online setting out courses
- → Site set-up support and/or ongoing mentoring and support for young engineers
- → Support with selecting and rolling out new software and equipment across your organisation
- → Specialist training such as machine control
- → Training needs analysis for your site engineers
- → Developing educational resources such as videos, written materials and online communities
- → Working with you to create a bespoke handbook or induction pack for your engineers
- → Support for your recruitment department or anyone involved in interviewing site engineers
- → Speaking at board meetings, company events or roadshows on a range of topical issues
- → Research on behalf of training groups, professional bodies and sector skills councils
- → Mastermind group for socially conscious construction company leaders

CHAPTER 1

INTRODUCTION

1.0   WHAT IS SETTING OUT?

Setting out is the process of measuring and marking out of construction works including buildings, houses, structures, roads and civil engineering projects. It involves taking the correct information from the design drawings and physically marking out the correct positions and levels on site, allowing the works to be constructed correctly. Points must be set out in the correct position and orientation in relation to existing features and to the correct dimensions. Examples of features which must be set out are:

→ the positions and levels of excavations

→ corners and levels of concrete slabs

→ position, alignment and verticality of reinforced walls and columns

→ centrelines and kerb lines of roads

→ the various stages of highways construction

→ positions and verticality of structural steel columns

→ manhole positions, levels and internal arrangements

→ locations of piles

→ earthworks for cuttings and embankments

→ plinths and bases for mechanical and electrical equipment

→ reinforced concrete sub-structures and super-structures

→ internal and external wall positions

→ cable and duct locations

→ landscaping features

1.1 WHERE DOES SETTING OUT FIT INTO THE BIGGER PICTURE OF A CONSTRUCTION PROJECT?

Setting out is only one element of the whole construction process, however it is an element which is pivotal to the success of the project. Not only must the different parts of a project be positioned correctly in relation to each other, but even within each road, structure or building, the component parts must fit correctly together so that the finished works look and function as per the design and conform to the project specification. The accuracy and speed of the setting out, and the quality of the records kept will impact on:

→ the costs of construction

→ the ability of the Contractor to collect payment from the Client

→ the project keeping to the programme and avoiding delays

→ the potential for using the most economically viable construction methods

→ the quality of the finished product

→ the aesthetics of the finished project

→ structural integrity of temporary works and finished structures

1.2 WHY IS IT IMPORTANT TO KEEP GOOD RECORDS?

It is not only the accuracy of the setting out that can make or break whether a project makes money or is completed on time and within budget, but whether the correct records are kept throughout the process. The engineer's field book could potentially be the only record of the reduced levels that steel reinforcement in a slab was set to, or the date that a certain structure was set out and which revision of a drawing was used. The Contractor's ability to invoice for work completed depends on the engineer's records and as-built surveys. When the quantity surveyor is reconciling the volume of concrete ordered versus the amount estimated in the bill of quantities, it is the engineer's records that will shed light on why they do not match. When the sub-contractor is insisting that a concrete pour is ready to go ahead, it is the engineer who will pick up on the fact that there is insufficient cover to the soffit of a slab which could lead to expensive remedial work. The quality, integrity and aesthetics of the finished products depend on the setting out being correct.

Anyone who has any responsibility for making sure things are built in the correct position and to the correct dimensions is carrying out setting out. Joiners, steel-fixers, concrete operatives, steel erectors and bricklayers are all carrying out setting out as part of their job, albeit that the engineer is responsible for ensuring that each feature is in the correct relative position and for checking the internal accuracy of tradespeople's work.

Upon completion of the works, some things will be hidden underground or inside concrete pours. In certain cases, the site engineer's records are the only way of knowing if something is structurally sound or not. When things go wrong, they could be a result of:

→ poor quality of concrete

→ poor structural design

→ poor temporary works design

→ method statements not being followed

→ setting out

Even if all your setting out is correct and you haven't made any mistakes, your records are needed to verify this, and so that other root causes can be investigated. The onus is you as the site engineer, and contractually on the Contractor to prove that all setting out is correct.

1.3 WHAT'S THE DIFFERENCE BETWEEN SETTING OUT AND SURVEYING?

Broadly, surveying is the measurement and mapping of features which already exist whereas setting out is the reverse of surveying. It is the process of taking information from design drawings and ensuring it is constructed to the correct level and position.

In a construction context, the term 'surveying' can have several different meanings. It can mean:

→ **land surveying** - measuring, monitoring and mapping of existing buildings, structures, urban features and large parcels of land

→ **site surveying** – measuring and marking out of lines and positions and recording as-built positions, as and when needed, on a live construction site

→ **engineering surveying** – an umbrella term for land surveying and site surveying

→ **building surveying** - assessing and analysing existing buildings

→ **quantity surveying** - the management of costs and risks of a construction project

Building surveying and quantity surveying are unrelated to engineering surveying and are completely different job roles and skill sets.

The term 'site surveying' encompasses the whole process from establishing the control points from which the works are referenced, to setting out the final positions and levels of the construction elements. It includes keeping the correct records, carrying out necessary checks to ensure the equipment is working as it should be and gathering relevant dimensional data and conveying the information to whoever needs it in the appropriate format. The equipment used for site surveying includes the:

→ automatic level

→ laser level

→ pipe laser

→ tape measure

→ total station

→ robotic total station and

→ GNSS equipment.

The term 'land surveying' tends to refer to activities such as structural monitoring, mapping and topographical surveys and installation of control points for large scale projects. In addition to the equipment used in site surveying, it may require specialist equipment such as UAVs (drones) and 3D scanners which can measure the 3D coordinates of millions of points in just a few minutes.

In the context of this book, setting out means site surveying.

1.4 WHAT IS GOOD PRACTICE IN SETTING OUT?

This book is based on good practice. In the context of setting out, good practice means following standardised methods and procedures of setting out, booking and checking, which are tried, tested and recognised.

There are various other sources of information detailing good practice including good quality reference books, company policies, quality management documents and procedures and British Standards and Eurocodes.

Examples of good practice include:

→ cross-referencing drawings

→ building in self-checks

→ checking work using a different method to the one used to set it out

→ seeking independent checks

→ carrying out and recording regular checks on equipment

→ recording work in a standardised layout

→ following the correct procedures for installing and maintaining control points.

1.5 WHY USE GOOD PRACTICE?

Making a habit of using good practice can save time, money, energy and resources. It can:

→ improve productivity

→ increase profits

→ improve accuracy

→ reduce the likelihood of costly mistakes

→ flag up errors before it is too late

→ reduce stress and anxiety for the engineer

→ help maintain good relations and harmony with project colleagues

→ contribute to the smooth running of the programme

→ minimise wastage

→ enable accurate reconciliation of materials

→ help to support claims or compensation events.

1.6 WHY ISN'T GOOD PRACTICE ALWAYS USED?

In an ideal world, all setting out would be done using good practice, but unfortunately it isn't. Some of the barriers include:

→ **the size and complexity of the project** - the larger and more complex the project, the more difficult it is to ensure that good practice is consistently being followed

→ **procedures not being in place** - some sites do not have a set of written-down procedures for setting out activities

→ **procedures not being followed** – even if a site has comprehensive written procedures, they are no good if nobody knows about them or follows them

→ **poor company culture** – where poor company culture exists, most engineers are not aware of what constitutes good practice and even those who do may feel they do not have 'permission' to do things the right way

→ **engineers or managers taking a short-term view** – in some cases engineers either choose, or feel they are expected, to take shortcuts to solve immediate problems. Over the duration of the project, this approach will cost much more time, money and energy

→ **the project being under-resourced** – when engineers are over-stretched, they may be under so much pressure and stress that they miss out vital checks or take shortcuts

→ **there is inadequate support for engineers** – when engineers don't have an experienced mentor they can ask for technical help, they need to muddle through

→ **managers and supervisors do not have relevant knowledge or experience** – some managers and supervisors have not had any setting out experience or training. They are unable to provide technical advice, are not able to ensure that the correct procedures are being followed and may have unrealistic expectations of engineers

→ **engineers do not have the right training** – it is often assumed that trainee and graduate engineers have learnt everything they need to know about setting out as part of their college or university course. This is often an incorrect assumption as many courses are academic rather than employer-focused. The content may not be relevant or comprehensive, group sizes may be so large that it's impossible for all students to achieve the required level of confidence and the lecturer may themselves have no industry experience whatsoever

→ **new technologies are evolving quickly** - the speed of education and training isn't keeping up with the roll-out of new technologies

→ **technology is mistakenly equated with accuracy** – there is a misunderstanding that high-precision equipment necessarily leads to higher accuracy. However, the instrument is only as good as the person using it. If the engineer is not following good practice, then this will be reflected in the results.

1.7 WHAT ARE THE PROBLEMS WITH BAD PRACTICE?

If good practice is not adhered to, it can lead to gross errors or unacceptable cumulative errors. These can in turn lead to:

→ **remedial work** - when remedial work is needed there is a costly process involved in identifying and verifying the error, deciding what to do about it, organising and waiting for remedial action and then carrying out the necessary remedial works

→ **redesign** – in some cases, the time implications and cost of remedial work may be so significant that it is cheaper and more practical to leave the error in place and have the subsequent works redesigned to accommodate it. When this happens, the lengthy and costly process involves identifying and verifying the exact nature and wider implications of the error, deciding at site level what action to take, seeking the relevant approvals on the chosen course of action, waiting for the redesign (which will most likely impact on subsequent stages of the works), the cost of the redesign (someone has to pay for it!), and copious email exchanges between members of the site team and between all parties. Once the redesign is complete, there may be a further drain on resources and costs down the line, in working out which party is contractually responsible for the cost. There will inevitably be complex legal disputes and contractual claims to be resolved and settled

→ **strained relationships** - when the engineer makes frequent mistakes or doesn't keep the correct records, relationships between the main contractor and sub-contractors, clients and designers can become strained. Regular errors can lead to a loss of trust, a blame culture or have a demoralising effect on the engineer and the project team generally

→ **damaged reputation** – if site engineers are not competent and do not consistently use good practice, the resulting poor-quality work and the negative impact on other parties will reflect badly on the contractor

→ **delays** – when one element of the work is delayed due to a setting out error, it can have a significant knock-on effect on subsequent activities

→ **hidden costs** – for a variety of reasons, setting out errors are not always allocated as such in the cost reconciliation

→ **poor efficiency** – when engineers are slow at setting out, the more engineers you need to do the same amount of work. One efficient and competent engineer could cover the workload of two or three who are not fully trained in good practice

→ **penalties** – when construction activities are held up by the setting out not being ready, this can lead to inflated equipment and labour costs and/or liquidated damages or other time-related penalties

→ **wastage** – setting out not being completed on time, or a setting out error being discovered at the last minute can lead to wasted materials. This is particularly relevant in the case of time-critical materials such as concrete. If the concrete arrives on site and the pour cannot go ahead, the concrete needs to be used or dumped elsewhere on site, or if that is not possible, it needs to be disposed of appropriately which itself incurs additional costs

→ **a poor-quality finished product** - it is not economically viable for some errors to be corrected which results in the finished product looking bad or not functioning as it was intended to. This can result in financial penalties to the contractor and/or damage to the contractor's reputation. In some cases, the mistake may be visible in the finished works, reflecting badly on the contractor forever. Even where remedial works are completed it can result in lower quality finished product than if it had been constructed correctly in the first place

→ **poor culture** - if site engineers are using bad practice, they will be passing it on to younger engineers, perpetuating the cycle and continuing to incur the associated costs

→ **repeated mistakes** - if proper records aren't kept, it is difficult to identify the source of an error. This makes it difficult to prevent the same mistake happening again in the future

→ **unsuccessful claims** - if proper setting out records aren't kept, it weakens or removes the chance of making a successful claim or compensation event

→ **inefficient handovers** – handovers between engineers are a fertile ground for errors. If an engineer does not keep the correct records and does not follow the required procedures, it makes it difficult to trace exactly what they have done and what source information they have used. This increases the likelihood of errors rooted in handover should they leave the company, move to another job or take annual or sick leave

→ **inaccurate reconciliation of materials** - if the proper records aren't kept, it makes it difficult to reconcile material usage

→ **an expensive snagging (defects correction) phase** – even if there are no major setting out disasters or setting out-related delays over the course of the works, poor setting out can lead to a complicated and costly snagging phase. Concrete defects such as honeycombing, insufficient cover to rebar or lack of fit of mechanical and electrical (M&E) elements can all be prevented if good practice is followed in all setting out.

1.8 WHAT ARE THE FINANCIAL COSTS OF SETTING OUT ERRORS?

Contractors are in business to make money. It can sometimes be perceived that managing with fewer or lower paid engineers can be a cost saving, but in most cases, it would make more sense to adequately resource a project with properly trained people. The costs incurred from a single setting out error due to lack of resources or poor training could easily equate to the salary of a good engineer.

Whenever a setting out error is made, there is a cost associated with it. Even if the error is detected before any construction work is done, at best there is still the cost of time wasted by the engineer in redoing the setting out correctly. At worst there could be substantial costs associated with standing plant, human resources and possibly discarded concrete.

The direct costs of setting out errors are often not transparent or accurately reported in project costs. They may be covered up or 'lost' in other parts of the job, either deliberately or inadvertently. This may be either with the intention of protecting the originator of the error or because it is difficult to quantify the direct and knock-on costs.

Costs may stem from gross errors, where the simple misreading of a single number can lead to extensive costs, or from a series of cumulative inaccuracies.

The main causes of setting out errors are:

> → inexperienced or poorly-trained engineers
>
> → engineers not doing proper checks
>
> → poor record-keeping
>
> → poor communication between engineers
>
> → engineers knowingly taking short cuts or risks due to time pressures

Even when doing everything correctly, there is a still the risk of error, however it is dramatically reduced if good practice is used.

If the true costs of setting out errors were quantified, it might highlight that the cost of proper training of site engineers and those overseeing setting out activities would be a drop in the ocean compared to the corner-cutting costs that might be saved within a few weeks, let alone over the lifetime of the project.

1.9 WHAT ARE THE CHALLENGES FACED BY ENGINEERS?

In the course of their setting out tasks, engineers are typically required to work and act upon large amounts of complex information. The role of a site engineer often involves a range of other factors including:

→ working long hours under high pressure

→ communicating with designers, some of whom do not fully understand the issues faced because they themselves may have had little experience of the practical side of the construction process

→ working for managers or supervisors who may not recognise the effort required to achieve the necessary standards of accuracy and precision

→ dealing with aggressive or unsympathetic people

→ explaining complex problems and potential solutions to lay people

→ facing pressure to make decisions on issues that are not in their remit, rather than wait for designer input.

1.10 HOW CAN I COMMUNICATE TECHNICAL ISSUES TO LAY-PEOPLE?

When explaining complex technical issues to people who do not have a specific knowledge of the subject you are explaining, it can help to:

→ assume that they don't know anything – give them some background information if necessary. They won't feel insulted. It is much more likely that they will be grateful that you have given them the full picture

→ remember to explain what you mean if you need to use any words they may not be familiar with

→ draw a diagram as you explain

→ give them plenty of opportunities to ask you questions

1.11 WHAT IS THE ROLE OF THE ENGINEER'S ASSISTANT?

The engineer's assistant is the person who assists the engineer with their duties, either by providing an extra pair of hands for tasks which require two people or by taking on elements of the workload, freeing up the engineer to work on other tasks. The engineer and their assistant can work most efficiently together when they are able to gel as a team and adapt to each other's preferred methods of working. Ideally, the assistant will be appropriately skilled and able to assist the engineer with a range of tasks. Often, however, the engineer is not afforded the luxury of a full-time assistant and will need to work with a site operative who happens to be available at the time. Depending on their skill level, availability and the nature of the job, the assistant may be required to do any of the following:

→ set up the equipment

→ assist the engineer in performing calibration checks on surveying equipment

→ help to move equipment around the site

→ hold the levelling staff or prism for taking measurements

→ move the levelling staff or prism as directed by the engineer and mark levels and positions accordingly

→ hold the zero end of the tape measure at the correct point

→ carry out independent checks on levels and dimensions

1.12 WHY IS AN ASSISTANT SOMETIMES REFERRED TO AS THE CHAINMAN OR CHAINBOY?

The term 'chainman' originates from the time when chains were used in surveying. Chains of fixed lengths were used to measure long distances accurately, since they were not susceptible to stretching. The chainman's job was to hold one end of the chain in position.

The term chainman or 'chainboy' is still used, however the gender-neutral term, assistant, is becoming more common.

1.13 HOW AND WHY SHOULD I TRAIN MY ASSISTANT?

Different situations will call for different styles and levels of training, but no matter whether you have a properly trained assistant working with you all the time or you have 'borrowed' a site operative from somewhere else on the job for a quick five minute task, it is essential that you communicate the information they need to do the job properly.

As well as improving accuracy and productivity, investing the time and energy in training your assistant can have significant benefits. It may:

→ improve accuracy

→ improve productivity

→ reduce the chance of errors

→ improve their motivation and reliability

→ create a strong working relationship

→ enable their development and progression

You don't need to design a complicated training scheme. You may decide that it is appropriate to address topics in a logical order or you may choose to more informally spend time explaining why things must be done in a certain way as new subjects arise in the course of the work. Here are some actions you can take:

→ **provide your assistant with resources** - refer them to this book and any other useful books

→ **encourage questions** - encourage them to ask questions or to keep a notebook with their questions in to ask you at the end of each day

→ **involve them in the work** - give them genuine tasks, for example setting up equipment for you

→ **ask for their perspective** - invite their opinion and involve them in your thought process when solving problems

→ **check if you have explained well** - ask them questions to check their understanding of what you have asked them to do

→ **set challenges** - set challenges for them, for example, to complete a two-peg test in under five minutes or set up the total station over a point in under three minutes

→ **provide practice opportunities** - give them opportunities to practice using different bits of equipment, for example let them have a go at checking some of the work you have set out

→ **teach the theory** - help them to understand the maths and theory involved

→ **assert yourself if necessary** - not all assistants are quick learners or happy to be there. You are responsible for the quality of their work so if they are doing something incorrectly, it is up to you to provide them with feedback and communicate clearly what is required of them

→ **remember safety** - educate them about the relevant safety issues. Don't ask or allow your assistant to do anything unsafe. You are the more senior person, so you are likely to be held responsible if they are injured or cause harm to others whilst under your guidance.

CHAPTER 2

WHEN YOU START ON A NEW SITE

2.1 WHAT DO I NEED TO THINK ABOUT WHEN I START ON A NEW JOB?

When you start working with a new company or are relocated to a new site, you will generally find yourself in one of two situations. Either:

→ it is a brand-new site and work has not yet started, or

→ it is a site which is already up and running.

In both situations, your actions in the first few days will impact on how smoothly the project runs, the cost of the project, your personal performance and your personal reputation.

On sites where the proper systems are in place and there is a culture of using best practice in setting out, this is an opportunity to show that you too work to the required high standards.

Where a poor site or company culture exists in terms of setting out, when you first arrive on a project is the critical time to set yourself apart from it. There are many complex reasons for poor setting out practice and you may be in a position to positively influence setting out procedures and practice. If you do not feel you are able to influence others, keep in mind that there is never any benefit in lowering your own standards it and could result in your reputation being irreversibly damaged.

You may find you are the only person on site with responsibility for setting out, in which case it's all up to you.

2.2 THE SITE IS UP AND RUNNING. WHAT SHOULD I DO FIRST?

If you are placed on a site which is up and running, some of the setting out and construction work will already have taken place. The first job on your list is to get an understanding of the relationship between what has been constructed so far and what is still to be set out.

If you don't ask the right questions at the outset, and you wait passively until the information drip-feeds its way to you, you'll find out most of what you need to know (eventually!) However, rather than gathering the information on a need-to-know basis, you will inevitably be a better-quality, more productive, more collaborative and more credible engineer if you are proactive from the start.

2.3   INFORMATION YOU NEED:

→ the master quality management file (inspection test plans, concrete pour check sheets, pro-formas)

→ the Survey Management File (see section **3.2**)

→ Method Statements and Risk Assessments

→ the Drawing Register

→ the drawings ('For Construction' issue)

→ the specification

→ relevant contract documents

→ services information

2.4   QUESTIONS TO FIND OUT THE ANSWERS TO:

→ What are my daily duties and responsibilities?

→ Who is the most senior person on the job if I have any problems?

→ Who puts the operatives to work?

→ What is the process for ensuring the latest revision of the drawings is being used?

→ Who is responsible for maintaining the Drawing Register? Where is it kept? What are my responsibilities in this area?

→ Who is responsible for the care, maintenance and security of surveying equipment?

→ What about an equipment checking regime? Who is the person responsible for ensuring regular equipment checks are carried out and recorded? (If no-one, at a minimum, set up your own file for your own equipment and if appropriate, offer to gather results of other engineers as well.)

→ Where can equipment be kept securely? Is it insured if it is kept in a van or vehicle?

→ Who is responsible for the control points? How were they established? How reliable are they? Where is the evidence?

→ What if I find an error in the control points?

→ Which is more critical for future setting out: primary and secondary control points or the relationship between elements which have already been constructed and elements which are still to be constructed?

→ Who is the contact at the equipment supply company?

→ If I am given conflicting information by different people, which one should I go with?

ACTIONS TO TAKE:

1. Walk the whole site, even the areas which you are not responsible for

2. Verify the existing control points (see section **18.22**)

3. Determine the accuracy of random as-built points in relation to each other and to the control points

2.5 I AM STARTING ON A BRAND-NEW SITE. WHAT SHOULD I DO FIRST?

If you are placed on a brand-new site and have responsibility for setting out, getting everything running smoothly can seem a daunting task. A methodical approach and confidence in your own ability are required.

INFORMATION YOU NEED:

→ the master quality management file (inspection test plans, concrete pour check sheets, pro-formas)

→ the Survey Management File (see section **3.2**)

→ Method Statements and Risk Assessment

→ the Drawing Register

→ the drawings ('For Construction' issue)

→ the specification

→ relevant contract documents

→ services information

ACTIONS TO TAKE:

1. create the setting out plan

2. set up the Survey Management File if one doesn't already exist

3. walk the site. Check the site boundaries and existing features reflect the drawings

4. identify whether there are survey stations you can use as your primary control points. If yes, plan the job in relation to these. If not, decide where to strategically locate your primary control points

5. analyse how services may impact on your setting out plan. How will they affect the staging of the works? Where are connections needed? Is there any conflict between the services information and the contract documents?

6. check the drawings.

CHAPTER 3

QUALITY MANAGEMENT AND DRAWINGS

3.1 WHAT DOES QUALITY MANAGEMENT MEAN IN THE CONTEXT OF SETTING OUT?

Quality management can be thought of as a set of processes and procedures which ensure that work is carried out to a specified standard and for which the evidence which can demonstrate the quality is recorded.

In practice, this takes the form of adhering to company policies and procedures and completing the relevant documentation.

There are a number of benefits to having a quality management system in place. It can:

→ provide transparency and traceability

→ avoid confusion

→ ensure records are kept which will support claims or compensation events

→ reduce the chances of errors occurring

→ monitor wastage and inefficiencies

→ prove the finished product is of the required quality

→ prevent accidents

→ keep track of what is working well and what is not

→ determine how future projects can make money

→ protect your personal reputation and your company's reputation

→ facilitate good site level relationships

Examples of forms and documentation include:

- → your yellow field book (which belongs to the project you are working on, not you personally!)
- → drawings
- → site diaries
- → Inspection Test Plans (ITP)
- → Method Statements and Risk Assessments
- → calculation sheets
- → Drawing Register
- → Technical Queries (TQ)
- → Request for Information (RFI)
- → Confirmation of Verbal Instruction (CVI)
- → Non-Conformance Report (NCR)

3.2 WHAT IS A SURVEY MANAGEMENT FILE?

The Survey Management File (sometimes referred to as the Setting Out File) is a hard-copy file which contains:

- → control point data, including a description of the methods used to install the control points, copies of any handwritten field data, spreadsheets, hand calculations, traverse data recorded in the total station, reports from CAD programmes used to process and adjust the raw data, results of least squares adjustment calculations
- → raw data gathered when installing secondary control points including information about the instrument set-up procedure and the primary control point which were referenced
- → the survey grid system
- → instrument quality control
- → instrument specifications
- → information about who is responsible for equipment
- → information about monitoring regimes

3.3 WHAT SHOULD THE SURVEY MANAGEMENT FILE CONTAIN?

The Survey Management File is the single source of information for all engineers on the project. At a minimum it should contain:

- → Quality control documents for equipment checks including required frequency
- → Inspection Test Plan pro-formas for any activities which require setting out, such as concrete pours, pipe runs and manholes
- → the required frequency of checking and maintaining control points
- → a summary of relevant accuracies and tolerances for the project
- → any critical information that engineers should have available to them
- → letter code and colour code conventions for marking up profile boards and batter rails.
- → Depending on the size, complexity and duration of the project it could also include:
- → file management and numbering systems for storing information in total stations
- → references to relevant contract documents, books, standards or company policies
- → procedures for installing control points
- → procedures for checking of setting out
- → information about the control points and how they were installed
- → standard offset distances to be used in setting out lines and positions.

3.4 WHAT IS A SETTING OUT PLAN?

A setting out plan is a strategic plan which is created at the outset of a project and considers site-specific issues relating to all setting out activities over the full duration of the project.

3.5 WHAT SHOULD A SETTING OUT PLAN COVER?

The setting out plan should take account of factors such as the:

- → complexity of the project
- → size of the project
- → nature of the project
- → resources available

The size of the plan and the level of detail will depend on the site. Every setting out plan will be different, however, broad headings should include:

→ **Overview** - layout plan showing how the sections will be split between engineers

→ **Site Specific Issues** - access arrangements, daylight working hours

→ **Health and Safety** - a detailed Risk Assessment covering all setting out activities

→ **Engineering Resources** - the number of engineers required, how the workload will be split, paying particular attention to interfaces between sections, provisions for sick and holiday cover

→ **Tools and Equipment** – a list of the equipment and tools required, secure storage arrangements, what to do if equipment is damaged, how to request and order additional equipment

→ **Staging** - how the staging of the works will affect the location and visibility of control points and how it might affect the setting out of the works

→ **Responsibilities** - who has overall responsibility within the site and within the company; who is responsible for setting up and maintaining the Survey Management File; who will re-order consumables such as spray paint, nails, timber boards and posts

→ **Lines of Communication** – for example, the point of contact for communicating with third parties such as designers and sub-contractors.

CHAPTER 4

DRAWINGS

4.1 INTRODUCTION TO DRAWINGS

Correct setting out is dependent on the correct interpretation of the construction drawings. Being able to understand and interpret drawings correctly is as important a skill as being able to use the equipment and record your work correctly. In addition to interpreting drawings, it is essential to ensure that the drawing you are using is the correct drawing, is the latest version and ties in with the information given on other drawings.

4.2 WHAT IS A DRAWING REGISTER?

Each company will have its own set of written procedures relating to how drawings should be documented and controlled for quality management purposes. One of the key documents in this process is the Drawing Register. There should be a named person responsible for ensuring the Drawing Register is kept up to date. For larger projects, there may be a dedicated Document Controller, but for smaller projects, the person responsible may be the Site Manager, Site Engineer or Section Engineer.

There should be a controlled list of people who must be notified when a new drawing or a revision to an existing drawing is issued. It is the responsibility of the end user to ensure that they are using the most up-to-date revision of a drawing.

4.3 HOW DO I CHECK THE INFORMATION ON THE DRAWING IS CORRECT?

Prior to commencing any setting out on site, it is essential to check the whole drawing package using a methodical approach. Each drawing should be inspected and cross-referenced with other drawings, the specification and any other relevant data. Check that:

→ the vertical distance between two points ties in with the difference between the corresponding reduced levels

→ the calculated distance between coordinated points ties in with dimensions given on road alignments, pipe details and building layouts

→ the sum of the intervals between piles, columns, gridlines etc. adds up to the corresponding cumulative distance

→ the levels and dimensions given on one drawing match the corresponding levels and dimensions given on other drawings showing the same information

→ all interface details are provided

→ all construction details are provided

→ the information on construction details ties in with the information shown on other drawings

→ the notes in the Notes column correspond with the information given on the drawings

→ any information missing from the drawings which you need to complete the setting out is available elsewhere.

When checking drawings, scrutinise each drawing thoroughly and compile a list of queries. You may find that as you go through the drawings, you find the answers to your initial questions. Once you have compiled the full list, ensure that you cannot answer any of the questions yourself by checking the drawings again or consulting the specification or other relevant information. Check with the other members of the site team that they cannot answer your questions. Each member of the site team should compile their own list of queries and the final list of questions should be issued to the Designer in the form or a Technical Query (TQ) or whatever your company's relevant quality management procedure is.

4.4 WHAT DIFFERENT TYPES OF DRAWINGS ARE THERE?

The different types of drawings in a typical construction package include:

→ the general arrangement (the overall layout drawings)

→ standard construction details

→ drainage layout, details, sections and elevations

→ roads layout, details, sections and elevations

→ structural steel layout and details

→ steel reinforcement layout and schedules

→ existing services

→ pipework layout

→ proposed services

→ reinforced concrete layout, details, sections and elevations

→ earthworks drawings

→ piling layout and details

→ temporary works drawings

4.5 HOW DO I INTERPRET DRAWINGS?

You should ensure you have a full understanding of all the information and detail of the drawings you are using for setting out. If there is even one small aspect of the drawings which you do not understand, it means you have a gap in your knowledge which could lead to a gross error. Go through each drawing and make sure that you fully understand the:

→ annotations

→ abbreviations

→ notes

→ terminology

→ relationships and interfaces between different sections

→ construction details

→ technical specifications

→ references to other documents and drawings

4.6 IS IT OK TO SCALE FROM A DRAWING?

A drawing is 'to scale' if the proportions on the drawing are all correct in relation to each other. If a drawing is marked as being 'to scale', it is usually to a convenient scale which corresponds to a scale rule, such as 1 in 100, or 1 in 250. A drawing is technically to scale even if it is to an awkward scale such as 1 in 231.3.

Some drawings are marked 'DO NOT SCALE'. This could indicate that even though the drawing was originally created to scale, the method of printing means it could lose its original size and proportions. It may also mean that the Designer has issued the drawings with indicative dimensions only and should not be used for setting out.

Some designers provide a 'scale bar' or reference on the drawing to allow you to calibrate the print to check it is correct.

In certain cases, it may be appropriate to scale from a drawing. Use your engineering judgment to decide if you should seek written confirmation from the Designer for any critical measurements.

4.7 HOW DO I USE A SCALE RULE?

Even if the scale is shown on the drawing, e.g. 1:200 or 1:50, always check its accuracy against a given dimension. This is because the drawing may have been created to the right scale but may have been distorted during printing. Also, you may have an uncontrolled copy of the drawing which may have lost some accuracy during copying.

Before scaling from a drawing, first check its accuracy by selecting the correct scale on the rule and measuring a distance on the drawing of which the dimension is shown. The longer this measurement, the more accurate, so select a dimension which is as close as possible to the full length of the ruler. If the dimension measured agrees with the dimension shown on the drawing, then you can proceed to take further measurements from the drawing.

If your scale rule does not have the scale you need, but a multiple of it, you can take the measurement and then convert it to the correct scale. For example, if your scale rule has a scale for 1:2500, but your drawing is to a scale of 1:250, you can take the measurement and divide the answer by 10. For example, if you measure 195m at a scale of 1:2500, this will convert into 19.5m at a scale of 1:250.

4.8 HOW DO I SCALE FROM A DRAWING WITH A NORMAL RULER?

If the scale is not indicated on the drawing, the actual scale can be calculated by comparing a known dimension on the drawing to the corresponding ruler measurement and calculating the ratio. The calculated scale can then be used to determine the dimensions of other elements on the drawing.

For example, if a dimension shown on a drawing is 70m and its corresponding ruler measurement is 0.178m, the scale is 1: (70/0.178), which is 1:393.3. If you then want to determine the dimension of a line whose ruler measurement is 0.283, you multiply it by the scale factor of 393.3, giving a dimension of 0.283 x 393.3 = 111.3m.

Use this method only when its accuracy is appropriate to the task you are doing. Use the longest dimension line provided as your calibration line and always verify the results against several other dimensions provided.

EXERCISE

1. If the scale on a drawing is 1:250, and the measurement along the side of a building is 21.6cm, what is the actual length of the building?

2. There is no scale on a drawing. You know that the length of 10 parking bays No. 3 to No.12 is 25m. When you measure the parking bays of the drawing with a ruler you find that it is 193mm.

    a. What is the scale of the drawing?

    b. If you then measure the distance on the drawing between two manholes, MH3 and MH4, to be 287mm, what is the actual distance between them?

4.9 IS IT OK TO PHOTOCOPY DRAWINGS?

It is not always practical to take all the relevant drawings out on site with you. Often, it is more practical to photocopy an A4 or A3 extract of an A1 or A0 drawing, showing only the information relevant to the specific task you are working on. In this case, clearly mark up the extract with the following information:

→ "UNCONTROLLED COPY!"

→ the date

→ the drawing number and revision

→ the purpose of the task

Each day you refer to the extract, check the Drawing Register to confirm that there have not been any revisions to the drawing.

If you hand a copy or an extract of a drawing to anyone else on site, for example the Foreman or Joiner Supervisor, ensure that they understand that the drawing is only valid for that day (or specified amount of time) and that they should discard it or seek to have the date extended at the end of the time period. Depending on the document control procedures in place, it may be necessary to obtain a signature.

4.10 IS IT OK TO USE ELECTRONIC VERSIONS OF DRAWINGS?

Unless there is special provision in the contract, the hard copy or PDF copy of drawings issued to the Contractor with the status 'For Construction' are the only contractually approved drawings. Electronic versions of CAD files are not necessarily contractually binding unless they are officially issued to the Contractor following the correct quality management procedures and with the correct correspondence in place. There will usually be a clause in the Contract stating that the hard copy drawings take precedence over the electronic files, and that the Designer is not responsible for any information extracted from the design files if it does not correspond with the information shown on the hard copy drawings. This means that even if the design information taken from the electronic drawing is incorrect, the Contractor is liable for any errors in the information. It is also worth noting that even in the absence of any written disclaimer, the Contractor is still liable if the Contract states that the 'For Construction' hard copy or PDF drawings are contractually binding. Unless it is a requirement of the contract, the Designer is not obliged to provide the Contractor with co-ordinates or electronic design information. If they do so, technically it is at the Contractors risk unless there is correspondence in place to say that the Designer accepts liability. Designers are often reluctant to issue co-ordinates or share electronic design files with the Contractor as it could potentially expose them to risk.

Examples of when electronic design data is issued to the Contractor include large highways, rail or earthworks projects. This may take the form of road string information or a digital terrain model.

CHAPTER 5

ACCURACY, ERRORS AND TOLERANCES

5.1 WHAT IS THE DIFFERENCE BETWEEN ACCURACY AND PRECISION?

Accuracy is the closeness of the measured value in relation to the true value. Precision is the closeness of a set of measurements to each other.

If you were to repeat a measurement several times, the results may be sufficiently accurate but not precise. Or they may be precise, but not sufficiently accurate. This concept is illustrated in Fig. 5-1.

Fig 5-1: Accuracy vs Precision

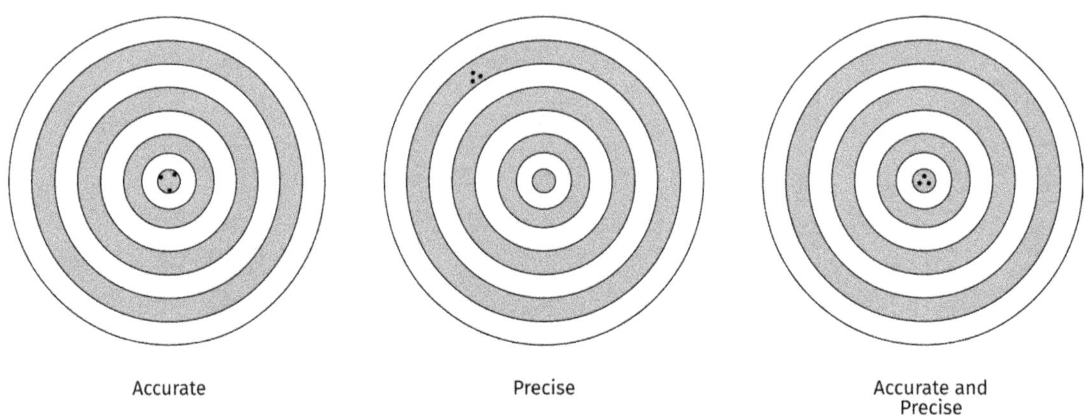

5.2 HOW CAN A MEASUREMENT BE PRECISE BUT NOT ACCURATE?

If a measurement has a high degree of precision, it means that you could repeat the same measurement over and over again and get the same answer every time. However, if there is a gross or constant error, then all of the measurements are incorrect even though they agree.

An example of this would be if the zero end of a tape measure had been damaged and then repaired causing it to be 70mm shorter. In this scenario, four measurements are taken between two fixed points with the faulty tape measure giving results of 15.775m, 15.776m, 15.775m and 15.773m and an average of 15.774m. This relatively small spread of 3mm may lead the measurer to believe that all is well even though the true distance between the two points is 15.704m, 70mm different to the measured value.

In this example, the variation between the readings is due to random errors such as the positioning of the tape against the mark or the user's interpretation of the tape reading.

Precision relates to how consistently you can get the same result using the same method and equipment. Precision can be broken down into two further elements: repeatability and reproducibility.

Repeatability is the likelihood of an individual person to repeat a measurement and get the same answer each time, whereas reproducibility is the likelihood of different users with different equipment to follow the same process and get the same results each time.

In the context of setting out, an example of a reproduceable activity would be measuring a round of angles with a total station. Consider a scenario in which Engineer 1 sets up their total station over the point and measures the angle between the two points as 41^13'27" taking only a left face reading. Engineer 2 comes along the next day using the same total station and sets up over the point and measures the angle between the two points as 41^13'29", again taking only a left face reading. The difference between the two results is only 2", however, if the total station has a collimation error of 20" and the true angle is 41^13'08", then the results are precise, but not accurate.

5.3 HOW CAN A MEASUREMENT BE ACCURATE BUT NOT PRECISE?

If a measurement has a high degree of accuracy, it means it is close to its true value.

An example of this would be if, on a windy, rainy day, a tape measure is used by an inexperienced engineer to take four measurements between the same two fixed points as in the example in section **5.2**, giving results of 15.698m, 15.709m, 15.701m and 15.708m. Even though there is a relatively large spread of 11mm, the results give an average of 15.705mm, which is only 1mm away from the true value of 15.704m.

The results in this scenario are not as precise as those in the previous example given in section **5.2**, but they are more accurate.

Consider a similar scenario to that considered in section **5.2**. An Engineer sets up their total station over the point and measures the angle between the two points as 41^13'05", taking only a left face reading. The same Engineer then comes back the next day using a different total station and sets up over the point and measures the angle between the two points as 41^13'11", again taking only a left face reading. The difference between the two results is 6", however if the average of the two readings is 41^13'08" (the true angle), then the results are coincidentally accurate, but not precise.

Accuracy improves with tools that are calibrated correctly and when the users are experienced and well-trained.

5.4   WHAT IS INTERNAL ACCURACY?

Internal accuracy refers to the accuracy of the dimensions within a single structure, as opposed to that structure's relationship with other structures, buildings, roads, drainage or existing features. Consider a brand-new building in the middle of a field. Even if the overall position (the external accuracy) of the building was not important, the internal accuracy is critical since the structural steel columns, beams and other fixed length elements must still fit together correctly.

5.5   WHAT IS AN ERROR?

In the context of setting out, an error is the difference between a measured value and its true value, or the difference between the true position of an element and its as-set-out position. Errors can occur in level (Fig. 5-2), position (Fig. 5-3), distance (Fig. 5-4) or angle (Fig. 5-5)

Errors fall into one of three categories:

> → random errors

> → systematic errors

> → gross errors.

Fig. 5-2: Level error

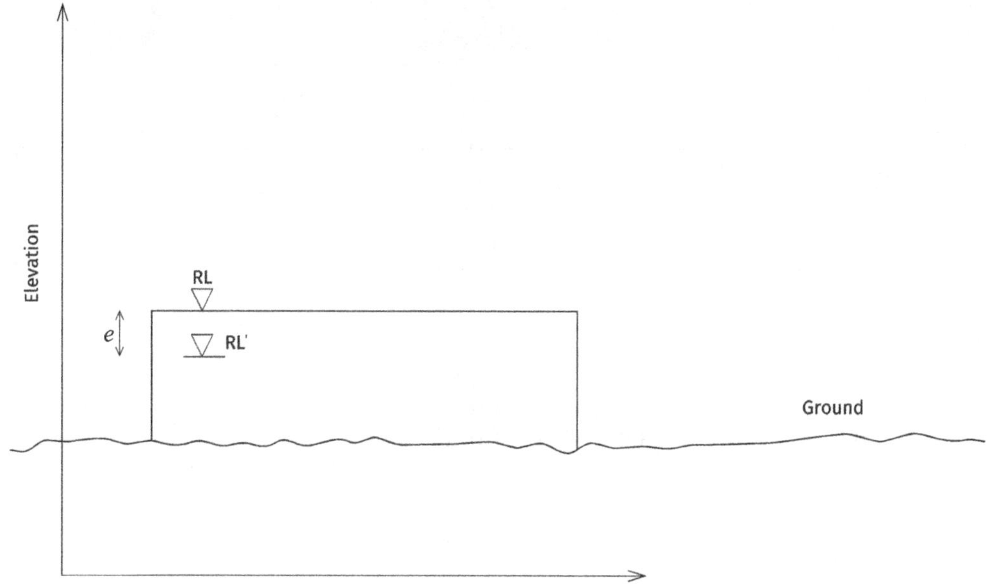

Fig. 5-3: Positional error

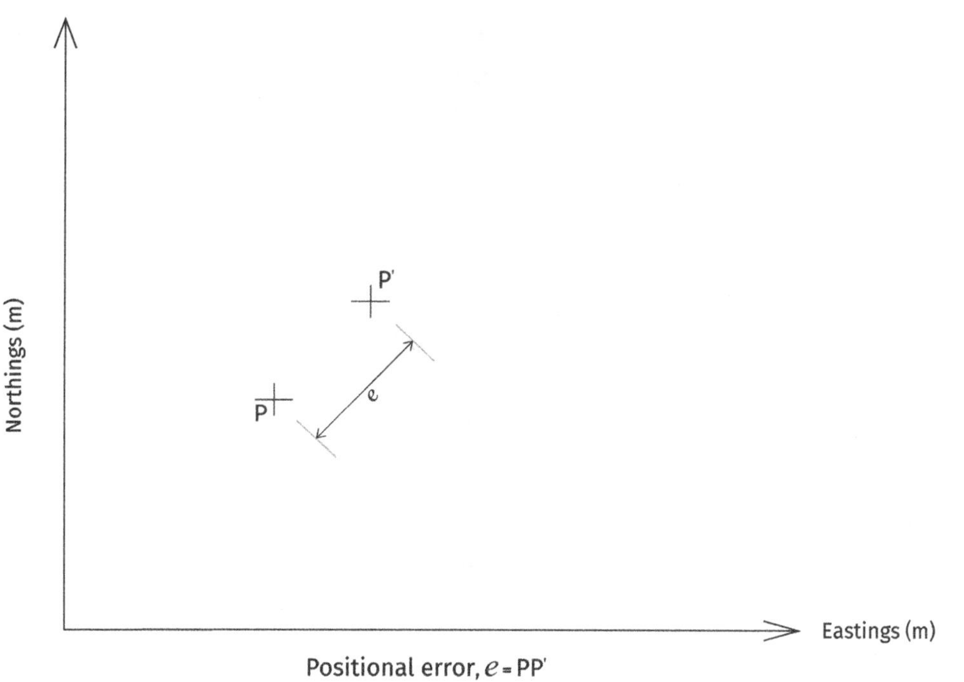

Positional error, e = PP'

5-4: Distance error

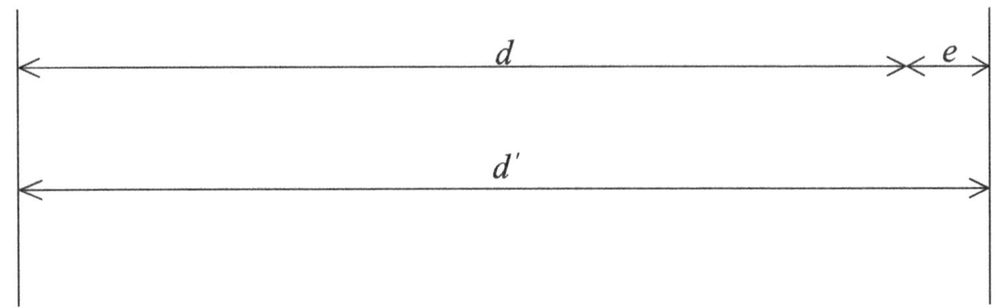

Dimensional error, $e = d' - d$

5-5: Angular error

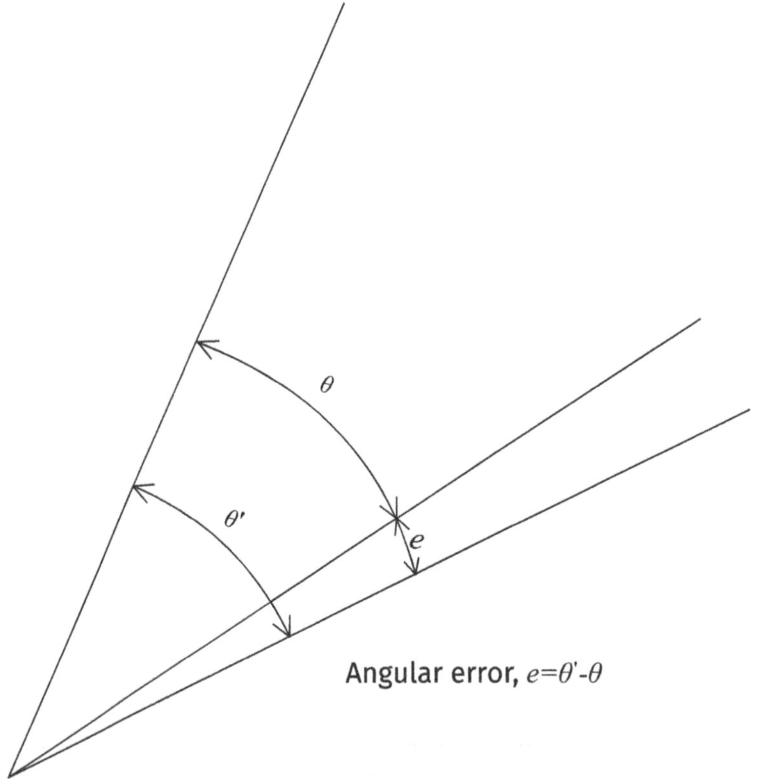

Angular error, $e = \theta' - \theta$

5.6 WHAT ARE RANDOM ERRORS?

Random errors are not mistakes but small, unavoidable errors which are present within the equipment or in the user's interpretation of the equipment.

Random errors vary from measurement to measurement, are randomly positive or negative and vary in magnitude and likelihood of occurrence. They comply with laws of probability and therefore follow a normal pattern of distribution. The effect of this type of error is reduced when measurements are averaged. The more times a measurement is repeated, the greater the accuracy when all the answers are averaged. This may have a cancelling effect and random errors can generally be assumed to be negligible.

An example of a random error is when sighting to a target with the Total Station. Even when the highest attention is paid to lining up the cross-hair, sometimes it will be a tiny amount off to the left and sometimes it will be a tiny amount off to the right, these errors being undetectable to the human eye. Taking the average of several readings will improve the accuracy of the final reading and bring it closer to the true value. Other examples of random errors are reading the levelling staff and taking tape measure readings.

Random errors can be minimised by taking repeated readings and calculating the average. Note that any erroneous results should not be incorporated into the average.

It is worth noting that for small sample sizes, this is not necessarily the case. For example, in levelling, a 1mm positive error and a 1mm negative error between intermediate sights taken to two different points will have a cancelling effect. However, if both errors are positive this will result in a 2mm error.

5.7 WHAT ARE SYSTEMATIC ERRORS?

Systematic errors are consistent in magnitude and direction and have a non-zero mean. This type of error is not reduced when readings are averaged. The error is cumulative when measurements are repeated end-to-end.

An example of a systematic error is a tape measure with a damaged end. If the start of the tape measure was 10mm but mistakenly assumed to be zero mm, then each measurement would be incorrect by a magnitude of -10mm. If five readings were taken, the starting point of each measurement being the end point of the previous measurement, the total distance would contain an error of 5 x -10mm, i.e. -50mm.

Other examples are scale factor errors, prism constant errors and collimation errors.

Systematic errors can be minimised by proper care of equipment, accurate setting up of instruments and regularly checking that instruments and equipment are in good adjustment.

5.8   WHAT ARE GROSS ERRORS?

Gross errors are the effect of human mistakes or oversights. They can be positive or negative and there is no limit to their magnitude or nature.

Examples of gross errors include misreading a drawing, using incorrect values for control points, and calculation errors.

The risk of gross errors can be dramatically reduced by incorporating the relevant checks, following site quality management procedures and by having critical tasks independently checked by another Engineer.

5.9   WHAT IS A TOLERANCE?

A tolerance, sometimes called a permitted deviation, is an allowable error. In the context of setting out, it means the maximum variation from the specified or true position or level before it would be classed as a defect.

A tolerance may be defined in terms of absolute accuracy (e.g. the allowable error in relation to the coordinate system for the whole site) or relative accuracy (e.g. the allowable error in the distance between column centres).

A tolerance can be given for a length, angle, position or level or for the acceptance criteria for instrument checks.

The tolerance given is for the final as-built position, rather than your setting out mark, so the accuracy of your setting out should allow for any additional error in the final construction.

5.10   HOW ARE TOLERANCES DETERMINED?

The tolerance specified depends on a combination of the following factors:

→ how the components physically fit together and connect

→ the cost and time involved in achieving a certain level of accuracy compared to the cost of deviating from the specified tolerance

→ the effect on the strength, durability and aesthetics

5.11   HOW DO I KNOW WHAT TOLERANCES TO WORK TO?

The tolerances for your project may be found:

→ in the specification for the contract you are working on

→ in the relevant company policies

→ in the relevant British Standard (which will be referenced in the Specification or other contract documents)

→ on drawings

5.12 WHAT ARE TYPICAL CONSTRUCTION TOLERANCES?

The following are indicative typical values for guidance only. It is your responsibility to establish the tolerances for the specific project you are working on:

→ level of concrete base slab for universal columns +/- 5mm

→ relative difference between the level of steel columns within a building +/- 5mm

→ distance between centres of bolt groups (with cones) +/- 10mm

→ distance between centres of bolt groups (cast in without cones) +/- 3mm

→ distance between bolts within a bolt group (with cones) +/- 5mm

→ distance between bolts within a bolt group (cast in without cones) +/- 3mm

→ cover to steel rebar +/- 5mm

→ variation in blinding level for a single base +/- 7mm

→ blinding level in relation to site datum +/- 5mm

→ position of wall kickers +/- 5mm

→ plumb of a RC wall or column +/- 5mm

→ plumb of a steel Universal Column (UC) +/- 5mm

→ level of formation for manhole +/- 20mm

→ level of formation, top of bedding, Invert Level (IL) and crown for concrete pipe +/- 20mm

→ level of formation for excavation for concrete slab +/- 20mm

→ level of bridge bearings +/- 2mm

→ position of bridge bearings +/- 3mm

→ centreline of masonry wall +/- 5mm

→ level of masonry wall +/- 5mm

→ position of cast-in pipes to tie into prefabbed pipes +/- 3mm

→ box-outs in RC walls +/- 3mm

→ thickness of roads – capping layer/subgrade/base course/wearing course levels at channel and crown +/- 5mm

CHAPTER 6

TOOLS, MARKERS AND EQUIPMENT

6.1 WHAT SHOULD MY TOOLKIT CONTAIN?

As a minimum, your toolkit should include:

- → strong rucksack / tool bag
- → claw hammer
- → lump hammer
- → string line
- → chalk line
- → nails for timber
- → nails for concrete / tarmac
- → spray paint
- → road marking crayon
- → French chalk
- → electrician's tape (red and white)
- → permanent marker pen
- → sharp-nib permanent marker pen
- → blunt / joiner's pencil
- → 30m measuring tape
- → 5m or 8m measuring tape
- → rubber eraser
- → pencil sharpener

→ survey book

→ calculator

→ spare powder chalk

→ retro targets

→ boat level

→ 1.5m spirit level

Some optional extras which might come in handy depending on the nature of your job are:

→ hand sanitiser

→ antibacterial kitchen / bathroom wipes

→ baby wipes

→ antiseptic wipes

→ sticking plasters

→ manhole keys

→ pinch bar

→ staff bubble

→ torch (for manholes, pipes, levelling in the dark)

→ dry J Cloths or paper towels for drying the 30m tape

→ G clamp

→ sledgehammer

→ retractable knife

→ plumb bob

6.2 HOW DO I MARK MY SETTING OUT?

There are many ways to create the physical marker which defines the position or level of the point you are setting out. The method you choose depends on the surface you are marking on, the nature of the element you are setting out, the accuracy required and the length of time the mark needs to remain. Marks may be temporary, semi-permanent or permanent.

Setting out markers include:

→ **carpenter's pencil or blunt pencil** – to create a temporary mark on light-coloured, smooth, dry surfaces such as finished concrete or timber shuttering

→ **chalk line** – for flat, dry, horizontal or vertical surfaces such as blinding concrete, structural concrete, timber or steel shutters

→ **electrician's tape** – for marking levels or positions on rebar. The correct point on the tape can be marked with a permanent marker or the edge of the tape can indicate the correct position. The tape should be marked with an arrow pointing to the correct edge

→ **a patch of road marking crayon with a pencil mark scribed in it** (Fig. 6-1) – for concrete, tarmac or rebar

→ **marker pen** - on concrete, shuttering and profile boards

→ **spray paint** – on soil, stone, concrete or tarmac. To spray a straight line, pull a string line taut between the two end points and spray over the top of it

→ **Hilti nail or survey nail** – in concrete or tarmac (Fig. 6-2)

→ **spray paint over string line** – to create a sharp line on concrete or tarmac. Lay the string line on the ground, have it pulled taut and spray paint over the top of it. When the string line is removed there will be a clear line where the string had been

→ **spray paint using a stencil** – for smooth dry surfaces such as concrete or tarmac

→ **protruding nails** – on pegs, profile boards or timber shutters so that a string line can be fixed in place when it is needed.

Fig. 6-1: Road-marking crayon with pencil mark

Fig. 6-4: Hilti nail or survey nail

6.3   HOW DO I USE A CHALK LINE?

A chalk line is a string line which is wound up into a chalk powder-filled casing. As the chalk line is pulled out, it is covered in a coating of chalk dust.

The chalk line is pressed to a flat surface and pulled taught at each end. As close to the midpoint of the chalk line as possible, the chalk line is lifted a few centimetres away from the flat surface and released (pinged!). The chalk line is then lifted away, leaving a neat, straight line a few mm wide.

Chalk power is available in different colours, usually blue or red, and is suitable for distances of up to about 10m-15m.

The casing of the chalk line is often shaped in a point so it can also be used as a plumb bob.

6.4 HOW SHOULD I TAKE CARE OF EQUIPMENT?

Surveying equipment is complex, fragile and expensive. Treating it well and maintaining it properly will not only save time and money in maintenance and repair costs but will ensure that the equipment is functioning correctly and providing the accuracy required for your setting out tasks. Repair bills for equipment such as a robotic total station can run into thousands of pounds.

Maintaining and caring for the equipment you use, whether it belongs to you or your employer, will show others that you are a credible and careful engineer. If you don't treat your equipment correctly, you may end up being blamed for damage or loss which is not your fault.

Most of what is involved in taking proper care of equipment is common sense. Here are some things to remember:

→ **never sit or stand on the equipment box** – don't stand on top of the box to reach your instrument. Apart from damaging the box, it is dangerous and may affect your accuracy

→ **keep the box closed** – the box is the protective case for your equipment. Keep it clean and dry. Once you have taken the equipment out of the box, close the lid to avoid dust, mud, debris or rain getting in

→ **open the box when wet** – if you have been working in the rain or damp conditions, when you bring the instrument indoors for storage, either take the equipment completely out of the box to dry or leave the box lid fully open. If you do not leave the box open, any moisture in the box will be forced inside the instrument as condensation and obscure your view through the telescope. If this does happen, there is a chance that the instrument may dry out naturally over time, however you may need to send the instrument to the workshop for repair

→ **store at room temperature** – avoid extremes of temperature

→ **store in a dry atmosphere** – avoid damp conditions

→ **use proper cleaning equipment** – for cleaning condensation or rain from the lens, use a lens-cleaning cloth. Use a puffer or air brush for tackling dust or grit particles

→ **be prepared for when things go wrong** – have a plan in place in case anything goes wrong with equipment. Who should you call for the right technical support? Trying to fix something without the right information and guidance can do even more damage

→ **don't bump the equipment** – when transporting the equipment in a car, wedge it securely in the boot of the car, or fasten a seatbelt round it. Never place it on the seat where it could fall off or on a flat surface where it could slide around. When carrying fragile equipment, don't swing it around or try to carry too much other equipment at the same time

→ **carry and store levels upright** – the compensator is more likely to be damaged if the level is tipped up on end, upside down or on its side

→ **keep up to date with calibration** - submit for calibration annually

→ **keep an equipment log** – when you take delivery of a piece of equipment, make a checklist of everything in the box including user manual, mini-prism , charger, rain cover, USB data stick, spare battery, Allen key, screwdriver, mobile phone charger lead, calibration certificate

→ **store spare parts safely** – if you think there is a risk of losing any of the parts, you may want to store them in a safe place until it is time to return the equipment

→ **keep levels upright** – whether they are in the box or on top of a tripod

→ **transport total stations in their box** – don't move the total station to a different location whilst it is still attached to the tripod.

6.5 WHAT IS THE DIFFERENCE BETWEEN TEMPORARY AND PERMANENT ADJUSTMENT?

Temporary adjustment is the set-up of the instrument including levelling it up and centring it over a point. Temporary adjustment includes adjustment to the:

→ foot screws

→ object focus screw

→ the cross-hair focus screw

Permanent adjustment is anything which requires a tool of any kind. Most permanent adjustments should be carried out in a specialist workshop, however, some basic adjustments can be carried out by the user on site. Permanent adjustments which can be carried out by the (properly trained) user include:

→ adjusting the line of collimation (automatic level)

→ adjusting pond bubble (automatic level, total station or backsight target)

→ adjusting the long bubble (total station or backsight target)

6.6 WHAT IS MEANT BY CALIBRATION AND MAINTENANCE OF EQUIPMENT?

Calibration of equipment is carried out in a specialist lab or workshop. The equipment is checked by trained technicians to ensure that the equipment is in correct adjustment. If necessary, permanent adjustments are made to the equipment.

Chapters **10** and **12** explain the checks that should be carried out by the user on total stations and levels. If any errors are found within the instrument, the user should not attempt to make any permanent adjustments themselves unless they are fully trained to do so. These days, it is common practice and relatively inexpensive to have the equipment calibrated and maintained by a specialist company. The only exception to this perhaps is adjustment of the line of collimation in the level, but even in this case, the user should only carry out the adjustment if they are confident and competent.

Typically, the calibration certificate is valid for 12 months, unless the casing of the instrument is opened up, which in some cases may invalidate the calibration certificate.

Maintenance in the form of an annual service is usually carried out at the time of calibration. This involves stripping the equipment down and cleaning and oiling the internal mechanisms.

The details of how to calibrate and maintain equipment are not within the scope of this book.

6.7 SHOULD I HIRE OR BUY EQUIPMENT?

When deciding whether to hire or buy equipment, you will need to weigh how often you will need it and what you need it for. If you need to use it every week or two weeks, even if it is for one small job, it will probably be more cost-efficient to own the equipment. If you only need it for a few weeks every few months, then it would probably be more ecomonical to hire it.

Things to consider when deciding whether to buy equipment:

- → The initial outlay vs the equivalent hire costs
- → Storage of the equipment when it is not needed
- → Insurance costs
- → The convenience of having it available at short notice without having to arrange hire

- → Calibration and maintenance costs

- → Breakdown and repair costs

- → Research the supplier's reputation in relation to after sales support

- → Check if any training is provided with the equipment

Things to consider when hiring equipment:

- → You are still responsible for insuring the equipment

- → You can chop and change between different types of equipment as often as you like

- → You will always have the most up-to-date equipment

- → If the equipment stops working, it will be replaced immediately

- → Depending on the length of the hire you may be able to negotiate long term hire rates

- → Hire purchase schemes are available

6.8 HOW MUCH DOES EQUIPMENT TYPICALLY COST?

The costs of purchasing and hiring equipment can vary dramatically depending on the quality, accuracy and functionality you require and your purchasing power. Here are some very approximate indicative costs:

	Buy (new)	Hire (per week)
Level, tripod and staff	£150 - £1000	£8 - £50
Laser level	£400 - £2000	£20 - £140
Mechanical total station	£3000 - £16000	£65 - £250
Robotic total station	£8000 - £20000	£200 - £500
GNSS	£17000 - £25000	£300 - £700
Pipe laser	£2000 - £3500	£20 - £50

6.9 HOW DO I CHECK IF A SPIRIT LEVEL IS IN CORRECT ADJUSTMENT?

Some spirit levels have a fixed bubble; however some have bubbles which can be adjusted. These can go out of adjustment over time, so should be checked regularly.

To check the horizontal bubble (Fig. 6-3):

→ Use the straight edge of the spirit level to draw a horizontal line on a wall.

→ Flip the spirit level over and draw a line on exactly the same line as before with the mid point of the spirit level at the middle of the line. If the spirit level is in perfect adjustment, then the lines will lie exactly on top of each other. If it is out of adjustment, then the lines will form a cross.

To check the vertical bubble, the process is the same as for the horizontal bubble except with a vertical line.

Fig. 6-3: Checking the adjustment of a sprit level

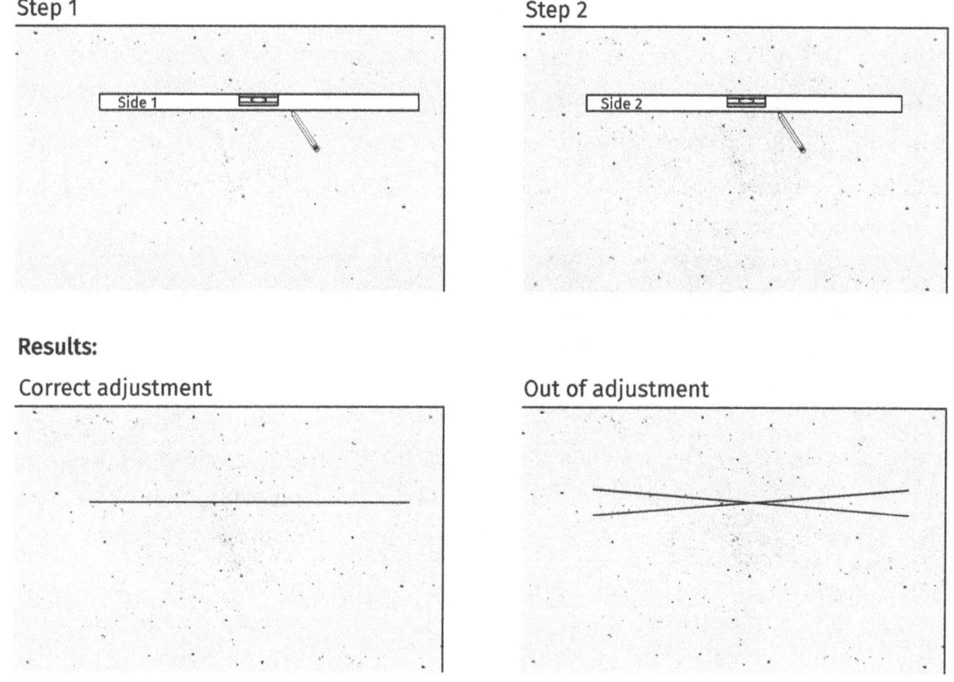

CHAPTER 7

SAFETY

7.1 WHAT SAFETY MEASURES SHOULD I TAKE?

Setting out activities are an integral part of any construction project. This section on health and safety is intended only as a broad overview in the specific context of setting out activities. It is not exhaustive and does not preclude the need for other health and safety training relevant to the specific environment you are working in. The information on the following pages is intended to assist in the preparation of a full and detailed risk assessment rather than as a comprehensive list:

Activity	Hazard	Suggested safety measures
Using a lump hammer or sledge hammer for banging in pegs	* banging fingers or thumbs * accidental letting go of hammer which hits someone else	* hard hat * goggles * proper training * supervision of inexperienced people * appropriate gloves
Using timber pegs	* splinters	* appropriate gloves when handling materials * goggles when nailing together/ dismantling
Nailing timber together and dismantling with a hammer	* cuts and scratches from nails * banging nail into skin * hitting thumb or finger with hammer * hitting head with hammer	* hard hat * appropriate gloves * supervision of inexperienced people * appropriate gloves

Activity	Hazard	Suggested safety measures
Banging pegs into the ground	* striking underground utilities services	* proper training * scan the area using a CAT scanner * obtain services drawings of the area * obtain a permit * follow company / site procedures
Weil's disease	* placing the levelling staff on the invert of drains or in areas where rats are present	* frequent cleaning of the staff with antibacterial wipes * wear gloves * wash hands before eating or smoking * learn about the symptoms to watch out for
Live traffic	* being distracted by the task you are doing	* always set up in a safe place or have a look out if you need to set up in an area which is not segregated from traffic
Levelling staff	* contact with or proximity to overhead electrical cables	* check the area before commencing work * use non-conductive levelling staff * don't have the staff extended any more than necessary * proper training for the person holding the staff * be aware that arcing can occur even if there is no direct contact

Activity	Hazard	Suggested safety measures
Carrying equipment around the site	* slips / trips and falls * feet / legs catching on levelling staff or tripod	* don't carry more than you can safely carry * don't carry equipment up ladders, use an alternative method * check all clips on tripods are working properly so they can't accidentally extend whilst being carried
Carrying equipment around the site	* muscular skeletal injuries	* manual handling training * check the proper carrying straps in working order
Stress / anxiety caused by high pressure work environment or lack of training	* short term or long term poor mental health * diminished concentration leading to unsafe actions	* proper technical training and a supportive environment can reduce stress and anxiety
Working in awkward spaces	* stooping / bending for long periods of time can cause muscular skeletal injuries	* stretch regularly * take regular breaks * don't continue if any aches and pains develop
Kneeling for long periods (e.g. marking out wall positions on concrete slabs)	* prepatellar bursitis	* avoid kneeling for long periods of time * take regular breaks * ensure surface is clear of sharp stones, nails etc. * wear good-quality correctly fitted knee pads when kneeling for long periods of time

Activity	Hazard	Suggested safety measures
Lasers	* damage to eyes (user or others)	* switch lasers off when not in use * when using laser dot, either turn laser off or point telescope away from people (e.g. towards the sky) in between sightings * check manufacturer's safety warnings and advice * don't look at laser beams * inform others on the site who may be affected
Looking into the sun	* damage to eyes	* don't look into the sun with any surveying equipment
Working alone	* unable to access support in the event of an accident or emergency	* avoid working alone where possible * ensure a full risk assessment is carried out
Moving equipment in vehicles	* loose equipment causing injury in the event of a sudden stop or crash	* heavy equipment should be secured in place if in the back of a vehicle, or should be placed in the back of a van

CHAPTER 8

LEVELLING

8.1 WHAT IS LEVELLING?

The principle of levelling is that the instrument, known as the level, creates a horizontal line of sight through a telescope. When we place a levelling staff vertically on physical point of which we know the height and take a reading to the cross-hairs, we are measuring the vertical difference between the known point and the instrument height. We can use this information to calculate the height of the line of sight relative to that known point. This gives us the instrument height, or height of plane of collimation (HPC). Once we have calculated the instrument height, we can then take staff readings to any other physical points and calculate the heights of those points in relation to our instrument height. Effectively, we are measuring the vertical distance between points, which we then convert into relative heights. The principle of levelling is illustrated in Fig. 8-1.

8.2 WHAT LEVELLING SKILLS SHOULD I HAVE?

You should be able to:

→ record work correctly and in the industry standard way

→ incorporate robust checks into all levelling

→ list the possible sources of error in levelling

→ carry out a level survey (existing features or as-built)

→ set elements out to a fixed level

→ measure the reduced level of ceilings and soffits

→ transfer a TBM (install a vertical control point network from scatch)

→ set up profile boards for level excavation

→ set up profile boards for drainage

→ set up batter rails for cut and fill.

Fig. 8-1: The Principle of Levelling

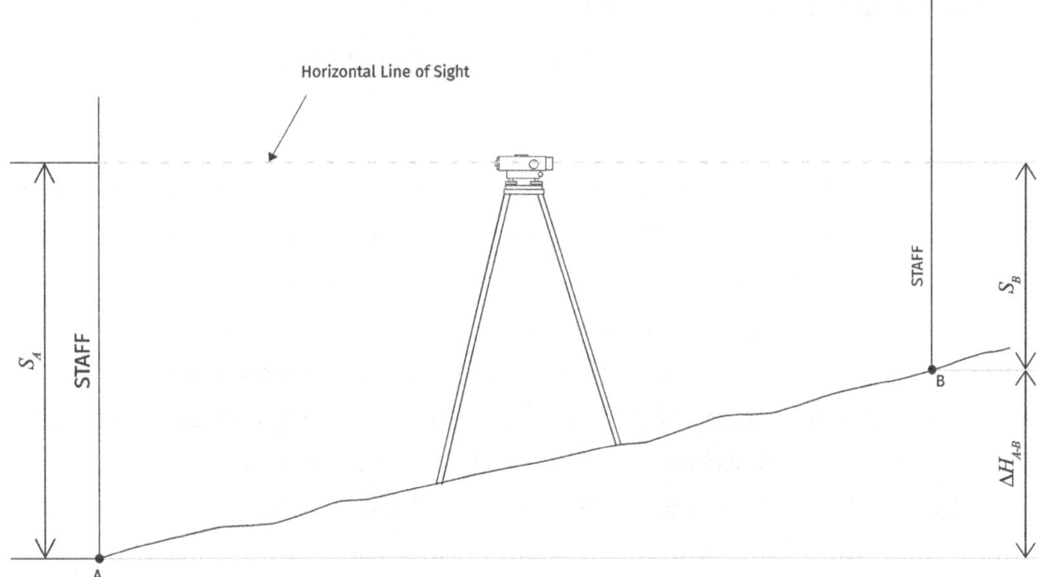

8.3 WHAT IS A LEVEL?

In the context of this book, the term 'level' refers to an 'automatic level': a precision survey-ing instrument which provides a horizontal line of sight coincident with the horizontal cross-hair of a reticule within a telescope. The level can be rotated horizontally through 360 degrees and therefore can point in any direction. There are several different types of level, all of which follow the same basic principles as the automatic level, which are described in more detail later in this chapter. They are:

→ laser level

→ digital level

→ precise level

A 'spirit level' is distinct from an automatic level.

8.4 WHAT IS LEVELLING USED FOR?

Levelling is used for setting out and measuring (surveying) the levels of pre-existing or as-built features, setting construction elements to their correct reduced level and estab-lishing and maintaining vertical control points.

8.5 WHAT ARE THE DIFFERENT PARTS OF THE LEVEL?

The different parts of the level are described here and can be seen in Fig. 8-2.

FOOT SCREWS

Three independent foot screws connect the tribrach to the trivet. They raise or lower the level as they are rotated either clockwise or anti-clockwise by the user. The foot screws are used to centre the pond bubble. This is known as temporary adjustment of the level.

On most levels, there is a line scribed at the mid-point of the range of each foot screw. It is optional whether you set each foot screw to its mid-point before starting the levelling process, however, it may be helpful to do this if you are having difficulties centring the bubble or if any of the foot screws are reaching the end of their travel and will not turn any further. If this does happen, do not force the screws as it will damage the instrument.

POND BUBBLE

The pond bubble, also known as the circular bubble or bulls-eye bubble, is the circular bubble on the base of the level, with a black circle scribed at the centre.

The pond bubble is multi-directional and reads in both the X and Y direction. The foot screws are adjusted until the small bubble is centred.

The movement of the bubble is prescribed by gravity, so if a foot screw is extended, or raised, the bubble will move towards it. If a foot screw is shortened, or lowered, the bubble will move away from it.

Most levels incorporate a mirror which is either adjustable in angle, or is fixed at 45 degrees to the bubble. This enables the user to obtain a birds-eye view of the bubble by looking horizontally.

The pond bubble can go out of correct adjustment and cause an error. See section **10.25** on pond bubble error.

THE FOCUS SCREW

This is found on the right-hand side of the level and is used for focusing the object you are sighting to, i.e. the levelling staff.

THE CROSS-HAIR FOCUS SCREW

This is the small screw on the eyepiece, closest to your eye, and is used for focusing the cross-hairs.

GUN SIGHT/OPTICAL SIGHT/OPEN SIGHT

The device on top of the barrel of the level which is used for approximate lining up of the instrument to the staff. This can take the form of a small tube containing an aperture which is only visible when the eye is in alignment with it, or two upstands with a 'V' shaped recess which come into alignment with each other when the eye is in the correct alignment. The user's eye should be about half a metre back from the level and moved about until the aperture can be seen. The level can then be manually rotated left or right until the aperture lines up with the staff. The user then looks through the eyepiece and uses the tangent screw for more accurate lining in.

THE TANGENT SCREWS

These are usually found towards the front of the level instrument and are used for moving the line of sight slowly to the left or right to position the line of sight correctly in relation to the staff.

THE TRIBRACH

The base section of the level which attaches onto the tripod head.

THE COMPENSATOR

The compensator is the internal device which automatically levels the instrument accurately. It is activated once the instrument has been approximately levelled manually.

It is a delicate self-levelling mechanism which utilises the effects of gravity to create a perfectly horizontal line of sight. It comprises a system of prisms and mirrors suspended by fine non-magnetic wires, an inverted pendulum and a damping mechanism. See Fig. 8-3 for an indicative diagram of the compensator mechanism.

Although the compensator is a great advance in levelling technology and helps to achieve excellent accuracy with a relatively quick and simple set-up process, it is susceptible to vibration caused by high winds and passing construction traffic. The effect is that the cross-hair visibly vibrates up and down and can make it difficult to take a staff reading.

To see the effects of the compensator for yourself, carry out the quick checks explained in section **10.19**

Fig. 8-2: Parts of the level (Image courtesy of Topcon Positioning Systems)

Fig. 8-3: The Compensator

Compensator level

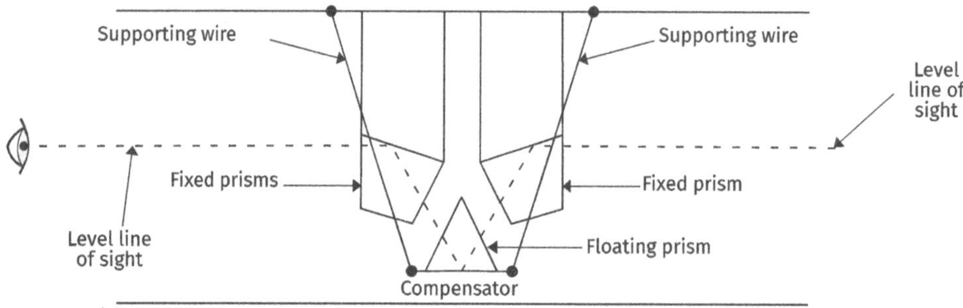

Compensator (Tilting forwards)

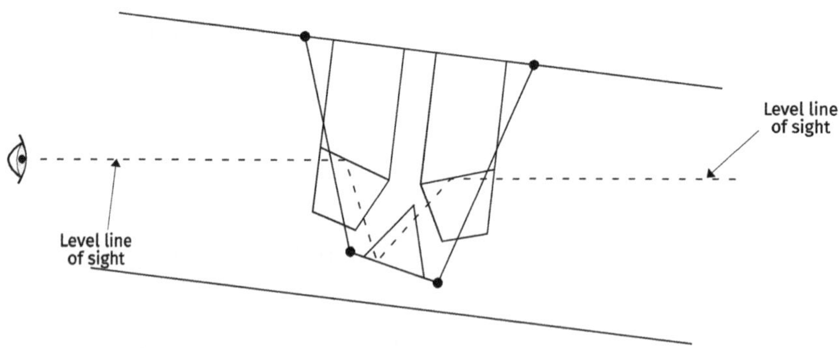

Compensator (Tilting backwards)

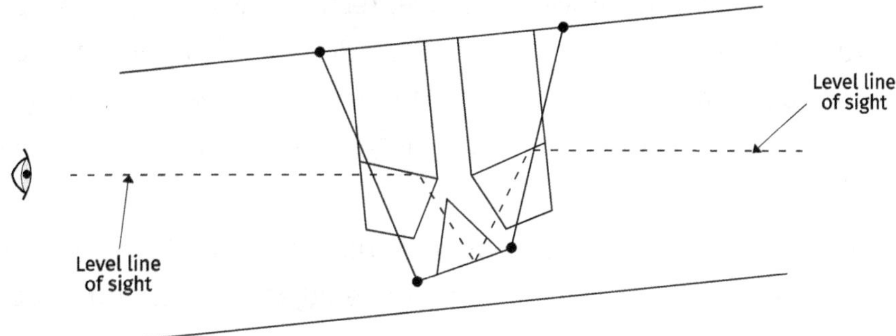

8.6 WHAT ARE THE DIFFERENT CROSS-HAIR ARRANGEMENTS?

The cross-hairs on the reticule of the level can take different formats. Some examples are shown in Fig. 8-4. No matter what the arrangement of the lines, you will always have a single horizontal line. Regardless of what the rest of the cross-hair arrangement is, it is always the single, horizontal cross-hair that you will take your reading to.

Fig. 8-4 Different cross-hair arrangements

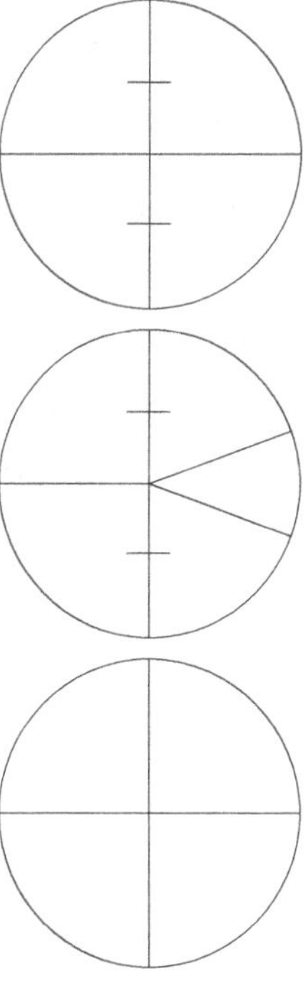

8.7 ABOUT THE TRIPOD

The purpose of the tripod is to provide a stable, relatively level surface to support the surveying instrument. It is the stand to which the instrument is attached. It comprises a head (a flat triangular surface with a round hole, through which the screw-fixing attaches to the base of the instrument) and three legs of adjustable length which are attached to the tripod head by hinges.

To enable the levelling of the tripod head, the tripod feet can be moved in or out or rotated laterally. The legs are telescopic and can be adjusted in length either by screw clamps, clip clamps or both.

If the clip clamps are in bad condition and at risk of slipping under the self-weight of the tripod and the instrument, once the instrument is set up, the screw clamps should be engaged.

The screw-fixing is a hollow tube with threads on the external surface and is attached to the tripod head on a moveable sliding connector. When setting up the instrument over a fixed point e.g. a nail, the lateral position of the instrument can be adjusted on the tripod head. The hollow tube enables the user to sight the nail with an optical or laser plummet.

There are two main types of tripod in common use. Timber or fibreglass legs are usually used for total stations. They are sturdier and have a larger head diameter to allow for lateral adjustment of the instrument when setting up over a nail.

Lightweight aluminium legs are adequate for use with levels, laser levels and backsight prisms.

Wooden and fibreglass legs can typically be adjusted in length from approx. 1m when fully closed to 1.7m when fully extended. Aluminium legs are usually, but not always, smaller and can be adjusted from approx. 0.9m to 1.5m.

The feet are typically made of die cast aluminium and taper to a point with a spike at the bottom. At the top of each foot is a small, ridged platform for pressing the feet into the ground.

Tripods are available with either flat or domed heads, however the flat head is the standard type found on construction sites.

8.8 LEVELLING TERMINOLOGY

Some common terms used in levelling are:

Above ordnance datum (AOD) – The vertical distance perpendicular to the earth's surface from the ordnance datum. Each territory has a specified point to which OD relates. In the UK, OD is the Mean Sea Level at Newlyn, Cornwall.

Backsight (BS) - the first staff reading taken to a point of known or assumed reduced level once the level has been set up in a new position. The backsight reading allows the height of collimation to be calculated.

Benchmark – a permanent reference mark on the ground, of which the reduced level is known or assumed. The reduced level may be given in relation to Ordnance Datum or a local datum. Historically, benchmarks took the form of a horizontal mark chiselled into the wall of a stone building. An angle iron was inserted into the point of the recess to create a bench for the levelling staff to sit on, allowing it to be accurately repositioned at the same point. Nowadays, a range of different marking techniques are used.

Change Point – a physical point in the ground to which a foresight reading is taken when all the survey points are not visible from one set-up position. The reduced level of the change point is calculated so a backsight reading can be taken to it from the subsequent set-up position. Ideally a change point is a reliable TBM, but can be any stable point with a well-defined high point for the staff to pivot on.

Datum – a horizontal plane of reference at a known or assumed reduced level.

Foresight (FS) - the final reading taken from a set-up position. The foresight is taken before completing a task or moving the level to a different position. It can be to a point of known reduced level for checking the closing error of a survey, or to a change point, in which case the reduced level of the change point is calculated, and the point is then used as the backsight from the subsequent set-up position.

Height of collimation / Height of plane of collimation (HPC) - the height of collimation, or height of plane of collimation (HPC), is sometimes described as the instrument height. It is expressed as a numerical value: the reduced level of the line of collimation.

HPC method – the standard method of recording levelling field data. This method uses the backsight reading to determine the height of collimation of the instrument. All points are measured, or for setting out, required staff readings are calculated, in relation to the height of collimation. This method involves fewer calculations than the rise and fall method and is more suitable when there are many intermediate sites.

Intermediate sight – any staff reading which is not a backsight or a foresight. It is the reading to the point being measured or the required staff reading to a point being set out.

Line of collimation - the line of sight coincident with the central axis of the telescope and the horizontal cross hair on the reticule. The line of collimation is assumed to be horizontal, however theoretically, there is always some error even though it may be insignificant.

Local datum – an arbitrary datum, specific to a site. The local datum is used for design and construction purposes and can be, but is not usually, coincident with the ordnance datum.

Mean sea level (MSL) – the level of the sea at the midpoint between high tide and low tide. Mean sea level is based on data gathered over time.

Ordnance benchmark (OBM) - physical markers which correspond to values given on Ordnance Survey maps. They are usually chiselled into a vertical surface, for example, the side of a building, however non-vertical surfaces may be used.

Ordnance datum – the datum used by Ordnance Survey. Reduced levels of spot heights and benchmarks on OS maps are given in relation to the ordnance datum. In the UK, ordnance datum is zero at Mean Sea Level (MSL), measured at Newlyn, Cornwall.

Ordnance survey (OS) – the government-owned national mapping agency which covers Great Britain. Ordnance Survey (OS) Limited was established in 1745 for military purposes.

Plane of collimation - the level instrument, and therefore the line of collimation, can be rotated laterally through 360 degrees. This creates a plane of collimation.

Reduced level (RL) – an elevation, height or altitude given in relation to a known or assumed datum. It is the vertical distance from the datum, measured perpendicular to Earth's surface. The reduced level is expressed as a numerical value and is typically indicated on section drawings by an inverted triangle, where the bottom tip of the triangle indicates the point to which the numerical value refers. On plan drawings, the RL of a point is marked next to a triangle with a dot at the centre or a small cross.

Rise and fall – an obsolete method of recording levelling field data. This method is more suitable for use in surveying rather than setting out, however it is rarely used nowadays as topographical surveys are generally carried out using a total station or Global Navigation Satellite System (GNSS) equipment rather than levelling.

Spot height – the reduced level of a point of which the position is recorded in relation to a coordinate system or existing features, but unlike a benchmark, there is no mark on the ground to indicate its location.

Temporary benchmark (TBM) – a benchmark which is utilised during a construction project, but which is destroyed, mothballed or no longer used once the project is complete. Fig. 8-5 shows various levelling terms.

Fig. 8-5: Levelling Terminology

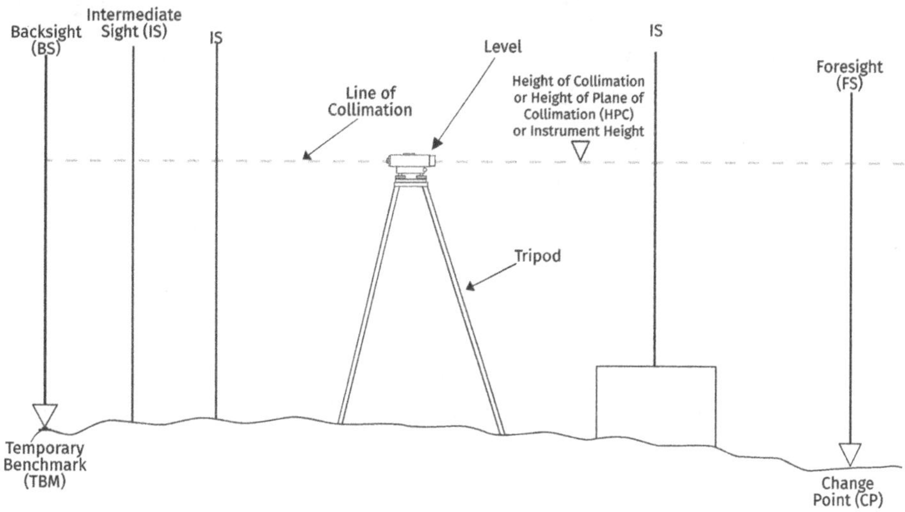

8.9 ABOUT THE STAFF

The staff is split into 1m sections, alternating between red and black (Fig. 8-6).

Each of these 1m sections is split into 100mm (0.1m) sections (Fig. 8-7).

Each 100mm section is split into 10 small blocks of 10mm (1cm) (Fig. 8-8).

The 'E' shape in the bottom half of each section comprises 5 blocks of 10mm, which is the first 50mm of that section. The top section of 5 blocks is another 50mm, making up the full 100mm. The 'E' shape is there to help you see the first 50mm at a glance so that for the blocks above it, you can start counting from the top of the 'E' instead of having to count all the way from the bottom each time.

Fig. 8-6 The levelling staff

Levelling Staff Detail (Main)

Levelling Staff Detail (Alternative)

Fig. 8-7 Staff Reading (Breakdown)

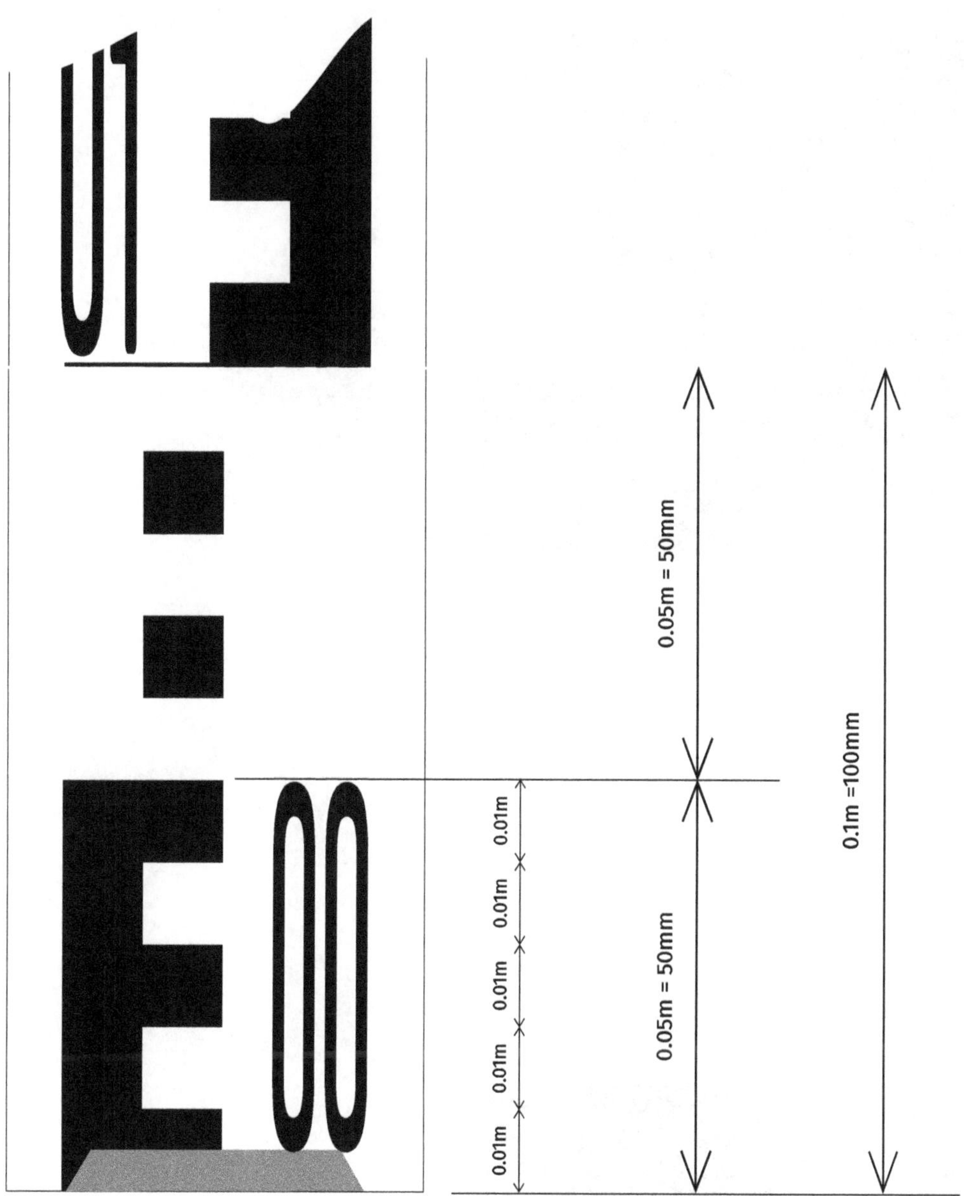

Fig 8-8: mm Readings On Staff

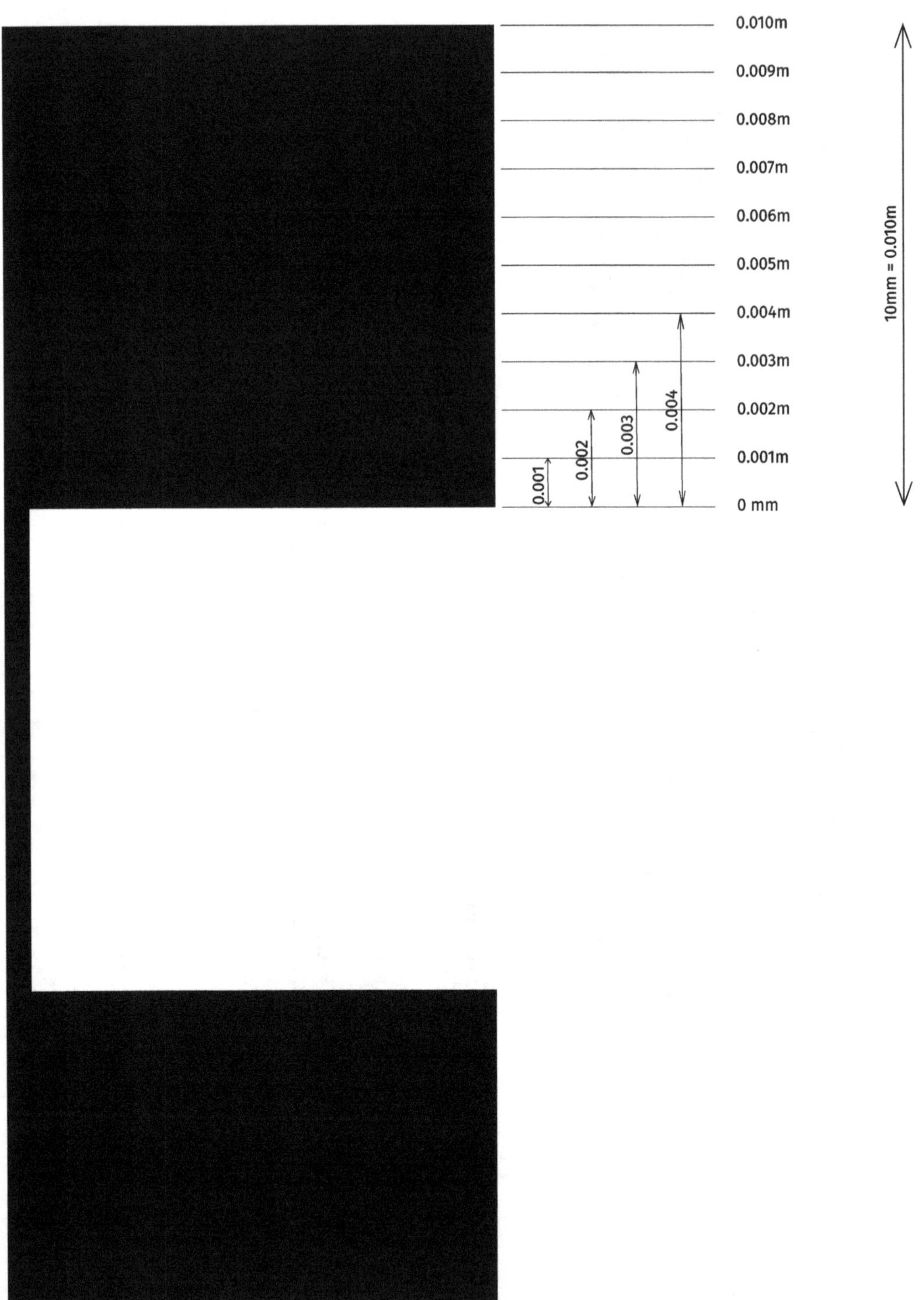

8.10 HOW SHOULD I LINE UP THE CROSS-HAIRS ON THE STAFF?

The single cross-hair should pass right through the staff. It should be the only line that is on the staff. The vertical cross-hair should be to the side of the staff and not touching it at all so that the stadia hairs are over to the side and not in contact with the staff. See Fig. 8-9

Fig. 8-9 Lining up the cross-hairs on the staff

Positioning of X-hair On Staff - Incorrect   **Positioning of X-hair To Staff - Correct**

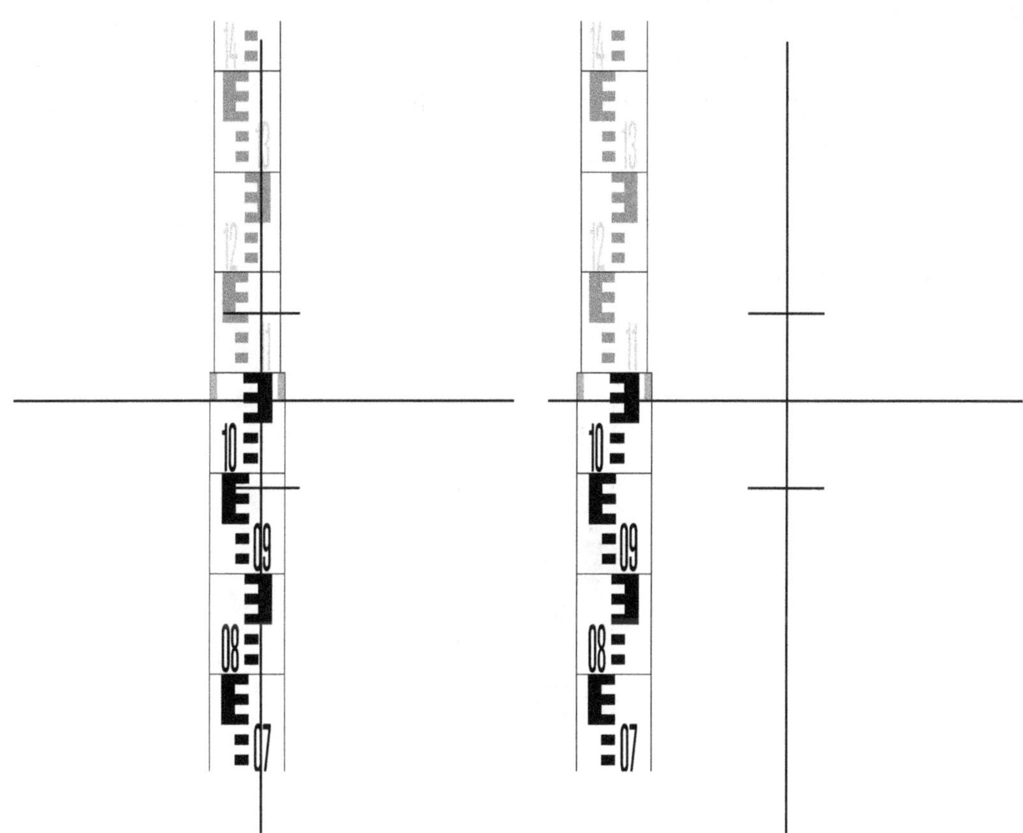

8.11 HOW DO I READ THE STAFF?

When you're learning how to read the staff, it's a good idea to look through it several different times and write in an extra digit each time. Using Fig. 8-10 as an example:

1. Sight the staff. From the cross-hair, look down the staff until you come to the beginning of the 100mm section. Write down the two-digit number. There are different types of staff. Some show a decimal point, and some don't. If your staff doesn't show the decimal point, add one in as you record it. So for example, 13 on the staff will be 1.3m.

2. Write down the number of full 10mm blocks between the bottom of the section and the cross-hair. For example, if there are four full blocks and then a part block, you will now have 1.34.

3. Finally, estimate the number of mm between the bottom of the 10mm block and the cross-hair. If you estimate that the cross-hair is 8mm from the bottom of the block, you will have 1.348.

Fig. 8-10: Reading the staff (Example)

Staff reading 1.348m

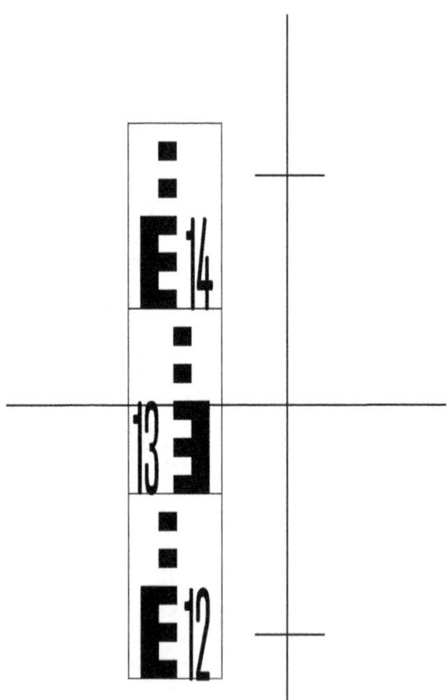

8.12 WHAT IS A LEVEL SURVEY?

A level survey is a survey which is carried out to calculate the reduced levels of points in relation to each other and/or in relation to either the Ordnance Datum or more commonly, in relation to a local design datum as shown in Fig. 8-11.

The positions and levels of the points can be shown on a scale plan or schematic diagram or may be given in list format, e.g. for chainages along the centreline of a road.

Fig 8-11: Level survey

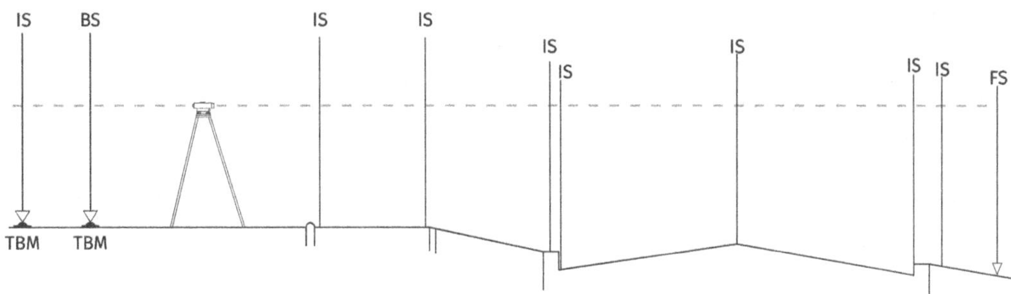

8.13 WHAT IS THE PURPOSE OF A LEVEL SURVEY?

There are many possible reasons for carrying out a level survey on a construction site. Here are a few examples:

→ the initial survey of the site (including locations of tie-ins such as road surfaces, drainage invert levels, building interfaces and pipework) to check that it ties in with the design information

→ when accurate information is required about existing features so that the detailed design can be finalised. Often, for costing and tendering purposes, the design is based on a survey of low accuracy, however some elements of the detailed design cannot be finalised until the works have commenced on site. This is common in 'design and build' or complex M&E projects

→ checking that one stage of construction is correct before moving onto the next stage

→ as-built survey to record the work completed, for payment purposes

→ as-built survey for use in claims/compensation events

→ as-built survey so that the correct quantity of materials can be ordered for the next stage

→ as-built survey for contractual requirements and completion of quality documentation.

8.14 HOW DO I CARRY OUT A BASIC LEVEL SURVEY?

The steps are as follows:

1. Set up your level where you have visibility to at least two reliable TBMs within 40m.
2. Take a BS to a TBM (TBM1)
3. Calculate the HPC
4. Take an IS to another TBM (TBM2)
5. Subtract the staff reading from the HPC to calculate the measured value of TBM 2 in relation to TBM1.
6. Compare the measured value of TBM2 to the expected value. If they agree within 2mm, continue to step 7. If they don't agree, do not proceed until you have identified and eliminated the error.
7. For each point in your survey, record the staff reading in the IS column.
8. Subtract the IS staff reading from the HPC to calculate the RL of the point.
9. Repeat steps 7 and 8 until you have completed your survey.
10. Take an FS to either one of your starting TBMs or to a separate reliable TBM.
11. Calculate the measured RL of the final TBM. If it agrees with the expected value within 2mm, then your closing error is acceptable, and your survey is complete.

An example of how to record a basic level survey is shown in Fig. 8-12.

Fig. 8-12: A basic level survey booking example

Date: 7 Feb 2019 Taken for: Greenfield Road/ Mansfield Road

From: TBM 23 To: TBM 14

BACK SIGHT	INTER MEDIATE	FORE SIGHT	COLLIMATION OR H.P.C	REDUCED LEVEL	DISTANCE	REMARKS
1.465			11.753	10.108		TBM 23
	1.361			10.212		TBM 14 (10.214 reqd, 2mm error ∴ OK)
	1.485			10.088		Chainage 150
	1.590			9.983		CH 160
	1.747			9.826		CH 170
	1.851			9.722		CH 180
	1.623			9.950		Manhole 1
	1.788			9.785		Gully 1
	1.990			9.583		Gully 2
		1.466		10.107		TBM 23 (12.500 reqd, 1mm error ∴ OK)

8.15 HOW DO I CALCULATE THE HPC?

The height of the plane of collimation(HPC), is calculated by taking a staff reading to a point of known reduced level (RL), i.e. a benchmark or temporary benchmark (TBM). The staff reading is added onto the RL of the TBM to give the height of collimation. Once the height of collimation is calculated, the RL of any other point can be calculated by taking a staff reading and subtracting it from the HPC.

EXERCISE:

1. You have been asked to determine the as-built levels at 4 corners of a small slab. The values of TBM12 and TBM 14 are 50.110m and 50.278m respectively. The staff readings are as follows: TBM12 (1.559), TBM14 (1.390), corner 1 (0.472), corner 2(0.468), corner 3(0.473) corner 4 (0.470), final reading to TBM 12 (1.557)

   a. Which column will each reading go into?

   b. What is the HPC?

   c. What is the error between TBM12 and TBM14?

   d. What are the RLs at the corner points?

   e. What is the closing error?

8.16 WHERE SHOULD I SET UP MY LEVEL?

When you are setting up your level, whether for a survey or setting out exercise, consider the following:

→ **sight lines** – you must be able to see at least two TBMs and the works area

→ **sight distances** – sight distance should be appropriate for the specific task you are working on

→ **live traffic routes** – for safety reasons, avoid setting up on or near live traffic routes

→ **the quality of the ground** – avoid soft, muddy or boggy ground

→ **shelter** – in high winds, try and set up in the most sheltered position

→ **sunlight** – when the sun is low in the sky, or shining at an awkward angle you may need to try several set-up positions to find the right spot.

8.17 WHAT IF I CANNOT VIEW ALL THE POINTS I NEED TO FROM A SINGLE SET-UP POSITION?

Sometimes it is not possible to set up in a position where you can carry out the full exercise from a single set-up. When this is the case, you can either use change points as shown in Fig. 8-13 and Fig. 8-14 or split the task into a number of separate exercises.

Fig. 08-13: Change points in a level survey

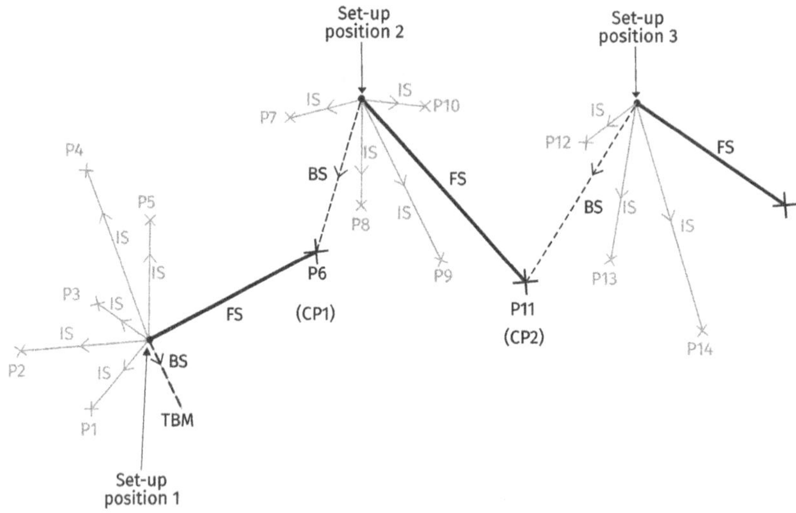

Fig. 08-14: Change Points in Elevation

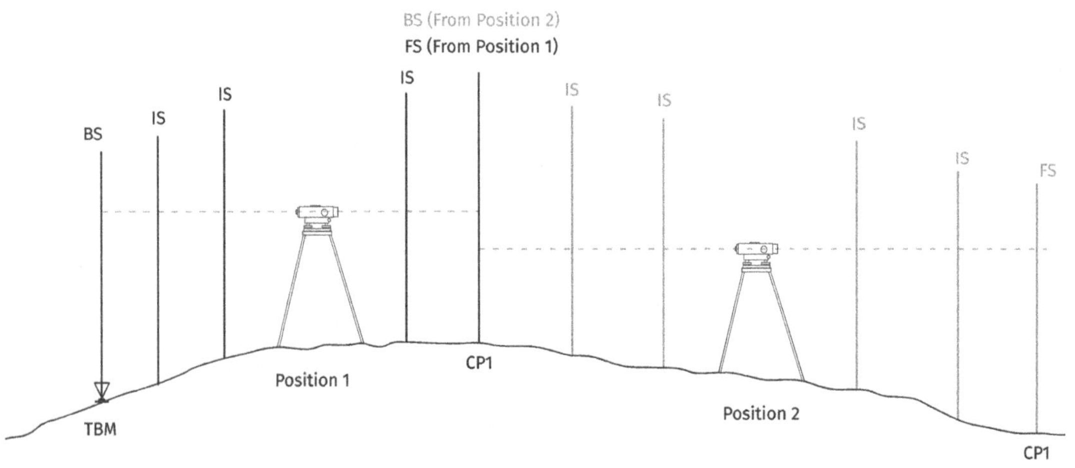

EXERCISE:

2. In the following survey, you measured P1 and P2 but needed to change your set-up position to sight P3, P4 and P5.

   a. Calculate the RLs of each point.

   b. What is the closing error of the survey?

BS	IS	FS	HPC	RL	Remarks
1.399				8.125	TBM1
	1.192				TBM3 (8.334 req)
	1.701				P1
	1.523				P2
1.538		0.275			CP1
	1.581				P3
	1.602				P4
	1.472				P5
		1.536			TBM5 (9.249 reqd)

8.18 HOW SHOULD I BOOK MY LEVELLING RESULTS?

When recording your levelling work in your field book:

1. each line in the book represents one staff position. You can have a foresight reading and a backsight reading to the same staff position

2. start each new job on a new page (even if you are just setting out one or two levels)

3. write down all relevant details including the date, drawing numbers, weather conditions and any other relevant notes

4. always write in pencil

5. if you need to switch the position of two numbers, write the number down in the new place before you rub out the old one

6. rub out or clearly cross out any numbers which are surplus to requirements

7. include a diagram if necessary

8.19 SHOULD I SET UP ON HARD OR SOFT GROUND?

There are advantages and disadvantages to both soft ground and hard ground. You should assess each unique situation and decide which is most appropriate.

Soft ground is more prone to ground movement, but the instrument is less likely to be knocked or blown over or the tripod tipping over.

On hard ground, there is no risk of ground movement, but there is a risk of the tripod feet slipping outwards.

8.20 HOW SHOULD I SET UP ON SOFT GROUND?

On soft ground, to obtain the maximum stability, the tripod feet should be spaced out so that they are at least 1m apart from each other. All three feet should be pressed into the ground as far as they will go by standing on the platform with all your body weight.

8.21 HOW SHOULD I SET UP ON HARD GROUND?

On hard surfaces such as tarmac or concrete, the feet should be placed close enough together that the weight of the instrument will not force the tripod to collapse if the feet lose their grip on the floor and slip outwards. Note that the spread of the tripod feet should not be so narrow that the tripod is unstable and can be easily knocked over sideways or blown over by the wind.

Once the tripod has been set up in position, before attaching the instrument, press down onto the head of the tripod with as much weight as possible to dig the points of the feet into the ground and to ensure that the feet do not slip outwards.

8.22 HOW SHOULD I SET UP ON A POLISHED SURFACE?

For setting up the tripod on a polished floor, there are various types of tripod stabiliser available. An example can be seen in Fig. 8-15. They may be rigid plastic or aluminium frames or fabric straps. Note that fabric straps are suitable for protecting your instrument from damage but are not necessarily sufficiently stable for preventing movement of the instrument for high accuracy purposes.

Fig. 8-15: Tripod stabiliser

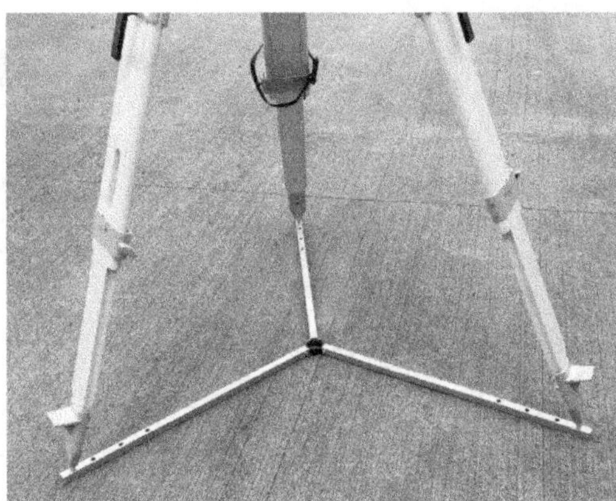

8.23 CAN I WORK IN THE RAIN?

Yes, levels and total stations are completely waterproof, so it is ok to use them in any weather conditions. After working in the rain, the instruments should be stored in a dry atmosphere with the box lid left open.

8.24 DOES THE CURVATURE OF THE EARTH AFFECT STAFF READINGS?

The line of sight is at a tangent to the surface of the Earth. Over short distances, we can assume that the line of sight is parallel to the datum. Sight distances in standard levelling (less than 50m) are short enough for this assumption to hold true.

For measuring level changes over any distance that would warrant Earth curvature corrections, a total station would be used.

8.25 WHY NOT USE THE SCALE ON THE BACK OF THE STAFF WITH MARKED GRADUATIONS?

The scale on the back of the staff is not intended for general levelling use. It might seem like it would be more accurate to use the graduations on the back of the staff, however the small lines are not detectable to the human eye over long distances. The back of the staff is more cluttered and difficult to read compared to the clear, bold, contrast red, white and black blocks. In addition, the arrangement varies from staff to staff: often the scale is inverted, with zero at the top of the staff rather than the bottom. An example of the graduations on the back of a levelling staff can be seen in Fig. 8-16.

Fig. 8-16: Graduations on a staff

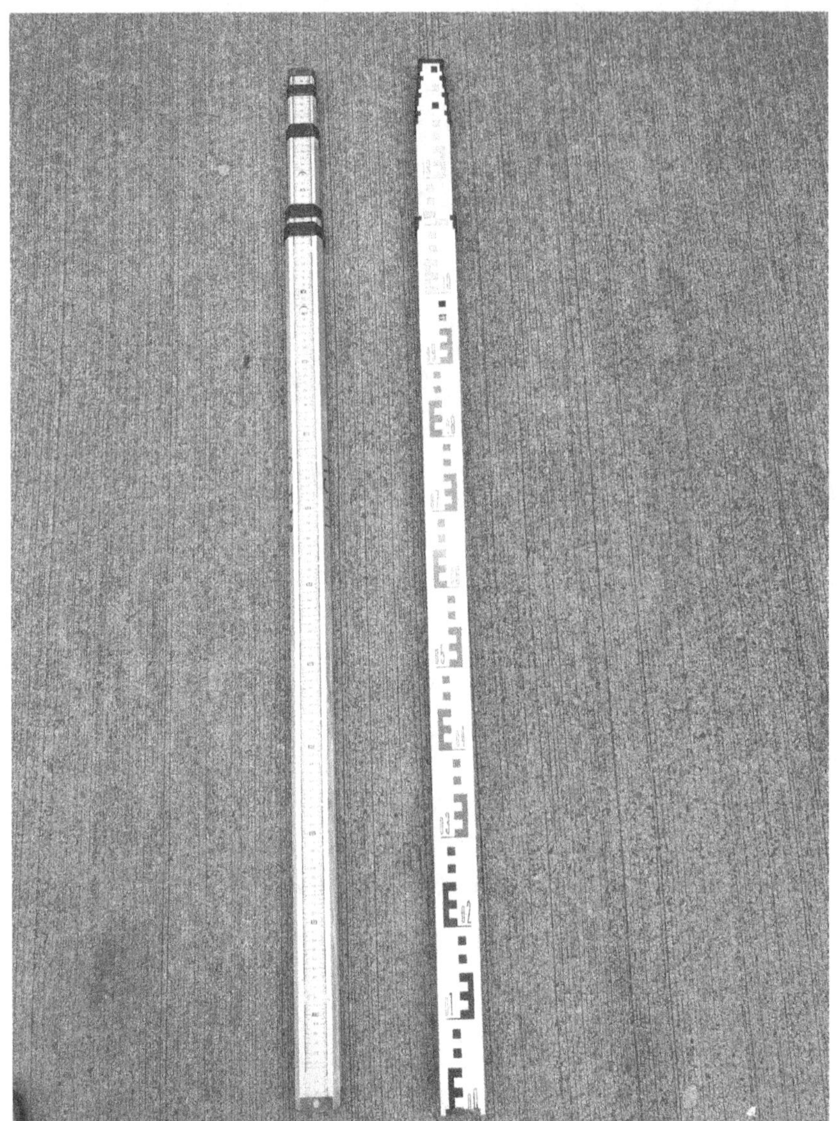

8.26 CAN I USE A TAPE MEASURE INSTEAD OF A STAFF?

In certain situations, it is appropriate to take a reading to a tape measure instead of a levelling staff, for example:

→ when the staff is too long and is obstructed by something overhead or below

→ when you are very close to the object you are sighting

→ when you have a vertical surface to rest the tape against, or verticality is not critical

→ when you are carrying out very accurate work, for example, setting out a cast-in pipe.

8.27 HOW DO I TAKE AN INVERTED STAFF READING?

An inverted staff reading, is a reading taken when the staff is inverted, in other words, held upside down with the zero end on the underside of a feature as shown in Fig. 8-17. This enables you to measure the reduced level of a point which is above the line of collimation for example, the soffit of a beam, the keystone of a bridge, or a ceiling. Using the same process as described in section **8.11**, Fig. 8-18 shows an inverted staff reading of 1.764m.

The bottom of the inverted staff can also be held flush with the top of a timber peg or slab, if you have accidentally set the height of collimation lower than the points you are measuring.

An inverted staff reading is taken in the same way as a normal staff reading and recorded in the same way as you would a basic level survey. The only difference is that you record an inverted staff reading with a negative sign. For an example of how to record an inverted staff reading, see Fig. 8-19.

Inverted staff readings should not be used as change points as it is difficult to rock the staff or to ensure that it has been positioned on exactly the right point.

To minimise staff verticality errors with inverted staff readings, have your assistant stand square onto the staff to the side of it rather than behind it. This means that they can visually check that it is vertical, and you can direct them to tilt it left or right to line it up vertically in the perpendicular direction.

Fig 8-17: Inverted Staff Reading

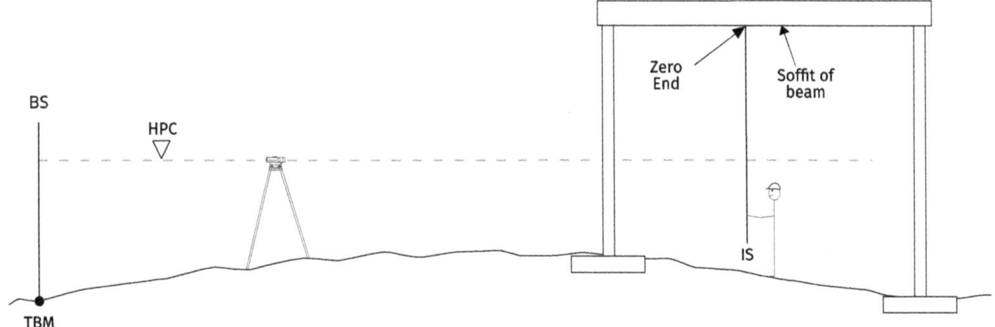

Fig 8:18: Inverted staff reading (Example)

Inverted Staff Reading 1.764m

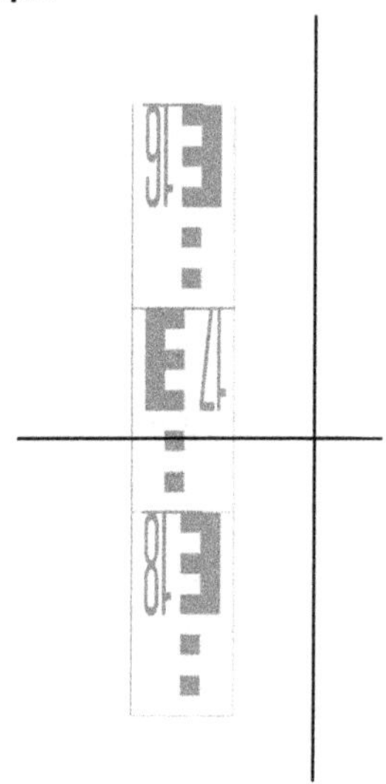

Fig. 8-19: Inverted Staff reading - Booking example

Date	21 August 2018		Taken for	Existing beam level	
From	TBM 13		To	TBM 7	

BACK SIGHT	INTER MEDIATE	FORE SIGHT	COLLIMATION OR H.P.C	REDUCED LEVEL	DISTANCE	REMARKS
1.526			104.802	103.276		TBM 13
	1.902			102.900		TBM 7 (102.899 reqd, 1mm error. OK)
	-2.488			107.290		Soffit of beam K1, approx. 1m from GLC
		1.525		103.277		TBM 1 (1mm error ∴ OK)

EXERCISE:

3. You have been asked to find the RL of the underside of the keystone of a masonry arch bridge. TBM1 = 32.155m, TBM2 = 32.205m. You take a BS to TBM1 of 1.765m, and IS to TBM2 of 1.714m. You then take an inverted staff reading of -2.597m. Finally you take a reading to TBM1 of 1.765m.

   a. What is the RL of the underside of the keystone?

   b. What is the closing error of the survey?

8.28 HOW DO I SET CONSTRUCTION ELEMENTS TO A FIXED LEVEL?

One of the most common tasks for site engineers is setting out levels to a given value. This is achieved by setting up the level and determining the height of collimation in the same way as a basic level survey. Then, instead of taking a staff reading to measure the level of something which exists, the staff is adjusted to the correct position to set out a new point. For an example of how to record your setting out in your field book, see Fig. 8-20.

The process is the same for the optical level and the laser level and is as follows:

1. Set up the level.

2. Take a BS to a TBM.

3. Calculate the height of collimation.

4. Take an IS to another TBM and check the measured value is the same as the expected. If so, continue.

5. Determine the required staff reading by subtracting the reduced level of the point you are setting out from the height of collimation. Record this in the IS column.

6. Direct your assistant to move the staff up or down against a vertical surface until the required staff reading is achieved.

7. Have your assistant make a mark at the bottom of the staff. This may be with a pencil on a flat vertical surface or with a strip of electrical tape round a steel bar. Ensure it is the point of the pencil in contact with the base plate as shown in Fig. 8-21 and not the side of the pencil as shown in Fig. 8-22. In the case of blinding, concrete or excavations where there is no vertical surface to make a mark, the staff reading is checked frequently until the correct level is achieved.

Here are some examples of levels which you might need to set out in this way:

- → cut-off heights for piles
- → formation level
- → blinding level
- → top of reinforcement
- → top of concrete
- → soffit level for formwork/falsework
- → cast-in pipes
- → box-outs for pipes
- → finished floor level

Fig.: 8-20: Setting elements to a fixed level

Date: 31 October 2018   Taken for: Pad founds A1 – H4, top of conc, 15.625

From: TBM 1   To: TBM 2

BACK SIGHT	INTER MEDIATE	FORE SIGHT	COLLIMATION OR H.P.C	REDUCED LEVEL	DISTANCE	REMARKS
1.189			16.689	15.500		TBM 1
	1.589			15.100		TBM 2 (15.100 reqd, 0mm error ∴ OK)
	1.064			15.625		Top of Base
		1.188		15.501		TBM 1 (1mm error ∴ OK)

Required staff reading

Fig. 8-21: Marking Under The Staff - Correct

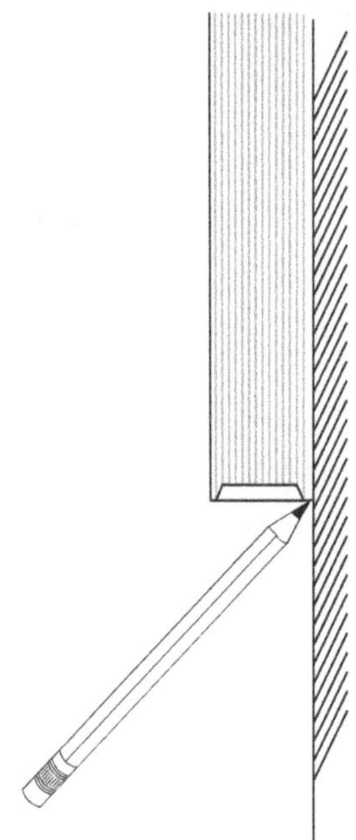

Fig. 8-22: Marking Under The Staff - Incorrect

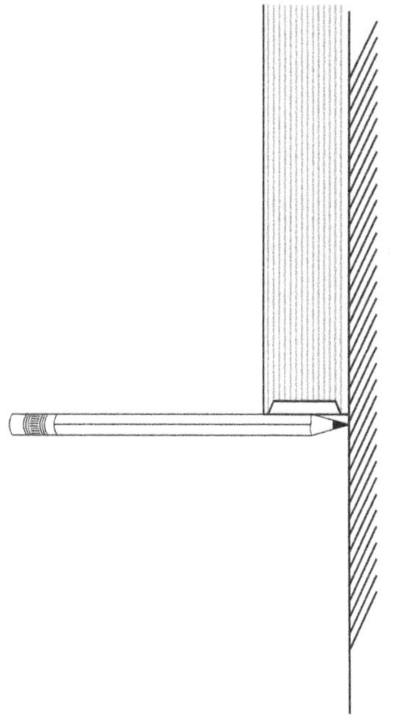

8.29 HOW DO I MARK A DATUM BY SETTING THE HEIGHT OF COLLIMATION AT THE CORRECT LEVEL?

Marking out a datum can be a quick and accurate way to set out levels. This involves making a series of marks or a continuous line at a fixed level and using it as a permanent Reference Line to measure from. This method is commonly used at the fit-out stage of buildings.

Once the datum has been marked, tradespeople such as ceiling fixers, floor fitters and joiners can then reference all their work from this line by measuring up or down from it with a tape measure. This method reduces the chance of gross errors and cumulative errors as all the different elements are set out from a single Reference Line.

You can set the datum by having the staff moved up or down until it is at the required staff reading, however, for multiple marks it may be quicker and more accurate to set the height of collimation in line with the datum and have your assistant make a mark in line with the cross-hair as follows:

1. Choose a datum level at a convenient height and round number, for example 1m above finished floor level.
2. Use conventional levelling to make a small clear mark at the correct level.
3. Set up the level by trial and error until the crosshair is on the mark made in step 2.
4. Have your assistant place the tip of the pencil on a vertical surface such as a wall or column, and direct them up or down until the tip of the pencil is in line with the horizontal cross-hair. Label the mark clearly, for example FFL + 1m.
5. Repeat step 4 at appropriate intervals along walls, allowing for the marks to be connected using a chalk line. On columns, make a mark on one face and if necessary, transfer the datum to the other faces using a spirit level.
6. An alternative to this method is to set up the level at a random height, calculate the HPC, make a temporary mark at this point and then use a tape measure to mark the required datum in relation to the mark.

8.30 WHAT ARE STADIA HAIRS?

There are a range of possible arrangements of the cross-hairs in a level, as shown in **figure 8-4**. Whatever the arrangement of the cross-hairs, there is always a horizontal cross-hair which extends the full width of the reticule as well as two short cross-hairs equal distances above and below it (Fig. 8-23). These short cross-hairs are known as stadia hairs and are designed for use in stadia hair tacheometry.

The purpose of these stadia hairs is to enable a horizontal distance measurement from the centre of the level to the staff position.

Fig. 8-23: Stadia hairs

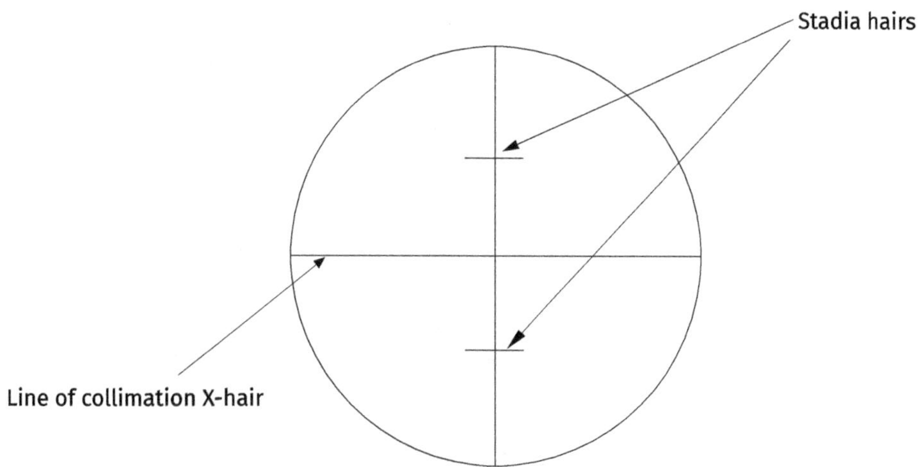

8.31 WHAT IS STADIA HAIR TACHEOMETRY?

Stadia hair tacheometry is an obsolete method of carrying out a topographical survey. The principle is that if the position of the level is known, relative angles can be measured from a reference object using the 360-degree scale on the vertical circle, distances can be calculated using the staff readings to the stadia hairs and the level can be measured by taking the reading to the main horizontal cross-hair.

8.32 HOW DO I MEASURE A HORIZONTAL DISTANCE USING THE STADIA HAIRS?

Due to the availability, convenience and accuracy of the total station, stadia hair tacheometry is no longer used for topographical surveys, however, the stadia hairs can be a handy way of quickly measuring a horizontal distance from the level to the staff. This method is useful if you don't have access to a total station and a tape measure is not suitable (for example, if the distance you are measuring is longer than your tape measure or if there is a void or obstruction between the points you are measuring).

To measure a horizontal distance using the stadia hairs (Fig. 8-24):

1. Take staff readings to the top stadia hair, S1, and the bottom stadia hair, S2.
2. Calculate the difference between them which is known as the staff intercept value, S, where S = S1 − S2.

3. Calculate the distance D, where D = stadia constant x S. In most levels the stadia constant is 100, however you should calibrate this for your own level against a baseline.

Fig 8-24: Measuring Horizontal Distance Using the Stadia Hairs

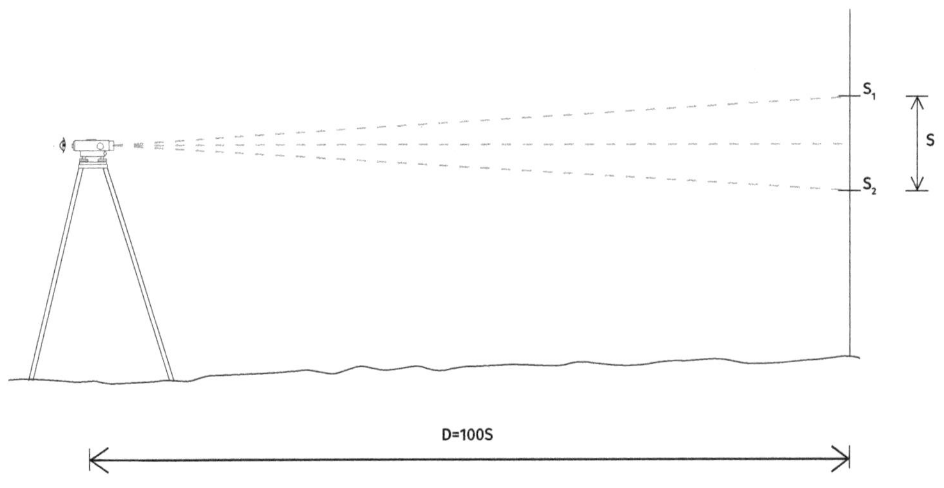

EXERCISE:

4. You need to know an approx. distance between two manholes. You set up over one of the manholes and have your assistant hold the staff on the other manhole. The staff reading to the top stadia hair, S1, is 1.883 and to the bottom stadia hair, S2, is 1.307. For a level which has a stadia constant, S, of 100, what is the distance between the manholes?

8.33 HOW ACCURATE IS STADIA HAIR DISTANCE MEASUREMENT?

This method is only accurate to about 20 or 30mm. Before you use this method, you should calibrate your accuracy against a baseline.

8.34 WHAT IS RECIPROCAL LEVELLING?

This is an obsolete method of levelling. Traditionally, this method would be used to determine the accurate level difference between two points where it is not possible to set up at the midpoint (because of an obstruction or void).

However, the generation of levels in common use today are, if appropriately checked and calibrated, sufficiently accurate for any standard site task. Today, total stations are readily available where greater accuracy is required over longer distances. See section **11.61** using the total station for levels.

8.35 WHAT DIFFERENT TYPES OF LEVEL ARE THERE?

In this book, the term 'level' refers to the optical type described above. You may also hear the optical level referred to as an engineer's level, dumpy level, site level, builder's level, automatic level or auto level. In broad terms, it refers to the instrument whose telescope you sight through to take a reading to a levelling staff.

8.36 WHAT IS AN AUTOMATIC LEVEL?

An automatic level is an optical level which incorporates a mechanism called a compensator. The compensator automatically sets the line of collimation to the horizontal once the user has centred the pond bubble.

8.37 WHAT IS A DUMPY LEVEL?

The dumpy level is an obsolete piece of equipment; however, the term 'dumpy level' has been retained. When someone refers to the dumpy level on site, it is safe to assume that they are referring to the automatic optical level.

You may come across the term 'tilting level', which is another obsolete piece of equipment.

8.38 WHAT IS A DIGITAL LEVEL?

The digital level operates on the same principle as the optical level, but instead of the user sighting through the telescope and manually reading the levelling staff, the instrument reads automatically to a barcoded staff.

Digital levels are not in common use on construction sites these days since automatic optical levels and total stations can be used to measure levels sufficiently accurately for most standard setting out tasks.

For setting marks or elements to required reduced levels, the optical level is more suited to the task as the level can be moved up or down until the correct reading is achieved.

8.39 HOW ACCURATE IS A DIGITAL LEVEL?

Although some digital levels achieve an accuracy similar to standard automatic levels, some do have the capability to read the barcoded staff to an accuracy of 0.1mm. Note that this reflects the accuracy of the of staff reading only and does not take into account any other errors such as staff verticality, collimation error or the positioning of the staff on the point.

8.40 WHY USE A DIGITAL LEVEL INSTEAD OF AN OPTICAL LEVEL?

Some drawbacks of the digital level, as opposed to the standard optical level are that they require battery power, which the optical level does not and that they are more expensive to hire or purchase. They are more suitable for use in measuring reduced levels than setting out and they can be difficult to use in low light or extremely bright conditions

In addition, unlike the total station, they do not provide a relative position which means a full topographical survey cannot be produced. For surveys with a digital level, it is necessary to sketch the layout of the points being measured or to add the levels to the correct locations on a pre-prepared plan or drawing, therefore a total station is more suitable for topographical surveys

8.41 WHAT ARE THE DRAWBACKS OF THE DIGITAL LEVEL?

Some drawbacks of the digital level, as opposed to the standard optical level are that they require battery power, which the optical level does not and that they are more expensive to hire or purchase. They are more suitable for use in measuring reduced levels than setting out.

Fig 8-25 shows a digital level and Fig 8-26 shows a barcoded staff. (*Images courtesy of Leica Geosystems.*)

Fig 8-25 Digital level

Fig 8-26 Digital level staff

8.42 WHAT IS PRECISE LEVELLING?

For levelling tasks requiring the highest order of accuracy, for example checking bridge bearings for flatness, a precise level should be used.

The precise level is a specialist piece of equipment which can achieve a double run value of 0.7mm on its own and 0.3mm with the use of a parallel plate micrometer.

A high-precision staff known as an 'Invar staff' is used in precise levelling. Invar is the tradename for the material used which is not susceptible to expansion and contraction. This type of staff has handles on the sides, a bubble to control the verticality and may also have legs for extra stability.

The precise level is used for specialist surveying rather than general construction-related setting out tasks.

8.43 WHAT IS A LASER LEVEL?

Rather than providing a horizontal line of sight through a telescope, the rotary laser level sends out a laser beam projected from a high-speed rotating disc. This high-speed rotating radial beam creates a laser plane.

A receiver comprising a visual display screen, emits a loud intermittent beeping sound. The frequency of the beep increases as the receiver moves closer to the line of collimation. An arrow on the display screen indicates whether the receiver is below or above the line of collimation.

There are several types of laser level available. The type referred to in this chapter is the robust and durable rotating laser level used outdoors on construction sites (Fig. 8-27) *(images courtesy of Leica Geosystems)*. It can be used in all weather conditions and is suitable for tasks such as:

→ earthworks

→ road construction

→ blinding concrete

→ drainage

The rotating laser level emits a laser beam rotating at a high speed, which creates a laser plane at the height of collimation.

A receiver attached to the levelling staff emits a beeping sound when it receives the laser signal. Different beep frequencies and an arrow on the receiver screen indicate whether the receiver is higher or lower than the height of collimation.

The receiver is attached to the levelling staff with a clamp and can be adjusted to the correct position on the staff. The receiver can be set to two different levels of sensitivity depending on the accuracy needed.

Most modern laser levels are self-levelling. The self-levelling mechanism works once the instrument is set up and levelled using the pond bubble by adjusting either foot screws or the tripod legs.

Most laser levels can be set to a horizontal plane or a fixed grade in a longitudinal or transverse direction.

Fig. 8-27 Laser level

Fig. 8-28 Laser level receiver

8.44 HOW DO I USE THE LASER LEVEL?

For a horizontal line of collimation, the laser level operates on the same principle as the optical level. The only difference is that instead of sighting through a telescope to take a staff reading, the receiver is moved up and down the staff until it is coincident with the line of collimation. In every other way, the laser level is, to all intents and purposes, the same as the optical level. It can be used for:

→ measuring reduced levels

→ setting construction elements to a fixed reduced level

→ measuring a relative height difference between two points

Any measuring or setting out carried out using the laser level should be recorded in your yellow field book in the same way as you would record any levelling carried out using the optical level. See sections **8.14** and **8.28**.

8.45 DO I NEED TO CARRY OUT A TWO-PEG TEST ON MY LASER LEVEL?

As with the optical level, the likelihood of your instrument developing an unacceptable collimation error depends on its quality and robustness. Unless your equipment comes with a guarantee that it will not go out of adjustment, you should carry out the two-peg test at intervals specified by your employer's policies, relaxing the regime only if you have solid engineering justification.

8.46 WHAT ARE THE ADVANTAGES AND DISADVANTAGES OF THE LASER LEVEL?

The laser level is an excellent tool when it is used correctly and for appropriate tasks. Like anything, it has its pros and cons. Here are a few to consider:

Advantages:

→ it is quick, convenient and sufficiently accurate if used by an appropriately trained person

→ it saves having to set up profile boards, screed rails, blinding pins etc.

→ it can be used by only one person and doesn't require an assistant

→ it can be used for gross checks on work which has been set out using an optical level.

Possible pitfalls:

→ there is scope for inaccuracy or gross error if the operator is not properly trained

→ the accuracy of the laser level is less than that of an optical level, so the range of tasks it is suitable for is limited

→ the laser beam can cause damage to eyesight

→ many people do not record their setting out in the correct way when using a laser level

→ untrained operatives may use an incorrect as-built level as a reference point for the next stage of work, leading to cumulative errors

→ untrained operatives may not be aware of possible errors such as staff verticality

→ there may be several laser levels set up on the same site and receivers pick up signals from all lasers.

8.47 CAN I USE A LASER LEVEL FOR STRUCTURAL CONCRETE SLABS?

A laser level can be used for large area concrete slabs with a tolerance of +/-5mm. The task should be supervised by someone who is properly trained and ideally, the laser level should be used in conjunction with marks on shutters and starter bars.

CHAPTER 9

ERRORS IN LEVELLING

9.1 WHAT ARE THE SOURCES OF ERROR IN LEVELLING?

Although the principle of levelling is relatively simple, there are many things that can go wrong if you are not aware of the potential pitfalls.

This section is designed to provide you with the information you need to:

→ develop good habits which will reduce the risk of error

→ become aware of your own particular areas of weakness

→ efficiently troubleshoot when things do go wrong (use the process of elimination to identify the source of an error)

→ train your assistant

→ mentor less experienced site engineers

→ speed up your learning curve

→ fully understand how the level works. It may be that there are some errors you personally may never make in practice; however, it is essential to understand where the possibilities exist and how you can minimise the risk of them occurring.

Sources of error in levelling include:

→ problems with the equipment

→ the equipment not being levelled correctly

→ misreading the staff

→ the staff not being vertical

→ movement of the level during use

→ using the wrong TBM or the wrong value for your TBM

→ calculation errors

→ recording errors

→ taking the wrong setting out information from the drawing.

9.2 WHAT ARE THE TYPES OF ERROR IN LEVELLING?

Errors in levelling can be broken down into three main categories:

→ equipment errors

→ operational errors

→ other errors

No matter what category an error falls into, if it goes undetected it is 100% due to human error. In theory, there is no error that cannot be brought to within acceptable limits by proper checking and due diligence. However, things do go wrong. By using best practice and building in the relevant checks, you will be able to detect errors at the earliest possible stage and rectify them before any harm is done.

9.3 WHAT ARE RANDOM ERRORS IN LEVELLING?

Random errors in levelling include:

→ human error in reading the staff

→ the positioning of the staff on the point

→ tiny movements in the clips of the staff

→ variations in the line of collimation between different set-ups

Reading errors are the random errors inherent in reading the staff. They occur because there are variations in the way the staff can be read. For example, you may be unsure whether the reading should be 1.536 or 1.537. Reading errors cannot be eliminated completely but can be minimised to within acceptable limits.

If left unchecked, random errors can accumulate and lead to unacceptable errors in your levelling. Here are a few tips to help minimise the effects of random errors:

→ **Practise** - the more experience and practise you have, the smaller this error will be.

→ **Calibrate your readings against known points** - check that you are getting the answers you expect against two known points. If you don't have any reliable TBMs available, choose any two points and take several sets of readings, setting up the level at a slightly different height each time. Practice until you consistently get the same difference between backsight and foresight each time.

→ **Don't round** - always read to the nearest mm as best you can. Don't be tempted to round your reading to the nearest 0, 5 or 10mm.

→ **Don't switch users** - the same person should take all readings from one set-up. If a user tends to always take the lower end of the possible range of readings or to always take the higher end of the possible range of readings, this will have a cancelling effect and the effects of reading errors will be minimised. The effect will be compounded if one user tends to read downwards and one tends to read upwards.

→ **Take more readings** - statistically, the more independent readings you take, the closer the average will be to the true value. For critical tasks, repeat the readings several times from different positions. Obtain an independent check from a third party.

→ **Avoid long sight distances** – for high accuracy work, keep sight distances lower than 30m.

→ **Minimise the number of change points** – the effects of staff reading errors are compounded at each change point. Consider the balance between the right number of change points vs the length of site distance.

→ **Avoid poor visibility** – rain, fog, lack of daylight, high temperatures and vapours from machinery can all impact on the quality of your staff readings.

9.4 WHAT IF THE BASE PLATE OF THE STAFF IS MISSING?

Through wear and tear, the base of the staff can become loose or detached altogether.

If the base plate of the staff is missing, then the outer edge of the staff should be in contact with the highest point of the nail as shown in Fig. 9-1 and not in contact with the surrounding ground as shown in Fig. 9-2. This consistency will cancel out any errors as it is the relative height difference between points which is being measured. In other words, it doesn't matter what the value of the starting point of the staff is, so long as the exact same point is used each time.

The recess in the bottom of the staff should not be placed over the mark.

Fig. 9-1: Staff Base Plate Missing - Correct

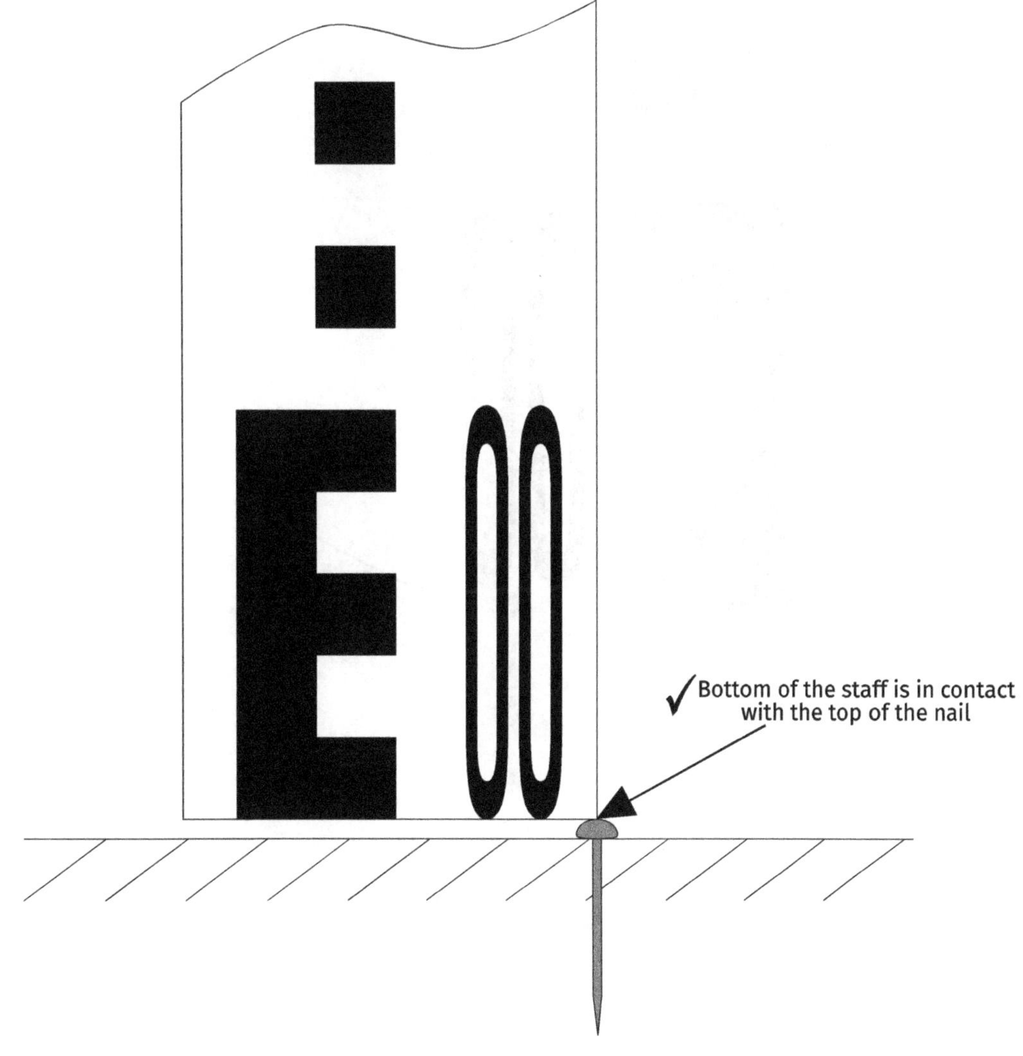

Fig. 9-2: Staff Base Plate Missing - Incorrect

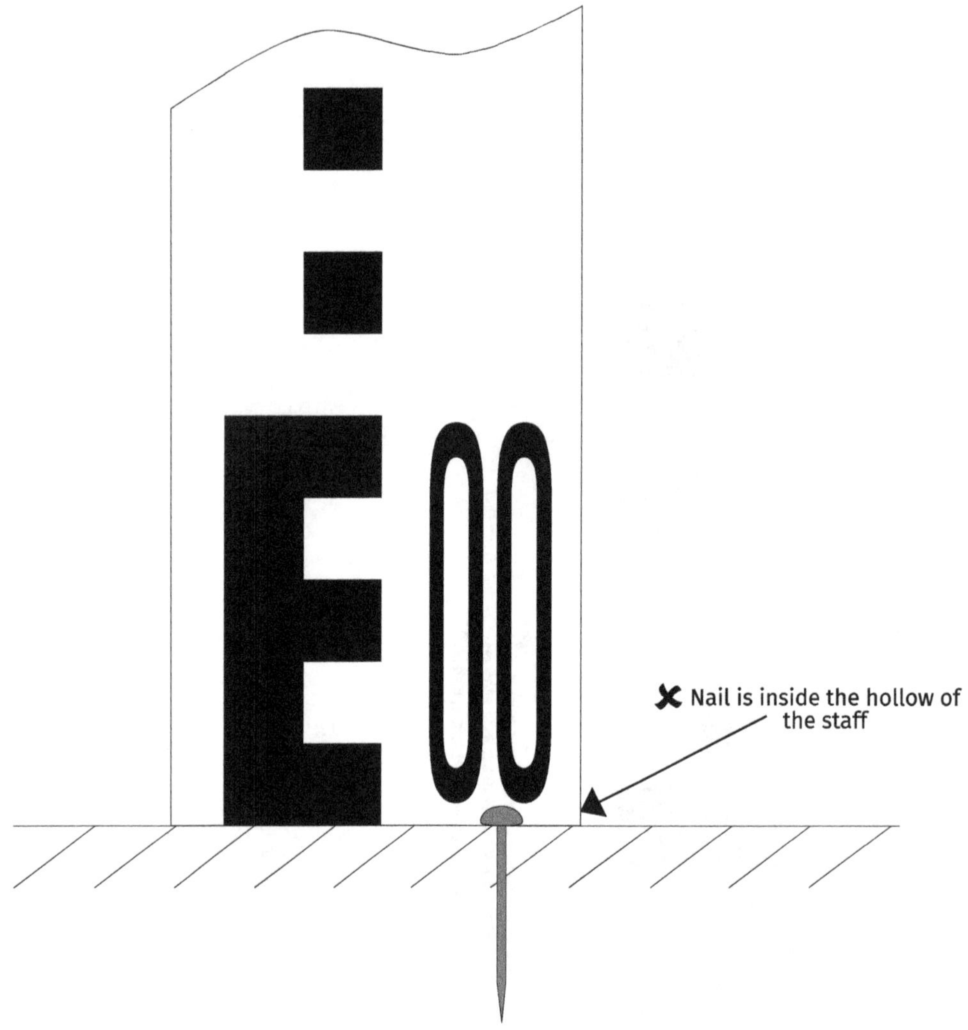

9.5 WHAT IF THE BOTTOM OF THE STAFF IS DAMAGED OR DIRTY?

The bottom of the levelling staff should be a secure, flat metal base and should be free from any dirt and debris.

Dried-on mud or concrete can become attached to the bottom of the staff. Any such dirt or obstructions will result in a differential error and should be removed before using the staff.

To minimise the risk of this error, train your assistant to check that the bottom of the staff is in good condition and is free from dirt or concrete before each reading is taken and cleaned before putting the equipment away.

9.6 HOW CAN I AVOID DAMAGING THE BASE OF THE STAFF WITH CONCRETE?

One of the most common causes of damage to the bottom of the staff, is the staff being used for the receiver of a laser level for setting the level of blinding or concrete base slabs. For this particular task, it is necessary for the base of the staff to come into contact with the freshly poured concrete in order to check that the concrete has reached the correct level.

One way to avoid this type of damage is to use a length of timber instead of a staff. Rather than placing the receiver on the staff at the correct position, measure from the bottom of the piece of timber and make a mark at the correct height on the timber where the receiver should be set. Once the pour is complete, the timber can be discarded, or the bottom sawn off to create a fresh surface for its next use.

If you do not have a suitable piece of timber, ensure that one of the operatives is responsible for frequent cleaning of the bottom of the staff so that concrete does not get a chance to set on the base plate.

9.7 WHAT IF THE CLIPS ON THE STAFF ARE DAMAGED?

The clips which lock the sections of the levelling staff into position can become worn out, damaged or clogged with dirt or concrete.

When the staff is extended, the clips should click into position and there should be no movement or 'waggle'.

Train your assistant to listen for the click and visually check that the clips have clicked properly into place as well as testing the staff for small movements.

If the clips break altogether, the staff will still work in the same way if you remove the bottom section and discard it. The staff will not start at zero, however this does not matter as levelling is based on relative measurements.

Don't necessarily throw the damaged section away as it might come in useful for something else e.g. the very quick two-peg test (see section **10.13**)

In the case of removing the bottom-most section of staff, there will be no flat base plate on the bottom. See section **9.4** on how to use the staff correctly in this situation.

9.8 WHAT IF THERE IS CONDENSATION WITHIN THE LEVEL?

If the level is left in a closed box after being used in wet weather conditions, condensation will be forced inside the instrument. If this happens, when you sight through the telescope at the staff, it will appear misty or foggy. This can make the staff either difficult or impossible to read.

To avoid risk of small reading errors or gross errors (it may not even be possible to read the numbers on the staff), do not use the level when it is in this condition. Take the level out of the box and leave it in a warm dry room for up to 24 hours until full visibility has been restored.

It may be worth keeping a spare level on site in case this does happen.

9.9 WHAT IS A STAFF MISREADING ERROR?

A staff misreading error is a gross error. It is basically a mistake made when reading the staff or recording the measurement in your field book. The misreading may be to one of the points you are surveying or setting out, or it may be in your backsight reading. If the error is in your backsight reading, every point you survey or set out from that set-up will contain the error.

Misreading errors can be caused by:

→ getting the numbers mixed up when you are reading the staff

→ reading the staff correctly but recording it incorrectly

→ not knowing how to read the staff properly

→ getting the decimal point in the wrong place

→ counting the wrong number of 10mm blocks

→ taking a reading to one of the stadia hairs rather than the line of collimation.

9.10 HOW CAN I REDUCE THE RISK OF MISREADING THE STAFF?

There are a number of precautions you can take to minimise the risk of misreading the staff:

→ **Start with two TBMs** – once you have taken your backsight to your TBM, take an intermediate sight to a different TBM. Calculate the reduced level of the second TBM and check it against its expected value. In effect you are comparing the measured difference in height to the expected difference in height between the two points as illustrated in Fig. 9-3. This will detect any gross errors in the backsight reading.

→ **repeat critical readings** - repeat critical measurements by setting up the level at a different height and repeating the readings.

→ **check readings in the bottom 100mm of the staff** – putting the decimal point in the wrong place tends to be much more common when taking readings at the very bottom of the staff. For example, an inexperienced engineer may read 0.008 as 0.080, or 0.046 as 0.460. Take extra care, or avoid taking readings in the bottom 100mm of the staff altogether. If you are unsure, verify the reading by checking it against another known point.

→ **do your calculations as you go along** – calculate your reduced levels as you go along. You will be much more likely to spot if you have misread the staff. Don't wait until you are back at your desk!

→ **carry out visual checks** – check that the reduced levels look right relative to the features you're surveying. For example, the crown of the road should be higher than the channel.

→ **perform random checks** - if you are prone to misreading the staff, carry out random checks every few readings by checking back to a reading you took a few minutes ago.

→ **double check** - once you have read the staff and recorded it in your book, look back through the telescope and check it again.

→ **place your eye close to the eyepiece** – this will reduce the chance of reading the wrong cross-hair.

→ **seek an independent check** - get another engineer to verify your results by carrying out the task from scratch and proving or disproving your answers.

Fig. 9-3: Checking Two TBMs

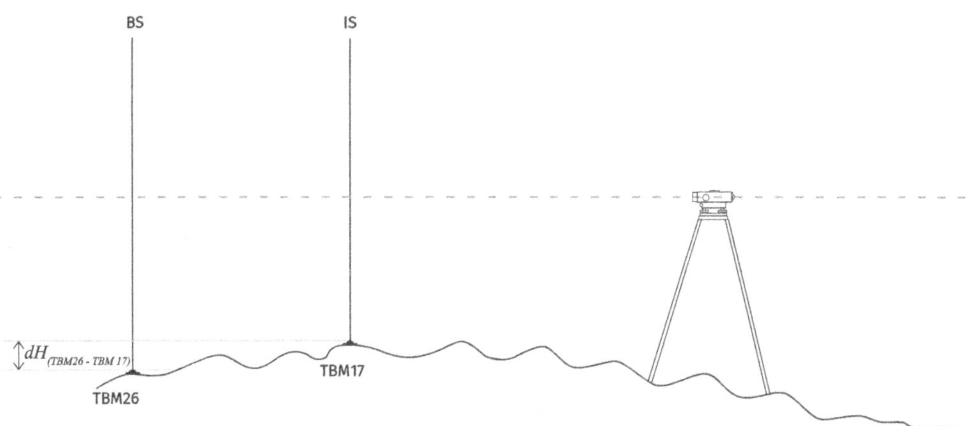

9.11 WHAT IS A PARALLAX?

Parallax is the apparent difference between the cross-hair and the true reading. When the eye is below the line of collimation, the cross-hair may appear higher up the staff and when the eye is above the line of collimation the cross-hair may appear lower down the staff as shown in Fig. 9-4. In theory, when the level is in perfect focus the effect of parallax should be zero. To minimise the effects of parallax, ensure that the cross-hairs are crisp, sharp and black and that the staff is in perfect focus. In addition, place the eye so it is virtually in contact with the eyepiece. The further back the eye from the eyepiece, the greater the effects of parallax. If you wear glasses, it is best to remove them before focussing the staff and cross-hairs.

Fig 9-4: Parallax - Eye Too Far Back

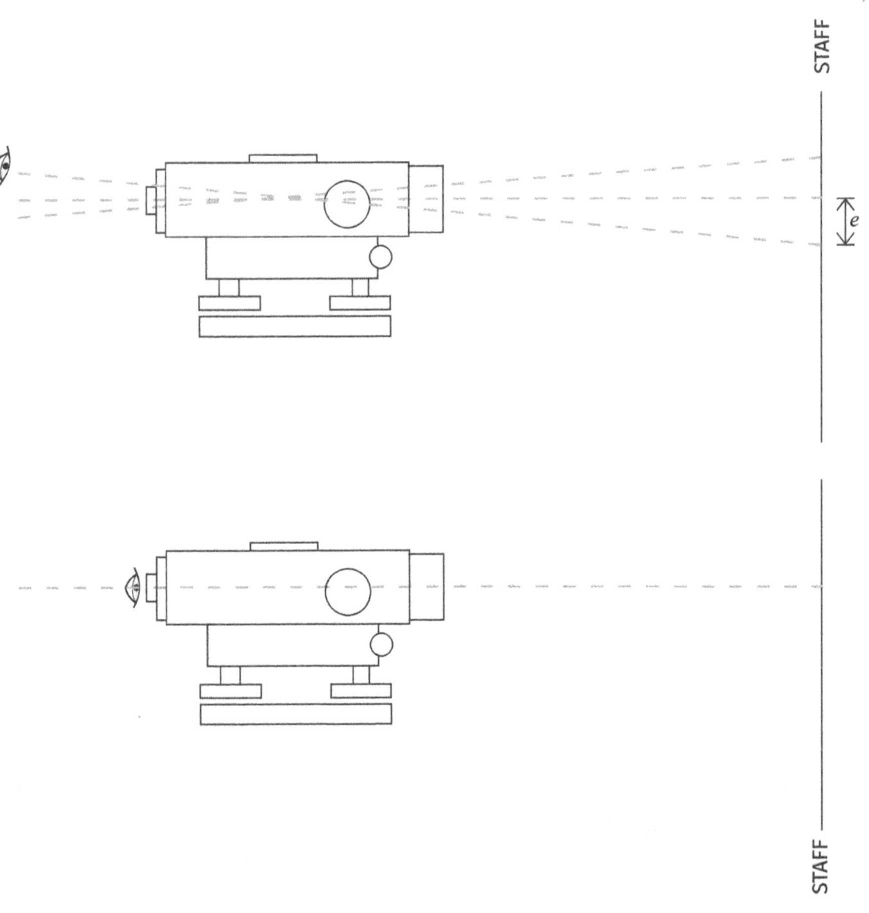

9.12 WHAT IS A STAFF VERTICALITY ERROR?

This section on staff verticality may seem disproportionately long for a seemingly simple aspect of levelling. However, the depth and detail in this section is warranted as staff verticality is one of the most common and largest sources of error. Lack of awareness or attention can mean a lot of hard work in eliminating all other sources of error is completely wasted.

The principle of levelling assumes that the staff is perpendicular to the line of collimation. If the line of collimation is horizonal, then the staff must be vertical.

If the staff is not vertical, then the staff reading will not reflect the true value; it will incorporate an error. When the staff is vertical, the distance from the bottom of the staff to the line of collimation is at its shortest possible distance. Any deviation from the vertical will mean the staff reading is larger than the true value, whether the staff is leaning forwards or backwards. See Fig. 9-5.

Since the assistant is standing directly behind the staff and the observer directly in front of it, even if the staff is significantly tilted backwards or forwards, neither will be able to tell.

Fig. 9-5: Staff Verticality Error

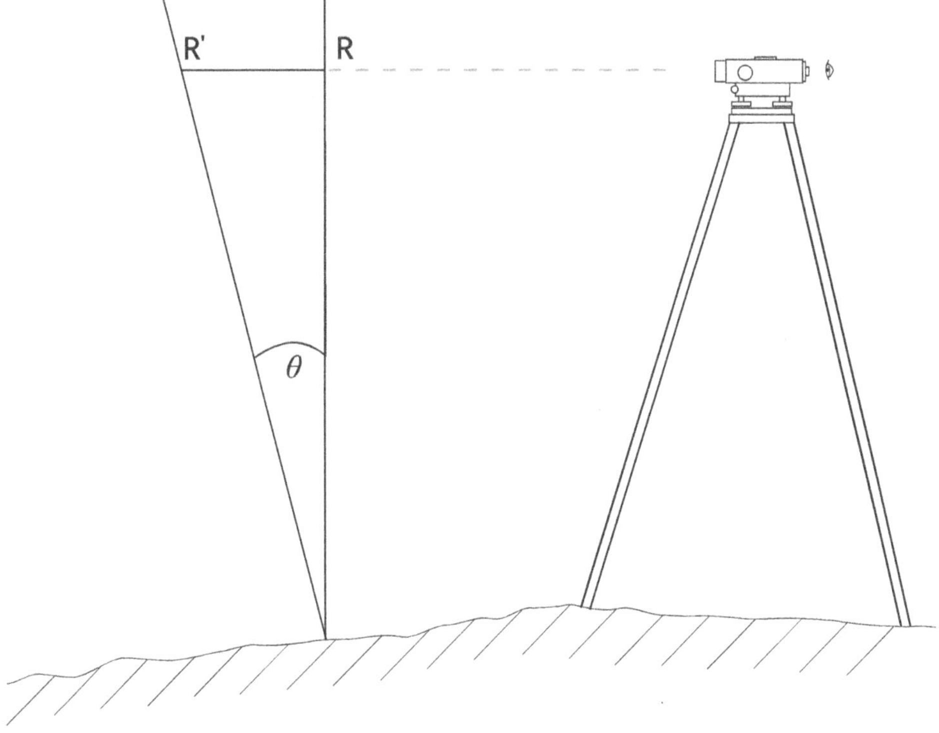

9.13 HOW CAN STAFF VERTICALITY ERROR BE ELIMINATED?

To eliminate the staff verticality error, the reading should be taken at the minimum point. The assistant should move the staff backwards and forwards by bending and straightening their arms. The staff reading will appear to increase and decrease. The maximum reading will vary depending on how far forwards or backwards the staff is tilted. The minimum reading however, will be consistent. It will be exactly the same each time the staff passes through the vertical position. This minimum reading, the lowest point on the staff that the cross-hairs go down to, is the value which should be recorded.

When your assistant is not ready for you to take a reading and the staff is not being rocked, they should turn the reading face of the levelling staff towards them and tilt it over so that there is no chance you can accidentally take a reading when they are not ready for you to do so.

9.14 WHAT IS THE LIKELY MAGNITUDE OF A STAFF VERTICALITY ERROR?

Mathematically, the limit of the staff verticality error is the top of the staff. The error varies depending on the tilt of the staff at the moment the reading is taken.

Realistically though, the staff is not likely to be much more than 5 or 10 degrees off the vertical without either the engineer or their assistant noticing that something is amiss.

If the staff is, say, 5 degrees off the vertical as per the diagram below, then a vertical distance of 1.658 would translate into a staff reading of 1.658/cos5 = 1.664. The error (i.e the difference between the true value and the value incorporating the error) is 1.664 − 1.658 = 0.006m, i.e. 6mm.

Bearing in mind that for tasks such as establishing control points, we are aiming to achieve an accuracy of less than 1mm per reading, it would be senseless to allow an easily preventable error of 6mm to wipe out any effort spent eliminating other sources of error.

9.15 WHAT IS THE CORRECT WAY TO ROCK THE STAFF?

It is up to you as the engineer to communicate to your assistant, the correct way to rock the staff. It is best to demonstrate this to your assistant rather than to explain it verbally. Explain to your assistant that the accuracy of the reading will be affected if they do not do it correctly.

Instructions are as follows:

1. Stand directly behind the staff.

2. Position the base of the staff on the TBM or feature to be measured.

3. Grip the staff with both hands at chest height. If only one hand is used, the staff may appear to wobble or wave from side to side which may affect the accuracy of the reading.

4. Place your feet such that your elbows are at your side and the staff is roughly vertical.

5. Tilt the staff back until it touches your forehead.

6. Tilt the staff forward until your arms are fully outstretched.

7. Repeat these two steps continually in a fluid motion, each repetition taking between 1 and 2 seconds.

When training your assistant, stand to the side of them to ensure that the full range of movement is achieved. Ensure that the staff is passing fully through the vertical and not stopping short of it. The movement should be exaggerated rather than small, otherwise to the person looking through the level, it is hard to tell whether the staff is being rocked properly or just wobbling about.

9.16 WHAT IF IT IS NOT POSSIBLE TO ROCK THE STAFF?

It is not always possible to rock the staff, for example if there is an obstruction behind the staff or the feature to be measured is above the line of collimation.

In this situation, have your assistant stand square on to the side of the staff. From this position they can see whether the staff is vertical along the line of sight. The person looking through the level should direct the assistant to adjust the top of the staff (or the bottom if it is an inverted staff reading) left or right. In this way the staff is plumbed in both directions.

When the staff is not being rocked, if you are unsure about the exact value of the reading, round down rather than up. This habit will mean your reading will always be closer to the true value. Be aware that it will not eliminate the error completely; the only way to do this is by rocking the staff through the vertical.

Brand new staffs usually come with a pond bubble which clips onto the back of the staff. If the bubble is available, this will help to ensure staff verticality when it is not possible to rock the staff.

9.17 WHAT IS REFRACTION?

Refraction is the phenomenon whereby a ray of light bends as it travels close to a parallel surface. In the context of levelling, the light ray is the line of sight and the parallel surface is the surface of the earth. See Fig. 9-6.

For general-purpose levelling on construction sites, it is safe to assume that the effects of refraction will be negligible.

As a general rule of thumb however, for levelling tasks which require high accuracy, aim to keep staff readings above 0.5m. Where this is not possible, keep sight distances balanced (equal distance to the backsight and foresight positions) and less than 25m in length.

Taking a staff reading close to the bottom of the staff is not a problem if it is on a raised feature such as the top of a long peg, fence post or outer edge of a concrete slab.

Fig. 9-6: Refraction

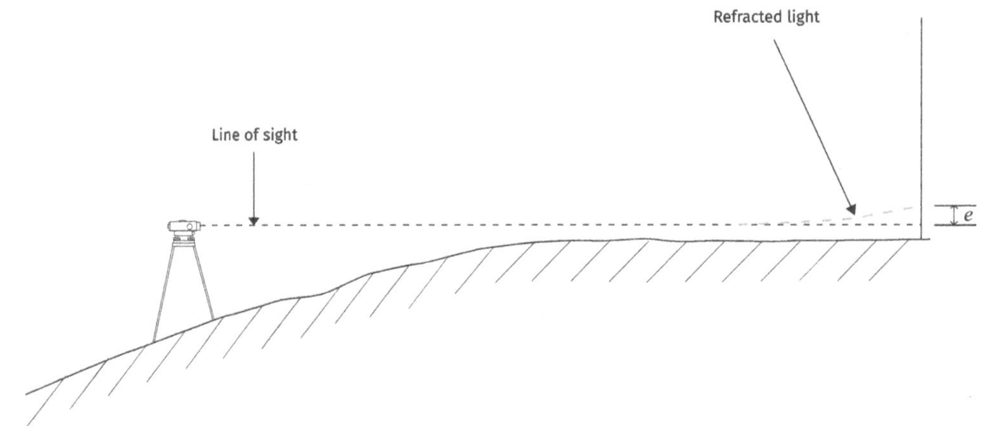

9.18 HOW CAN I MINIMISE THE RISK OF ERROR CAUSED BY THE LEVEL MOVING DURING USE?

There are several reasons why the level might move once it has been levelled:

→ ground movement

→ feet sinking into the ground

→ clips on tripods legs slipping

→ kicking or knocking the tripod legs

To increase the likelihood of detecting movement if it does occur, it is good practice to frequently check back to your TBMs. To minimise the risk of movement occurring, here are some tips:

→ **check the bubble before every reading** – get into the habit of checking the bubble before each reading, the same way you would unconsciously check in the wing mirror of your car before pulling out into traffic

→ **check that tripod leg clips are working properly** – set up the tripod on a hard surface and press down on the tripod head over each leg

→ **avoid soft ground** – avoid setting up on muddy or boggy ground, particularly if it is susceptible to vibration from construction machinery or traffic

→ **avoid tarmac on hot days** – in high temperatures tarmac softens. The self-weight of the tripod and instrument can create sufficient pressure on the pointed tripod feet for them to sink

→ **don't step near the feet or legs** - get into the habit of not stepping over the legs but taking exaggerated steps around them giving plenty of clearance. Ask other people to do the same. This not only reduces the risk of the ground surrounding the feet sinking under the weight of the person but also reduces the risk of the legs being kicked or knocked.

9.19 HOW CAN I MINIMISE THE RISK OF USING THE WRONG VALUE FOR A TBM?

Using the wrong value for your TBM means that no matter how accurate your readings, all your measurements from that set-up will be wrong. Here are a few scenarios which could lead to this happening:

→ you may be given incorrect TBM information

→ you may transpose figures when writing down the TBM in your field book

→ benchmark values may be reassigned to accommodate design changes or elements which have been set out incorrectly

To minimise the risk of using the wrong benchmark information, there are several measures you can take:

→ **start with two TBMs** – once you have taken your backsight to your TBM, take an intermediate sight to a different TBM. Calculate the reduced level of the second TBM and check it against its expected value. This will detect any gross errors in the TBM

→ **use the Survey Management File** - avoid passing TBM information directly between engineers. All TBM data should be held centrally in the Survey Management File. A proper system of communication should be in place for alerting engineers to any updates.

9.20 HOW CAN I MINIMISE THE RISK OF USING THE WRONG TBM?

The wrong TBM may be used if:

→ you are aware of the correct position but you do not communicate this clearly to your assistant and they hold the staff on the wrong physical point.

→ you misinterpret the location of the TBM and send your assistant to the wrong point.

To minimise the risk of the wrong TBM being used:

→ **start with two TBMs** – once you have taken your backsight to your TBM, take an intermediate sight to a different TBM. Calculate the reduced level of the second TBM and check it against its expected value. This will detect if the staff is being held on the wrong physical point.

→ **show, don't point!** – rather than pointing your assistant in the right direction, walk over to the benchmark with them. Show them the correct mark on the ground. If there are other TBMs nearby, draw them to your assistant's attention so that they can distinguish between them.

→ **label TBMs clearly** - if possible, mark the name of the TBM on the ground next to it.

→ **don't put TBMs close to each other** - avoid putting TBMs too close to each other as this can cause confusion. Before installing a new marker, check the setting out file and communicate with other engineers to ascertain if there is already a suitable point you can use in the vicinity.

→ **remove disused or incorrect markers** – avoid cluttering the site with nails and other marks which indicate they are a TBM. Remove obsolete TBMs which are not regularly used and checked. Remove or clearly indicate any reference marks on vertical surfaces which are either incorrect or finished with.

→ **use the Survey Management File** – communicate via the Survey Management File rather than passing TBM information directly between engineers.

→ **clearly define your TBMs** – in the Survey Management File there should be a diagram, description, photo and measurements from reference objects to clearly identify TBMs. TBMs which are not entered into the Survey Management File and are used for only one or two tasks should be removed straight after the task.

9.21 HOW CAN I MINIMISE THE RISK OF MISREADING DRAWINGS OR USING THE WRONG VERSION OF A DRAWING?

Misreading a drawing or using the wrong version is surprisingly easy and surprisingly common. To minimise the risk of using the wrong setting out information, here are a few precautions you can take:

→ **check to as-built points** – once you have set up your level and carried out a check to a TBM, check to an as-built feature which ties in with the element you are setting out.

→ **cross-reference drawings** - locate the information on at least two different drawings, for example the plan and the section (see section **4.3** for a methodical approach to this).

→ **check the drawing register** – make sure that you are using the most up-to-date version of the drawing. Be cautious when photocopying extracts of drawings to take out on site with you as the extract is an uncontrolled copy. Check for updates each morning and destroy any extracts as soon as you have finished the task to ensure the extracts are not used later.

→ **record the drawing number** – whenever you set something out, even if it is one single level, record the drawing number that the information came from in your field book and your site diary.

→ **seek a third-party check** – ask someone to independently determine the value of the feature you are setting out from the construction drawings.

9.22 WHAT IS THE EFFECT OF GROUND MOVEMENT AT TBMS?

Subsidence, heave, settlement and compaction can cause differential level changes in the ground as illustrated in Fig. 9-7. When this happens, your TBM remains the same in relation to the ground around it so there may be no visual evidence to suggest that your TBM has been affected. Here are some measures to help you reduce the risk of error caused by ground movement:

→ **start with two TBMs** – once you have taken your backsight to your TBM, take an intermediate sight to a different TBM. Calculate the reduced level of the second TBM and check it against its expected value. This will detect if there has been any differential movement between the TBMs

→ **choose TBMS which are not too close together** – the TBMs you use to check your set-up should be sufficiently far apart to register any differential movement. If they are on the same section of slab for example, the entire element may have moved downwards but the TBMs would be correct in relation to each other, giving a false impression that there has been no movement

→ **avoid placing TBMs close to excavations** – avoid installing TBMs close to the edge of deep excavations. When working within an excavation, regularly check the TBMs in the excavation against your primary control points

→ **use a range of TBMs for checking purposes** – don't rely on the same TBMs all the time. Varying the TBMs you use is a way of frequently cross referencing your control points, increasing the chance of detecting any differential movement

→ **plan your TBM locations** - at the outset of the project, strategically locate your primary control points as near to the extents of the site as possible and as far as possible from any of the areas of the site which are likely to be affected by ground movement.

Fig. 9-7 Differential movement of TBMs

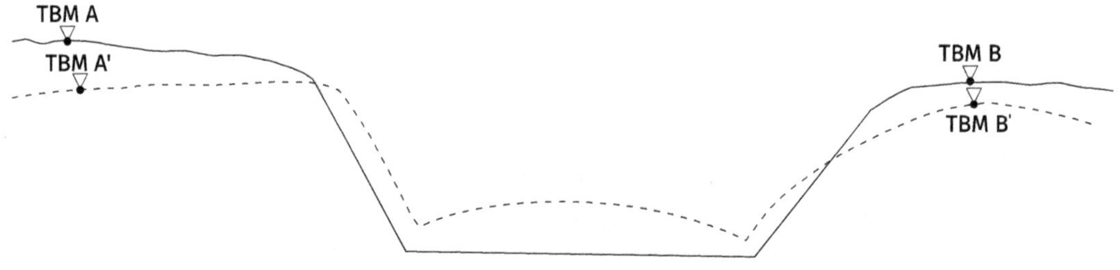

9.23 HOW CAN I MINIMISE THE EFFECTS OF CALCULATION ERRORS?

Calculation errors can occur at any stage of the setting out process, from calculating the required levels from drawings, to installing TBMs and eventually using them to do your final setting out on the ground. Here are some tips you can use to reduce the likelihood of calculation errors occurring:

→ **start with two TBMs** – once you have taken a backsight to your TBM, take an intermediate sight to a different TBM. Calculate the reduced level of the second TBM and check it against its expected value. This will detect any calculation error in your height of collimation which would subsequently affect all the other results from that set-up

→ **record your field data in a standardised way** – the industry standard methods taught in this book are logical, tried and tested and build in self-checks designed to highlight many different types of calculation errors. Develop the habit of recording all your work in this way and you will begin to easily spot when something is not right with your numbers

→ **use diagrams and sketches liberally** – create a visual representation of the task you are working on. For example, if you are calculating the level of formation, draw a section through the slab showing all the relevant information such as RLs at the top and bottom of the slab, the depth of the slab and the depth of blinding. Check that all the numbers add up. Include diagrams in your field book and in your site file

→ **use spreadsheets** – for large surveys or frequently repeated tasks, it may be worth setting up a spreadsheet for checking your calculations, especially if you do not have someone who can carry out independent checks on your work

→ **do your calculations as you go along** – calculate your reduced levels as you go along. This will increase the chance of detecting calculation errors at the time, minimising any confusion and further errors later down the line

→ **incorporate self-checks into your work** – this is an important skill and habit to develop. Use the best practice methods taught in this book

→ **use different methods of coming up with the same answer** – where possible, approach a problem from two different angles and you should end up with the same answer

→ **repeat all your calculations** – even for simple tasks like adding and subtracting, don't just accept the first answer that you come up with. Once you have completed a set of calculations, start from scratch and check that you get the same answers when you repeat it

→ **use your head as well as your calculator** – carry out quick checks in your head. For example, in a list of numbers, add up the millimetres column in your head and check that the final digit of the total sum of the numbers corresponds with the final digit of the sum of the millimetres column

→ **pay attention to minus and plus signs** – remember to record inverted staff readings as negative. Practise HPC and RL calculations until you can do them in your sleep!

→ **pay attention to units** – get into the habit of converting all levels and dimensions into meters

→ **carry out visual checks** – calculate the theoretical level difference between two points and do a visual check on the actual features. Does it look right?

→ **seek an independent check** - get another engineer to verify your results by carrying out the calculations from scratch and proving or disproving your answers.

9.24 USING THE WRONG EDGE OF THE ELECTRICAL TAPE

If you are using electrical tape as a marker for your reduced level, whether to guide the staff to the correct level or to indicate the level being set out, use a marker pen to mark the correct edge. When the tape is being used to mark the level of the top of concrete, the bottom of the tape should mark the correct level. This means that once the concrete has been poured, the tape is still visible. If the top of the tape is used, it is covered up as the concrete is poured and it's not possible to tell from the markers if the concrete level is too high.

9.25 HOW DO I MINIMISE THE EFFECTS OF WEATHER CONDITIONS?

In windy conditions it is difficult to keep the staff vertical so have your assistant rock it through the vertical. The higher the staff reading, the greater the effect of verticality errors so minimise staff readings where possible, especially in windy conditions. It is good practice to transfer TBMs down into the bottom of excavations (Fig. 9-8), rather than setting up at ground level and taking high staff readings to foresights within the excavation (Fig. 9-9).

Hot weather can also affect the quality of staff readings. Avoid taking readings close to the ground on a hot day as there may be a shimmering effect caused by heat rising from the ground.

Fig. 9-8: Reduced staff reading (TBM within excavation)

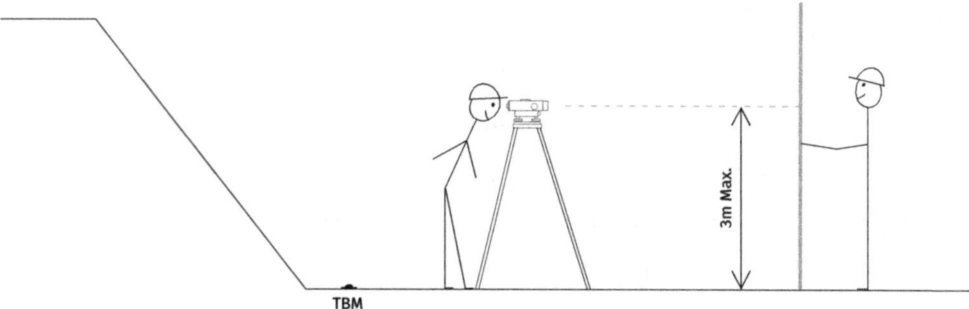

Fig. 9-9: High staff reading

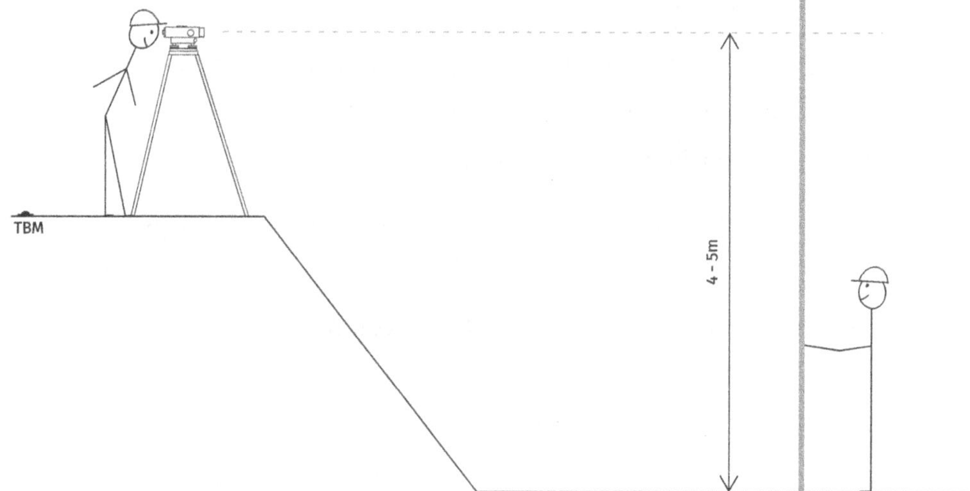

9.26 HOW CAN I AVOID READING TO THE WRONG CROSS-HAIR?

It is possible (and surprisingly easy!) to accidentally take the staff reading to one of the stadia hairs instead of to the horizontal cross-hair.

The risk of reading to the wrong cross-hair is increased when the eye is not close enough to the eye-piece of the level. This is because when the eye is set back from the eye-piece, it is possible to achieve a line of sight through the level even if the eye is either slightly above or slightly below the line of collimation. If the eye is above the line of collimation, the line of sight will be looking down towards the lower stadia hair and the middle horizontal cross-hair may be missing from the view altogether (Fig. 9-11). If the eye is below the line of collimation, only the upper stadia hair may be visible.

Even the most experienced engineers can make this error every now and again. Here are a few measures you can take to reduce the risk of this error occurring:

1. set up the level just below your normal eye level, so that you are just slightly bending your legs as you sight through the eye piece. If you set it up so high that you are standing on tiptoe or so low that you are bending down, your line of sight will be at an angle to the line of collimation rather than along it.

2. put your eye as close as possible to the eye-piece.

3. set the vertical cross-hair to the right or left of the staff so that the stadia hairs are not passing through the staff at all.

4. be extra careful when the staff is far away as the staff will be small in relation to your field of vision and the stadia hair may cross the full width of the staff making it less likely that you will notice the error.

5. make a habit of locating all three cross-hairs and saying to yourself, "I see the top, the bottom and the middle cross hair. I am consciously reading to the middle one!"

6. incorporate checks which will help to detect if you have made this gross error, e.g. checking to two benchmarks at the start of the task and calculating reduced levels as you go along rather than retrospectively.

7. if you find you are prone to making this mistake frequently, create an extra column in your survey book as a reminder to double check each reading.

8. practise! The more you practise, the more it will become a habit to check you are reading to the cross-hair rather than the stadia hair. Eventually it will become an unconscious habit, like looking in the wing mirror of your car before you change lanes.

Fig. 9-11: Wrong Stadia Hair-Eye Too Far Back

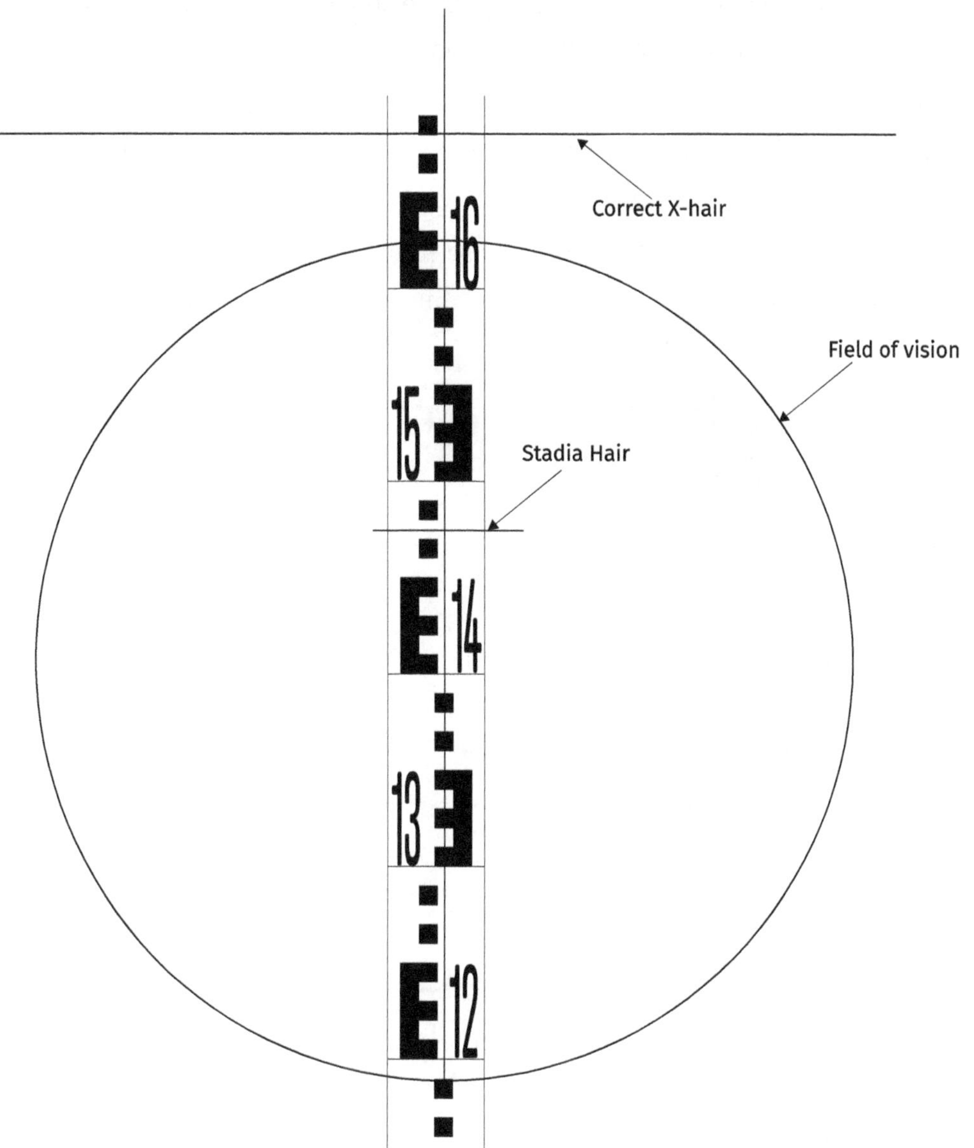

9.27 HOW CAN I AVOID FORGETTING TO LEVEL THE INSTRUMENT?

It sounds a bit ridiculous to point this out as an error, and the chances are that if you made this error, your standard checks would pick it up before it was too late. However, it's worth pointing out a few good habits which will reduce the chance of this happening to you:

→ centre the bubble as soon as you place the level on the tripod.

→ leave the level in the box until you are ready to start your task. Don't attach it to the tripod with the intention of levelling it up later.

→ if you are moving the whole tripod with the level attached from one location to another and plan to level it up at a later time, tilt the legs to make it visibly off level so there's no chance you could accidentally start taking reading without levelling it.

→ if you ever get interrupted and walk away from your level for any reason, always check the bubble and take a staff reading to your original backsight before continuing with your task.

9.28 WHAT SHOULD I DO IF THERE IS POOR VISIBILITY?

Sometimes visibility can be compromised, for example by:

→ fumes rising from plant

→ fog, rain or snow

→ a heat haze from the ground

→ condensation inside the instrument

→ vibration of the compensator causing a shaky cross-hair

If any of these things happen, you can either take measures to improve the situation or delay the task you are doing until conditions are better.

9.29 WHY ISN'T THE BUBBLE WORKING?

If you are having difficulty centring the bubble and have concluded that the bubble is not working properly, it is because you, the user, are doing something wrong. The bubble cannot be 'broken' unless gravity is broken!

It is more likely that:

→ the foot screws are damaged. If you are turning the foot screws but they are not extending or reducing, it could be that the foot screws have come detached from the base.

→ the bubble is hard up against the edge and you are turning the foot screws in the wrong direction. Try turning them in the opposite direction.

→ the bubble is hard up against the edge and even though you are turning them in the right direction the bubble is not moving away from the edge. This is because you have not turned the foot screws far enough yet.

→ you are turning the foot screws in the wrong direction.

If you are struggling to get the bubble centred:

→ Put the level back in the box

→ Level the tripod head by eye by adjusting the length of the legs or repositioning the tripod feet

→ Adjust all the foot screws so they are at the midpoint of their travel

→ Start again

9.30 THE FOOT SCREWS WON'T TURN ANY FURTHER. WHAT SHALL I DO? 🚗

If a foot screw comes to the end of its travel, there are two options:

→ adjust the length of the tripod legs or reposition the feet until the bubble slightly overshoots the centre circle. Then use the foot screws to move it back into the circle

→ ignoring the bubble re-level the tripod head by eye, set the foot screws to their midpoint and start again.

9.31 WHY ISN'T THE FOCUS WORKING? 🚗

If you are having difficulty focusing your object or your cross-hair, it is possible that the focus mechanism is damaged, however it is much more likely to be a user error.

Focusing problems can arise if:

→ either the object or the cross-hair focus have been set to one of the extremes

→ the object and cross-hair focusses are in too much contrast to each other

→ the staff is too close to the level

→ there is nothing to focus on. If you are sighting onto, say, a large painted wall, or a large area of sky, it's impossible to tell if it is in focus or not

→ you are not pointing where you think you are pointing

If you are having problems focusing the staff or if you cannot see any cross-hairs, ignore the staff for now and try the following:

1. Point to something at least 50m away.

2. Adjust the focus screw until you have a clear focus

3. Adjust the cross-hair focus screw until the cross-hairs are crisp and sharp

4. Line up on the staff using the sights on the top of the level

5. Now try focusing the staff and cross-hairs again. You will probably have no problem this time.

9.32 WHY AM I GETTING ERRONEOUS RESULTS?

There are many possible reasons for getting erroneous results. Use the following check list to eliminate possible causes:

→ Is the pond bubble centred?

→ Does the pond bubble remain centred when you rotate the instrument through 180°?

→ Have you stamped the feet right into soft ground?

→ Could the feet be slipping outwards if they are on a smooth, flat surfaces?

→ Did you check to at least two reliable TBMs at the beginning of your task?

→ Has your assistant held the staff on the right TBM?

→ Has your assistant held the staff correctly on the TBM?

→ Is your assistant rocking the staff right through the vertical? Or could they be stopping short?

→ Have you used the right values for your TBMs?

→ Have your carried out a two-peg test to rule out the possibility of a collimation error?

→ Have you checked that the compensator is not sticking?

→ Have you read to the wrong cross-hair?

→ Are your cross-hairs and staff perfectly focussed?

→ Have you checked that the clips on the tripod legs are not slipping?

→ Is the staff fully extended and locked in place?

→ Have you extracted the right information and from the right version of the drawing?

→ Could there be an error in your previous setting out?

→ Is the bottom of the staff clean and in good condition?

CHAPTER 10

INSTRUMENT CHECKS ON THE LEVEL

10.1 WHAT CHECKS SHOULD BE CARRIED OUT ON THE LEVEL?

There are 3 checks you should regularly perform to ensure your level is working correctly. They are the:

- → collimation test (two-peg test)
- → compensator check
- → pond bubble check

10.2 WHEN SHOULD I CARRY OUT INSTRUMENT CHECKS ON MY LEVEL?

You should carry out the checks your level:

- → when the instrument is first delivered to site
- → every week or two weeks depending on your company's policies
- → if you find an unexplained error when checking between control points or in the work you have set out
- → if the instrument is knocked, either in or out of the box
- → when you get it back after you have lent the instrument to another engineer

10.3 WHAT IS A COLLIMATION ERROR?

Although the level itself may appear simple and robust from the outside, internally it contains fragile working parts. The line of collimation is physically adjustable and, as a rule of thumb, anything that can be adjusted can go out of adjustment.

In any levelling activity that we carry out, we are making the critical assumption that the line of collimation is perfectly horizontal. If the line of collimation deviates from the horizontal, it is at an angle, θ, to the horizontal. There will be a difference between the theoretical staff (s) reading and the actual staff reading (s'). This difference is the 'collimation error'. See Fig. 10-1.

The magnitude of the error is directly proportional to the distance from the level to the staff, i.e. the further away the staff is from the level, the greater the error. The collimation error is therefore expressed as a vertical distance per horizonal distance, mm/m.

When a collimation error is detected which is outside of the acceptable tolerances, the instrument is said to be out of adjustment.

Fig 10-1: Collimation Error in a Level

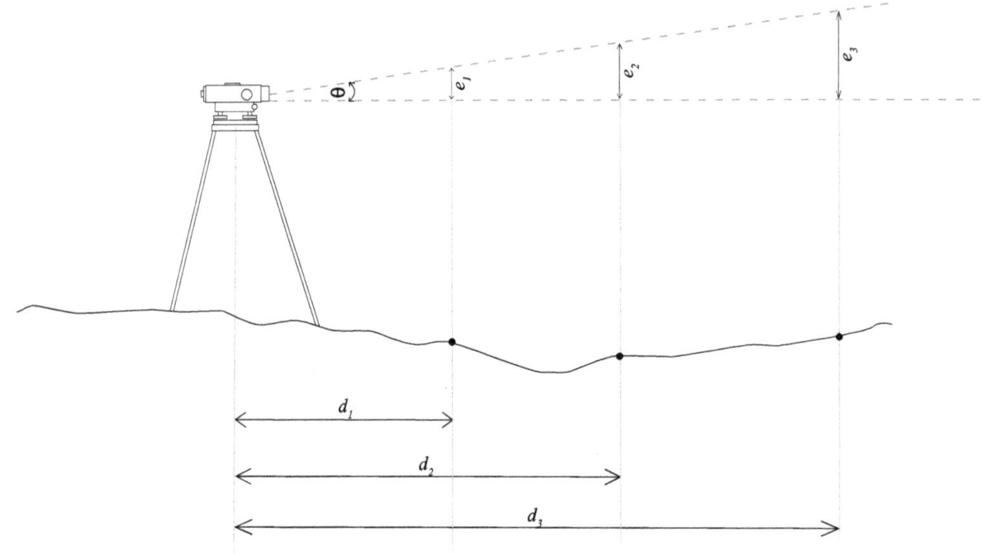

10.4 HOW DO I CHECK IF MY LEVEL HAS A COLLIMATION ERROR?

All levels contain a collimation error to some extent. In fact, it is a practical impossibility to have a level which is completely free of error. For the instrument to have no error whatsoever, the error would need to be 0mm over the distance infinity, which is clearly impossible. There is no expectation that the level should have zero error, rather that the magnitude of the error over a specified distance must be within acceptable limits.

The standard test for checking that your level is working correctly is call the two-peg test. The process will help you ascertain if the line of collimation is truly horizontal. If it is not, then your level contains a collimation error.

There are two reasons to carry out a two-peg test on your level:

> → as part of ongoing checks and quality management procedures

> → to rule out collimation error as a reason for other measurement or setting out errors

10.5 WHAT IS THE PRINCIPAL OF THE TWO-PEG TEST?

The principal of the two-peg test is that we measure the true difference in height between two points (H) and compare it to the difference in height between the two points incorporating the collimation error (H'). To calculate the collimation error, E, we subtract one result from the other. See Fig. 10-2.

To calculate the true difference in height between the two points, we set up at the midpoint. The angle, θ, is consistent whichever direction the level is turned, so when the level is set up equidistant between two points and the staff readings taken, the error (E) will mathematically cancel out. The difference between the two staff readings ($S_A - S_B$) will give the true difference in height (H) even if there is a significant collimation error inherent in the instrument.

To calculate the true difference in height incorporating an error, we set up close to one point and far away from the other. We set up the level close enough to one of the points that we can assume that the magnitude of the error is insignificant, i.e. 0mm. In the other direction, the error will be exaggerated over the longer distance. The staff reading will incorporate the exaggerated error over that distance.

The difference in staff readings ($S_A' - S_B'$) will be the difference in height incorporating the collimation error (H').

The collimation error $E = (S_A - S_B) - (S_A' - S_B') = H - H'$

Fig. 10-2: Two-Peg Test

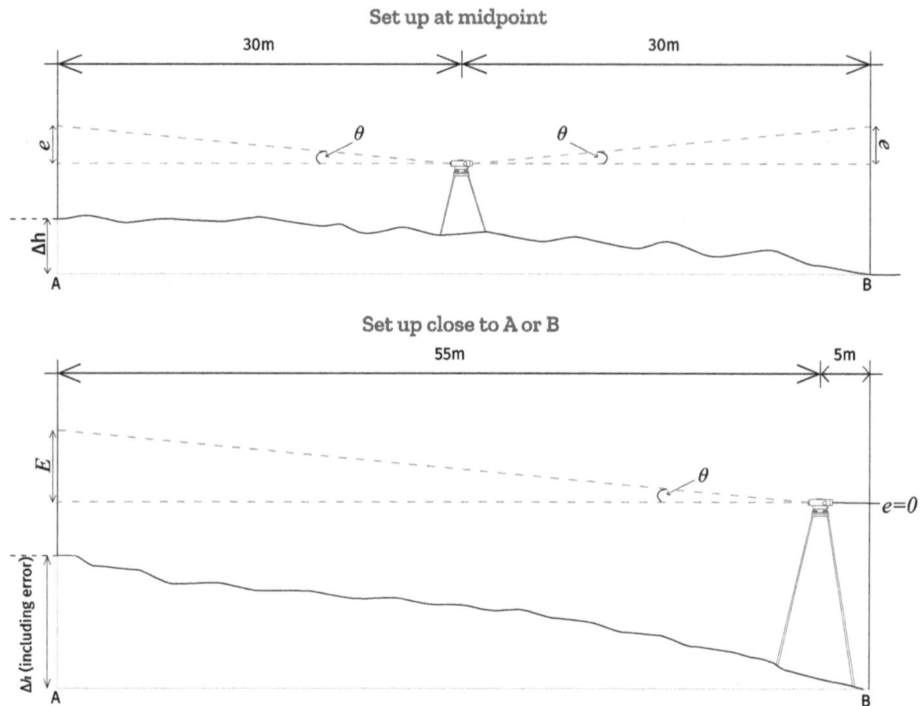

10.6 HOW DO I CARRY OUT THE TWO-PEG TEST?

The process is as follows:

1. Choose or create a suitable marker on the ground (Point A).

2. Pace out 60m in a straight line and choose or mark another point on the ground (Point B). See section **15.10** calibrating your paces.

3. Pace back 30m and set up the level.

4. Attribute an arbitrary Reduced Level of 10.000m to point A.

5. Take a backsight to A (S_A) and a foresight to B (S_B) and enter them into the appropriate column and row in your field book (Fig. 10-3). Calculate the Height of Collimation and the Reduced Level of B.

We have now calculated the Reduced Level at B in relation to A. Because the level is set up equidistant from the two points, any error will mathematically cancel out and the result is the true Reduced Level at B.

6. Pace 5m towards the centre from either of the points, it does n't matter which, and set up the level.

7. Take a backsight to $B(S_{B'})$ then a foresight to $A(S_{A'})$. Calculate the Height of Collimation for the new set-up position and from that, subtract the foresight to A to calculate the Reduced Level at A.

In theory, if there is 0mm collimation error in the instrument and we have eliminated all other sources of error, the second Reduced Level at A should be exactly the same as the first. The collimation error is the difference between the two. Record the error in the Remarks column.

8. Repeat steps 3 – 7 to obtain a full, independent set of results.

9. If E2 matches E1 within 2mm AND RL_{B1} matches RL_{B2} within 2mm, then the average of E1 and E2 is the final collimation error. If there is a discrepancy of more than 2mm between either of these, then the test must be repeated to identify the erroneous set of results. The erroneous set should be discarded completely, and the two good sets of results should be used to calculate the final collimation error.

Fig. 10-3: Two-peg test booking example

Date __8 Sept 2018__   Taken for __Two Peg Test (Serial no 23185)__

From __A__   To __B__

BACK SIGHT	INTER MEDIATE	FORE SIGHT	COLLIMATION OR H.P.C	REDUCED LEVEL	DISTANCE	REMARKS
1.653			11.653	10.000		A
1.322		1.032	11.943	10.621		B
		1.941		10.002		A (Error = 2mm/55m)
1.614			11.614	10.000		A
1.498		0.992	12.120	10.622		B
		2.116		10.004		A (Error = 4mm/55m)
						Collimation Error = 3mm/55m

EXERCISE:

1. You carry out a two-peg test on your level using the arrangements in Fig. 10-2 and the results are as follows:

→ From midpoint, BS to A = 1.629m, FS to B = 1.319m

→ From close to B, BS to B = 1.553m, FS to A = 1.861m

→ From midpoint, BS to A = 1.672m, FS to B = 1.361m

→ From close to B, BS to B = 1.489m, FS to A = 1.796m

   a. What is the collimation error in the level?

10.7 WHAT ASSUMPTIONS ARE WE MAKING WHEN WE DO THE TWO-PEG TEST?

When we carry out the two-peg test, we are making the critical assumption that we are minimising the effects of all other sources of error and we are treating them as 0mm. This assumption allows us to calculate the residual error inherent in the instrument itself, thereby isolating and measuring the collimation error.

It's not possible to eliminate random errors entirely, so the final collimation error calculated will incorporate these errors.

10.8 WHAT SHOULD I DO IF I DISCOVER A COLLIMATION ERROR IN MY LEVEL?

If your level has a collimation error this will result in the corresponding error being incorporated into your survey or setting out results. In other words, your results will be wrong.

If you complete the two-peg test and find that the collimation error in your level is outside of acceptable limits, you have several options:

→ continue to use the instrument for tasks which will not be adversely affected by the error. For example, if the collimation error is, say, 15mm and you need to set up profile boards for an excavation, you may deem that it is more cost-effective to accept the effects of the error than to keep the associated plant standing. The effect of the error will be maximised where backsights and foresights are not balanced, i.e. the distance from your level to your backsight staff position is much shorter or longer than the distance to your foresight staff position (Fig. 10-4). To minimise or cancel the effects of the error, ensure that backsight and foresight distances are equal (Fig. 10-5). If however, your collimation error is, say, 8mm and you are setting packer levels for steel column bases where a much higher level of accuracy is required, you might decide to choose one of the options below

→ decide whether it is appropriate to carry out the adjustment yourself. The pros and cons of this are given in section **10.11**. Follow the instructions in section **10.12** to adjust the line of collimation of your level

→ do not use the level. Send it away to be adjusted by a suitably qualified technician, e.g. the survey equipment hire company which the instrument is on hire from or the company which issues your annual calibration certificate

Fig. 10-4: Exaggerating effect of unequal site distances

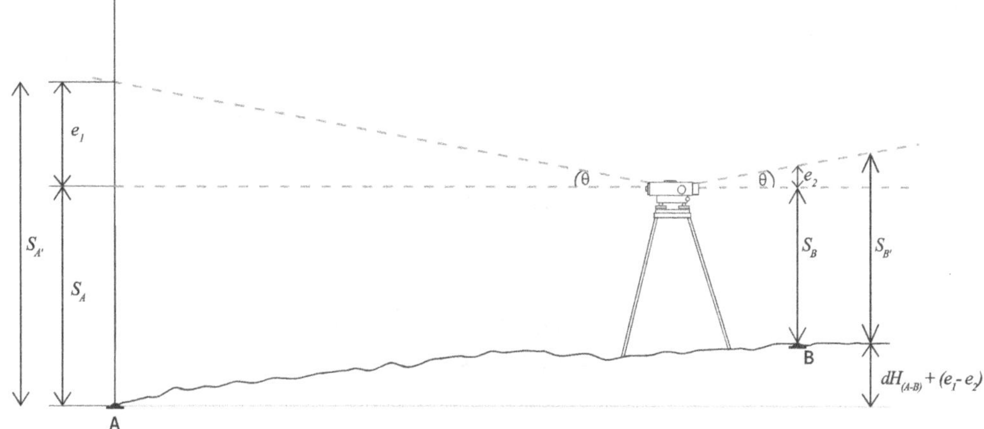

Fig 10-5: Cancelling effect of equal site distances

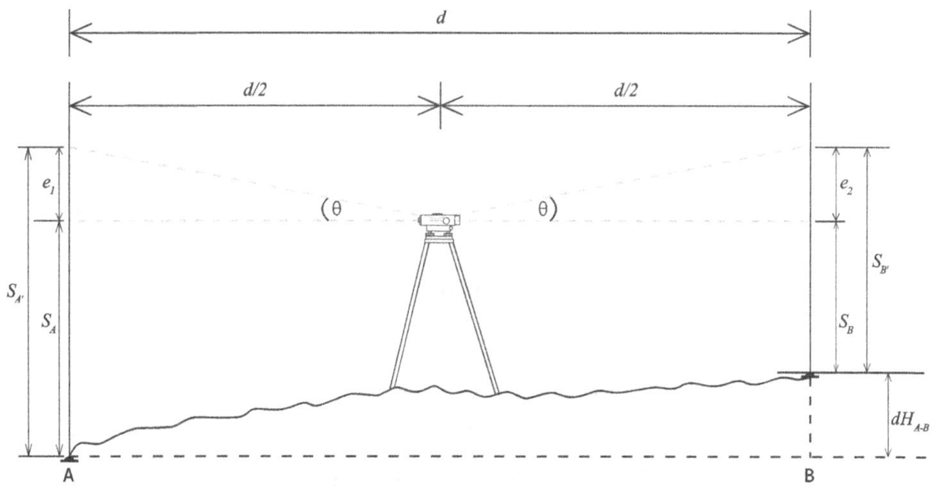

10.9 HOW CAN I MINIMISE THE EFFECT OF A COLLIMATION ERROR?

If you carry out the two-peg test and discover a collimation error in your level, you can still use it temporarily if the error is within acceptable limits for the task you are doing. For example, if you are setting out levels for an excavation, and the error in your level is, say, 5mm/55m and it is the only one you have, then it would make sense to go ahead and use the level rather than stop the works. If you do decide that this is necessary, remember that:

→ the error increases with increasing horizontal distance from the level to the staff, so keep sight distances as short as possible. For example, if your collimation error is 5mm/55m, over 20m it will only be 1.8mm.

→ equal sight distances will cancel out the collimation error so set up your level in a position which will give you approximately the same horizontal distance from the backsight position to the foresight position. In fact, it is good practice to do this wherever possible, in case your level has a collimation error which you don't know about.

10.10 WHAT ARE THE ACCEPTABLE LIMITS FOR COLLIMATION ERROR?

For the acceptable limits for collimation error of an automatic level, refer to your company's procedures or the procedures for the project you are working on. If you are unable to locate this information, use the following maximum error as a guide:

→ +/- 3 mm per 60m

→ +/- 1mm per 20m

If you find a collimation error greater than this, repeat the two-peg test several times and ideally seek an independent check by someone else to confirm that your results are correct.

10.11 CAN I ADJUST THE LINE OF COLLIMATION MYSELF?

Adjusting the line of collimation is classed as a permanent adjustment. In certain situations, it may be appropriate to carry out the adjustment yourself, for example if you are working in a remote location where a new instrument cannot be delivered quickly enough, or if you have an urgent task which needs to be completed before a new instrument can be delivered. If you are a freelance engineer, you might be thinking about avoiding the cost and hassle of having the instrument adjusted by a specialist.

There are several factors to take into consideration when deciding whether it is appropriate to adjust the line of collimation yourself:

→ check your company's policy. If your immediate line manager does not know the answer, in larger organisations there is usually a Head Land Surveyor, or Head Engineer who will know the answer. In smaller companies you may need to ask a Director or your Project Manager for guidance. If you are a freelance engineer with your own equipment, the choice is up to you.

→ if for any reason you do not feel confident or comfortable carrying out the operation or if you are unsure whether it is ok to do so, then do not attempt it.

→ if the instrument is on hire, check with the supplier. They may have a policy that the calibration certificate is invalidated once the cover has been removed.

→ if you decide to go ahead and adjust the line of collimation yourself, to prevent damage to the equipment, ensure you have all the necessary technical information at your disposal, including the manufacturer's manual.

10.12 HOW DO I ADJUST THE LINE OF COLLIMATION?

To adjust the line of collimation, the steps are as follows:

1. Carry out the two-peg test to establish the collimation error.

2. Set up 5m from either of the points and take a backsight to point A to calculate the height of collimation.

3. During the two-peg test you calculated the true RL of B. Subtract this from the height of collimation to calculate the required staff reading at B. Draw a pencil mark across the staff at that position.

4. Following the manufacturer's instructions for the specific instrument you are using, adjust the line of collimation until the horizontal cross-hair is over the pencil mark. This is usually done by removing the cap at the eyepiece to reveal a hidden capstan screw. Be aware that in older levels, there may be two screws, one at the bottom and one at the top. One must be released as the other one is tightened to keep the right amount of tension. If one screw is overtightened without releasing the

opposite screw, the wire can snap and cause damage which can only be repaired by technicians in the workshop.

5. Once the required staff reading is achieved, repeat the two-peg test, ensuring that you carry out two full sets of readings as described in section **10.6**.

10.13 HOW CAN I MAKE THE TWO-PEG TEST QUICKER?

There are a couple of tricks you can use to speed up your two peg test:

→ Method 1: If it is safe to assume that there has been no differential movement of points A and B, you can use the same two points each time you do the test. This means that you do not need to carry out the first part of the test in which you set up at the midpoint. You can go straight to part 2, setting up close to one of the points. Take one set of readings, adjust the height of the level and repeat. If the differences between A and B match within a few mm, take the average and calculate the collimation error based on the difference between the average and the true known height difference.

→ Method 2: To make the test even quicker, staff sections can be permanently set up at positions 60m apart (or thereabouts). You don't even need an assistant to hold the staff, or spend time walking from one point to the other. Simply set up, take the two readings, set up again, take the two readings. The whole process will take less than two minutes.

10.14 WHAT IS LIKELY MAGNITUDE OF A COLLIMATION ERROR?

The collimation error may be anything from zero up to about 50mm. Anything larger than this would indicate internal damage to the instrument.

10.15 WHAT ARE POSSIBLE CAUSES OF A COLLIMATION ERROR?

Small collimation errors, i.e. a few millimetres, can build naturally over time due to normal wear and tear, including being transported from place to place (either by being carried by hand or transported in vehicles), being picked up and put down whilst in the box, or vibration or movement whilst in use, e.g. tripod feet being pushed into the ground whilst the level is attached.

Larger errors are likely to be the result of the instrument being inadvertently banged or dropped.

10.16   DO I REALLY NEED TO CARRY OUT A TWO-PEG TEST ON MY LEVEL?

This is a common question I am asked, and I can understand why. In many companies, this critical task is overlooked, not because it is not important, but because there is a lack of awareness that instruments can be knocked out of adjustment and the effect that will have on the efficiency and quality of setting out activities. Even if company policies state the required time interval, the communication does not always reach the site engineers on the ground or their supervisors.

The majority of engineers who I come into contact with tell me that they rarely carry out the test and many are not even aware of its existence, let alone have been taught the process and how to record it properly and keep it as part of the site quality management records. The answer is 'yes', you do really need to!

10.17   ARE THERE ANY CIRCUMSTANCES WHERE I WOULD NOT NEED TO CARRY OUT THE TWO-PEG TEST?

It may be the case that you have a high-quality shock resistant level which specifies that no checking is required. If this is the case then:

→ check your company policy on whether it is ok to relax the checking regime, and/or

→ check it anyway, it only takes a few seconds

Any relaxations in checking regimes should be carried out from a position of conscious choice rather than from a position of ignorance and should be fully justified, documented and approved.

10.18   HOW DO I CHECK THE COMPENSATOR?

The compensator is a fragile mechanism and can get stuck and can also go out of calibration. Because it is hidden inside the instrument it is not possible to check visually. High quality instruments are less prone to this error than cheaper or older versions.

10.19 HOW CAN I CHECK THAT THE COMPENSATOR IS WORKING CORRECTLY?

There are two quick tests you can do to check that the compensator is working as it should:

→ **Test 1** - Once the instrument is set up and levelled, take a staff reading. Then press the compensator button if there is one, or tap the top of the level. If the compensator is functioning correctly, the horizontal cross-hairs will visibly vibrate up and down and settle at the same reading.

→ **Test 2** – Sight the staff and rotate one of the foot screws slightly. If the horizontal cross-hair stays on the same reading, the compensator is working.

10.20 HOW CAN I MINIMISE THE RISK OF A COMPENSATOR ERROR?

If your compensator is 'sticking' or damaged, it may be obvious when you sight to the staff. The line of sight may be pointing significantly upwards or downwards, which will be a reminder to press the reset button.

On the other hand, the error may be less dramatic and will only be detected by following the correct checking procedures.

To minimise the risk of a compensator error occurring, take the following measures:

1. press the compensator release button or tap the top of the level with your finger each time you set up. You do not need to do this for each reading, only when you have relocated the instrument

2. always carry the level horizontally and the right way up when it is in its box. Don't swing it around, jolt it or store it upside down, on its side or on end

3. avoid humidity. Leave the level box open indoors overnight when the instrument has been used in the rain

4. when repositioning the level whilst still attached to the tripod, place the legs each side of your shoulder. Don't close the legs together and carry it like a battering ram with the level pointing downwards.

To detect a compensator error, always measure to two reliable benchmarks before commencing your task. Check that they are both correct in relation to each other.

10.21 WHAT SHALL I DO IF THE COMPENSATOR IS NOT WORKING?

If you have reason to believe that the compensator is jammed, stuck or damaged in any way, you should return it to the workshop for repair. This is a job for a trained specialist and should not be attempted on site.

10.22 WHAT IF MY LEVEL DOES NOT HAVE A COMPENSATOR?

Generally, all levels on construction sites today will have a compensator. The use of older generation levels without compensators e.g. tilting or dumpy levels, is not in the scope of this book. Automatic levels are significantly more accurate, convenient and reliable and can be purchased for as little as £85 so it is worth considering replacing any older levels.

10.23 MY LEVEL DOESN'T HAVE A COMPENSATOR RESET BUTTON. WHAT DOES THAT MEAN?

If your level does not have a compensator reset button it may be that you have a high-quality shock-resistant level which does not need checking. Refer to the manufacturer's instructions to find out if you have a shock-resistant level. Alternatively, it may be the case that your level is very basic and simply doesn't have a compensator reset button. If this is the case, tap the top of the level each time you set up in a new position.

10.24 WHAT IS THE POSSIBLE MAGNITUDE OF A COMPENSATOR ERROR?

The magnitude of a compensator error depends on the sight distance. The further away the staff, the larger the error. A compensator error will normally be obvious and may even result in the cross-hairs pointing above the top of the staff.

10.25 HOW DO I CHECK THAT THE POND BUBBLE IS IN CORRECT ADJUSTMENT?

If the pond bubble is out of adjustment, then the line of collimation will not be horizontal even when the pond bubble is centred. This will result in errors.

To check that the pond bubble is in correct adjustment, follow these steps:

1. Use the foot screws to centre the pond bubble.

2. Rotate the telescope through 180 degrees.

3. If the bubble is still centred, then it is in correct adjustment. If the bubble moves off centre, the pond bubble is out of adjustment.

4. Decide whether it is appropriate to make the permanent adjustment to the bubble yourself or to return it to the supplier to be adjusted by a specialist. See section **10.27** for further guidance.

10.26 WHAT SHOULD I DO IF THE POND BUBBLE IS OUT OF ADJUSTMENT?

If you discover that the pond bubble is out of adjustment, you have four options:

→ use trial and error to find the position at which the bubble remains in the same postions as the level is rotated through 360 degrees.

→ carry out a permanent adjustment on the pond bubble. See section **10.28**

→ do not use the level until you have had the permanent adjustment carried out by an expert

→ use the instrument as a temporary solution over short, equal sight distances for low accuracy tasks such as earthworks. Make a note of the problem on each page of your field book.

10.27 SHOULD I ADJUST THE POND BUBBLE MYSELF?

The pond bubble is a relatively simple working part of the level. If the correct procedure is followed, there is a low risk of damage in attempting to adjust it yourself. The worst-case scenario is that you make a bad job of it and need to submit it to the lab for adjustment in which case you're no worse off than you were when you first discovered the error. If for any reason you do not feel comfortable making the permanent adjustment or are concerned that your employer would expect you to submit the equipment to the lab, then do not attempt it.

10.28 HOW DO I ADJUST THE POND BUBBLE?

To remove a pond bubble error, take the following steps:

1. Centre the pond bubble using the foot screws.

2. Rotate the level horizontally through 180 degrees.

3. Note how far off centre the bubble is.

4. Adjust the foot screws to bring the bubble half the distance back towards the centre.

5. Use the hex key to adjust the screws to bring the bubble all the way back to the centre.

6. Repeat steps 2 – 5 until there is no movement of the bubble when rotated through 180 degrees.

7. To leave the screws firm by removing backlash, the final turn of the hex screw should be clockwise.

CHAPTER 11

TOTAL STATION

11.1 ABOUT THE TOTAL STATION

The total station is a piece of hi-tech, high accuracy surveying equipment which can measure horizontal angles, vertical angles and distances between physical points. It comprises a telescope which can rotate horizontally and vertically, with a distance measurement function (Fig. 11-1). It is essentially a theodolite, which measures the horizontal and vertical angles, plus an Electromagnetic Distance Measurement (EDM) component , which measures distances, plus onboard software which processes the angles and distances to convert them into coordinates and reduced levels. Fig. 11-2 shows the axes of the total station in relation to each other.

Fig. 11-1: Total station (Image courtesy of Topcon Positioning Systems)

Fig. 11-2: Total station axes

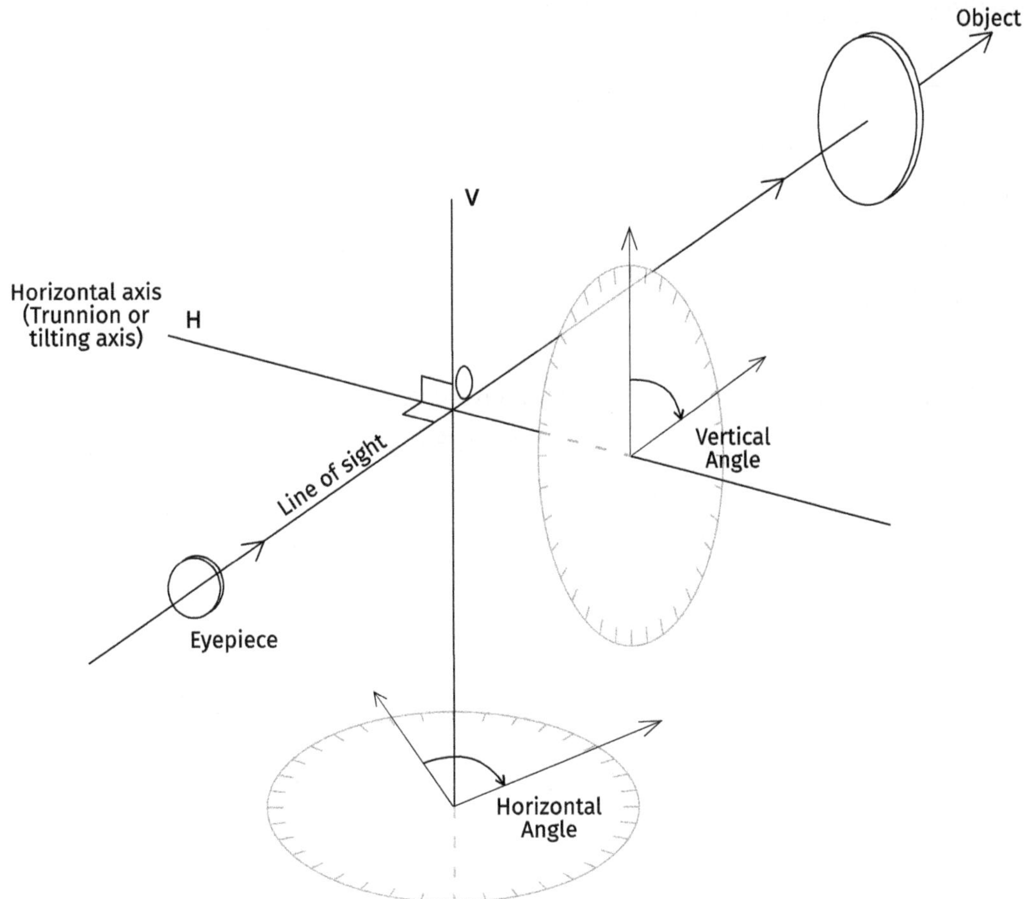

11.2 WHAT CAN TOTAL STATIONS BE USED FOR?

Total stations can be used for any surveying or setting out tasks, provided there is inter-visibility between the total station and the point being sighted. Total stations can be used for:

- → measuring and setting out angles, distances, coordinates and levels
- → measuring and setting out offsets from lines
- → measuring and setting out arcs and offsets from arcs
- → measuring irregular areas and volumes
- → processing measured and imported data

11.3 WHAT TOTAL STATION SKILLS SHOULD I HAVE?

You should be able to:

- → set up the total station over a point
- → carry out the relevant calibration checks (horizontal and vertical collimation error, trunnion axis, prism constant, optical/ laser plummet, diaphragm orientation)
- → list the possible sources of error when using a total station
- → use techniques for improving and checking your accuracy and precision
- → view, edit, add and delete data
- → install a network of primary control points from scratch (traverse)
- → install accurate secondary control points (retro targets) using the correct procedure
- → create a local coordinate system for a building on gridlines
- → use the correct procedures for measuring and setting out reduced levels with the total station
- → use a range of methods for plumbing columns and walls
- → set-up the position and orientation of the total station using the resection or occupied point programs
- → take a topographical survey and record the results systematically
- → measure the horizontal distance and level differences between two points
- → set out points of known co-ordinates using the stake out function
- → set out points in relation to a baseline using the reference line function
- → set out points at given chainages and offsets along a radius e.g. road centrelines using the reference arc function
- → measure irregular areas and volumes
- → transfer large amounts of data from the total station to the computer and vice versa
- → describe the capabilities of total stations and robotic total stations.

11.4 TOTAL STATION ERRORS

Sources of error in the total station include:

- → local ground movement
- → legs sinking into soft ground (including tarmac on hot days)
- → clips on legs not working properly and slipping down

→ sighting errors

→ recording errors

→ using the wrong prism constant

→ incorrect centring over the point of the total station or backsight (mini-prism or tribrach)

→ the total station not being levelled correctly (Fig. 11-3)

→ verticality of mini-prism or detail pole ('pogo')

→ poor visibility

→ using the wrong control points or incorrect values for the control points

→ taking the wrong information from the drawing

→ poor quality control points

Fig. 11-3: Total station detail not level

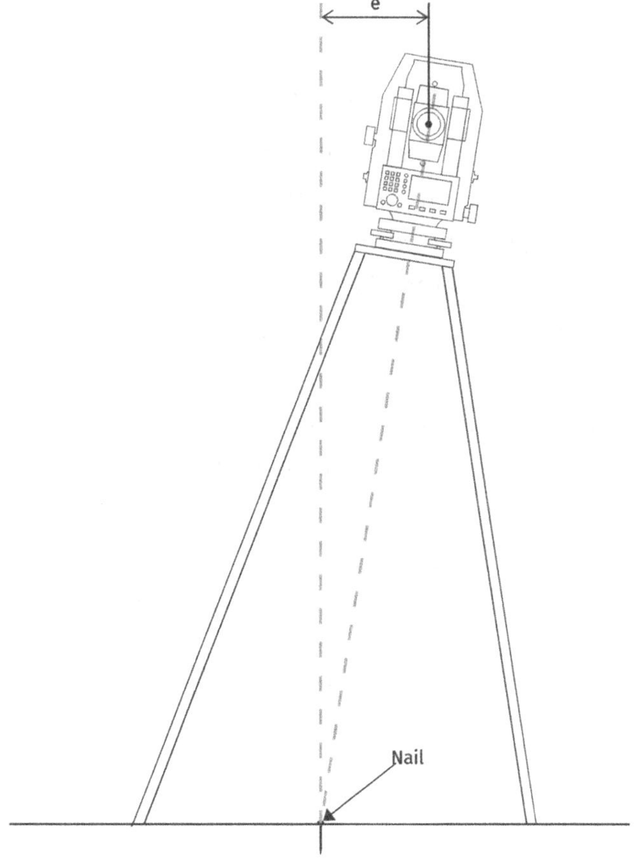

11.4 WHAT IS THE HORIZONTAL CIRCLE?

The horizontal and vertical circles are graduated circular scales within the total station. They can be thought of as 360-degree protractors.

The centre of the horizontal circle is coincident with the vertical axis of the total station. As the total station is rotated about the centre of the circle, the angle reading is displayed on the screen. The angle increases as the total station is rotated clockwise and decreases as it is rotated anticlockwise.

When the total station is initially switched on, zero degrees on the circle is at a random point. It is not pointing north as it has no way of knowing its orientation until you reference a known point as part of the instrument set-up process.

The function of the horizontal circle is to measure relative angles. To take a horizontal circle reading (HCR), or Hz, to a point, the single vertical cross hair should be lined up on the target. To calculate the angle between two points, it is necessary to record the HCR at both points and subtract one from the other.

11.5 WHAT IS THE VERTICAL CIRCLE?

The centre of the vertical circle is coincident with the trunnion axis. Unlike the horizontal circle, which is at a random orientation each time the instrument is switched on, the vertical circle is fixed in position. The vertical circle reading (VCR or V) is 0° when the telescope points vertically upwards (perpendicular to the earths curved surface), 90° when it is horizontal on face left, 180° when it is pointing vertically downwards and 270° when it is horizontal on face right.

The function of the vertical circle is to measure angles relative to the zenith (when the VCR is at 0°). To find the vertical angle between two points, it is necessary to record the VCR at both points and subtract one from the other.

The total station uses the vertical angle readings combined with slope distances (from the total station to the prism) to calculate horizontal distances and vertical distances.

11.6 WHAT ARE OPPOSITE FACES IN A TOTAL STATION?

Because both the horizontal and vertical circles can rotate through 360 degrees, it is possible to observe the same point from two diametrically opposed points on each of the circles. If you were to sight a fixed point and then rotate the horizontal circle through 180 degrees and flip the telescope over so it is pointing at the same fixed point, you can sight back to the exact same point, but this time on the opposite face. The opposite faces are called face left and face right, left face and right face or face 1 and face 2.

11.7   HOW DO I KNOW WHICH IS FACE LEFT AND FACE RIGHT?

The terms face left and face right date back to the use of traditional theodolites when face left indicated the vertical circle was to the left of the user when looking through the telescope. With modern total stations, it doesn't matter which you call face left and face right, or face 1 and face 2, so long as you keep the same convention for the duration of the task you are doing. One way to remember is to always use the same convention, for example face left is when the display screen is facing you and face right is when the screen is facing away.

11.8   HOW DO I CHANGE FACES?

To change faces:

→ Starting with the eyepiece of the telescope towards you, rotate the telescope through 180 degrees horizontally so the eyepiece is now facing away from you.

→ Rotate the telescope vertically so the eyepiece is towards you again. You have now changed faces.

11.9   WHY WOULD I NEED TO CHANGE FACES?

In certain circumstances it may be necessary to set out your points using both faces, for example when:

→ you have detected a collimation error in your total station and have no option but to use the instrument

→ the point you are setting out requires a high level of accuracy

→ you are forced to use short backsight distances with long foresight distances

When a collimation or circle offset error is present in the instrument, the mean of the left and right face will mathematically cancel out any error and give the correct position.

When measuring angles, the mean of the minutes and seconds of both faces gives the true answer. This is illustrated when taking left and right face readings for traverse observations, in section **19.4**.

When setting out, if a point is set out twice, once on each face, the midpoint between the two marks is the correct position. If the two marks are coincident, no collimation error is present.

11.10 HOW DO I SET OUT A POINT USING BOTH FACES?

Once you have set out your point, simply change faces by rotating the horizontal circle through approximately 180 degrees and flipping the telescope over. Set out the point again using this opposite face. You should find that your second mark is within millimetres of the original mark. If the two marks are not coincident, the mid-point between the two marks is the true position.

11.11 WHAT IS THE MAXIMUM SIGHT DISTANCE OF A TOTAL STATION?

The maximum sight distance of a total station depends on the type and quality of the instrument you are using. Increasing the number of prisms allows distances of up to 12km away to be measured.

Provided that the total station has the capability, it is possible to take a measurement even if the prism is so far away that it is not visible to the human eye.

When the coordinates of the target are not known, a large back board is needed to help you locate the target through your telescope.

When the coordinates of the target are known, the total station will guide you to the right location with arrows on the screen

11.12 HOW IS THE DISTANCE ACCURACY OF THE TOTAL STATION EXPRESSED?

The distance accuracy of the total station EDM function is expressed as two components:

Xmm + Yppm

The first component, Xmm, means that regardless of the distance to the target, there may be an error of up to + or − Xmm.

The second component, Yppm, means that in addition to the Xmm in the first component, there is an error of up to Yppm (parts per million), or Ymm for every 1 million mm of sight distance.

Example:

For a total station which has a distance accuracy of 2mm + 2pmm, the error would be:

→ over 10m, the error would be +/-(2mm + 2 (10,000/1,000,000)) = +/- 2mm

→ over 100m, the error would be +/-(2mm + 2 (100,000/1,000,000)) = +/- 2.2mm

→ over 1000m, the error would be +/-(2mm + 2 (100,000/1,000,000)) = +/- 4mm

11.13 HOW SHOULD I LINE UP THE CROSS-HAIRS ON A TARGET?

The horizontal and vertical cross-hairs should be treated as two completely separate tools for completely separate jobs. The single horizontal line is the correct tool for the job for measuring vertical angles and the single vertical line is the correct tool for the job for measuring horizontal angles. Fig. 11-4 shows a typical cross-hair arrangement for a total station.

Each time you take a measurement, before you start, be clear in your mind whether it is the horizontal angle which is critical to the task you are doing or the vertical. If you are measuring or setting out positions, i.e. eastings and northings, it is the horizontal angle which is critical. The vertical position of the vertical cross-hair will have no effect on the accuracy of horizontal positions. If you are measuring or setting out levels, it is the vertical angle which is critical and therefore the horizontal cross-hair. The horizontal angle will have no effect on the accuracy of the levels.

For low-accuracy tasks such as earthworks, it is ok to line up the small cross at the centre directly on the target.

Lining-up errors are minimised with a robotic total station because they lock onto theoretical focal point of the prism.

Fig. 11-4: Total station x-hairs

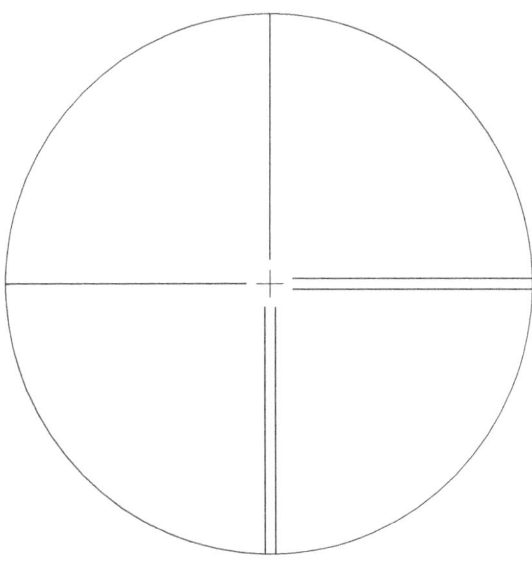

11.14 WHAT IS THE PROCEDURE FOR LINING UP THE CROSS-HAIRS FOR MEASURING OR SETTING OUT A POSITION?

To line up the cross-hairs when measuring or setting out a position:

1. Use the horizontal tangent screw to line the vertical cross-hair through the lateral centre of the PGM (Fig. 11-5), target arrows, tip of pole (Fig. 11-6 and Fig. 11-7) or the centre of the cross (Fig. 11-8). For a simple horizontal angle measurement (Fig. 11-9 and Fig. 11-10), Step 2 is not needed - record the angle.

2. Use the vertical tangent screw to line the horizontal cross-hair through the centre of the target arrows. Do NOT adjust the horizontal tangent screw even if the centre cross appears off-centre to the left or right.

3. Press the relevant key to take the measurement

Fig. 11-5: Lining up on backsight - horizontal measurement

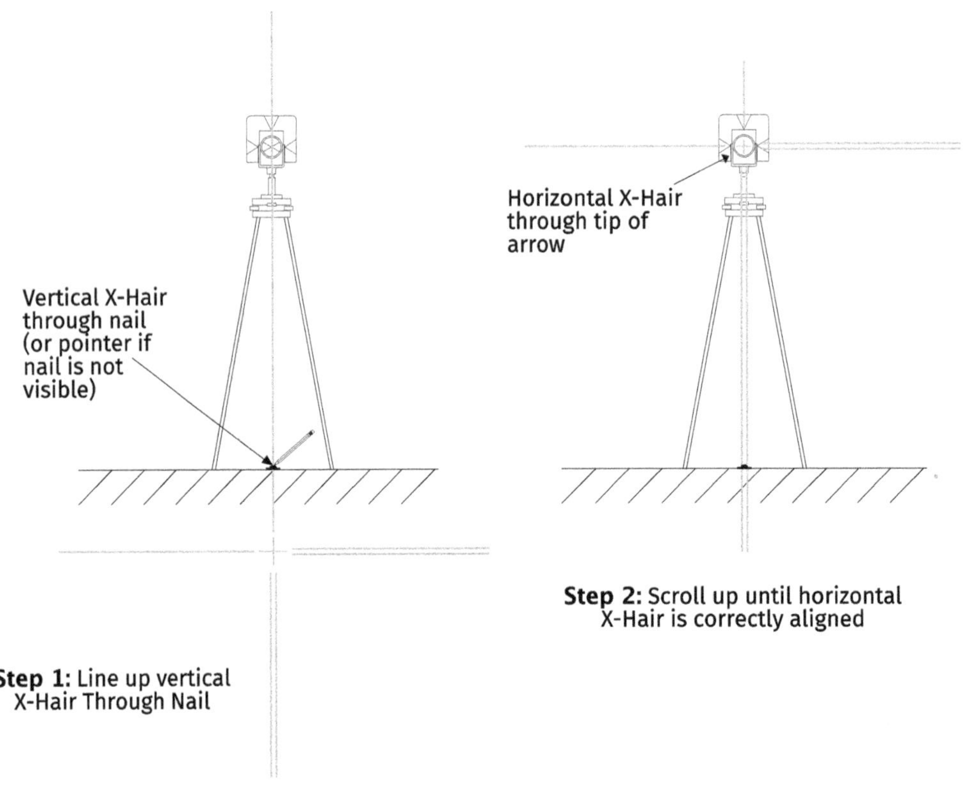

Fig. 11-6: Lining up on mini prism - horizontal

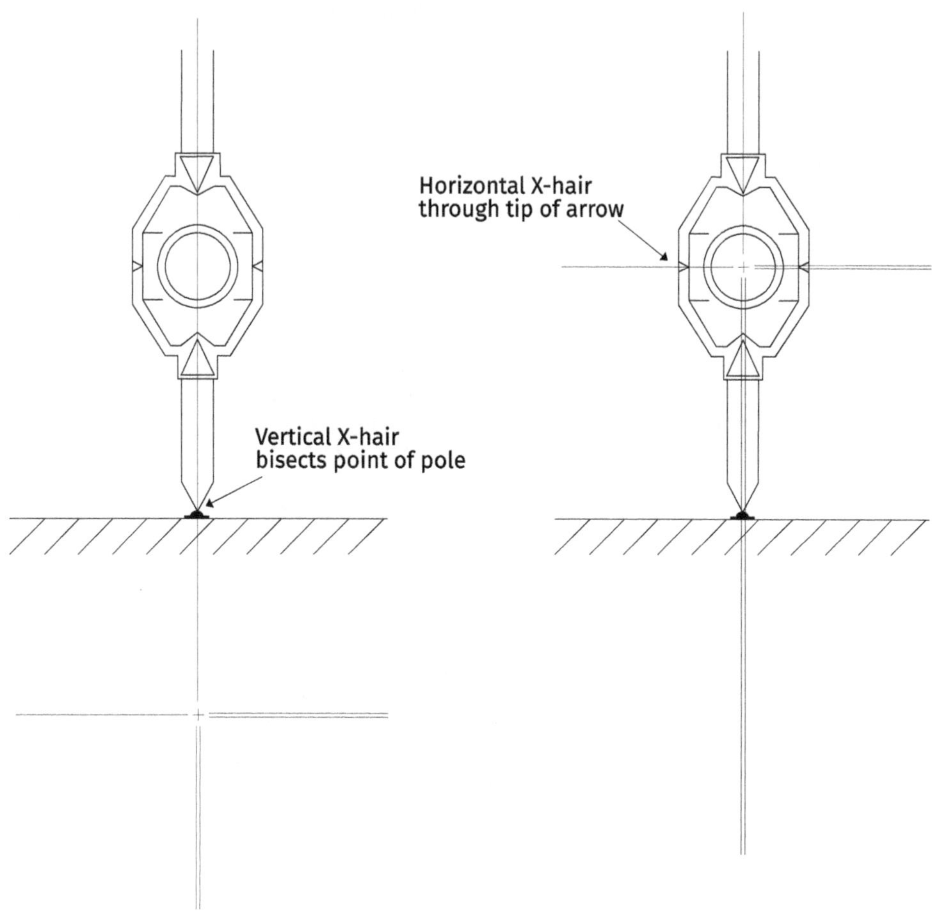

Step 1:
Line up the vertical x-hair

Step 2:
Scroll up until the horizontal x-hair
is correctly aligned.

Fig. 11-7: Lining up on mini prism - wonky

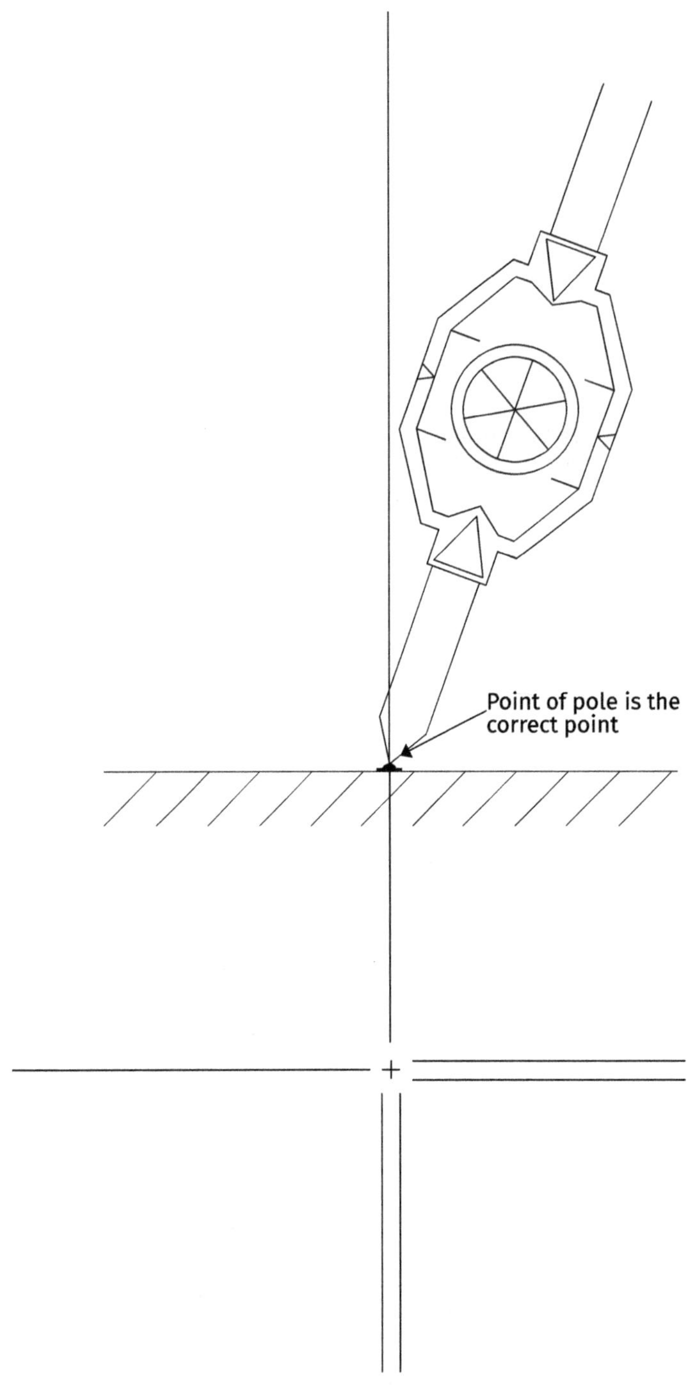

Fig 11-8: Lining up on retro for horizontal measurement

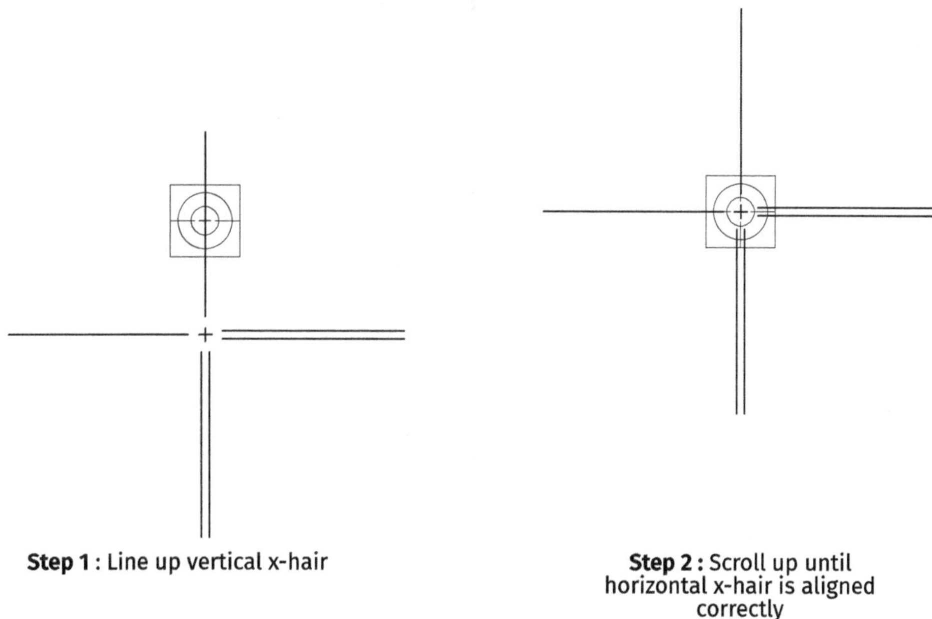

Fig 11-9: Lining up on nail for horizontal measurement

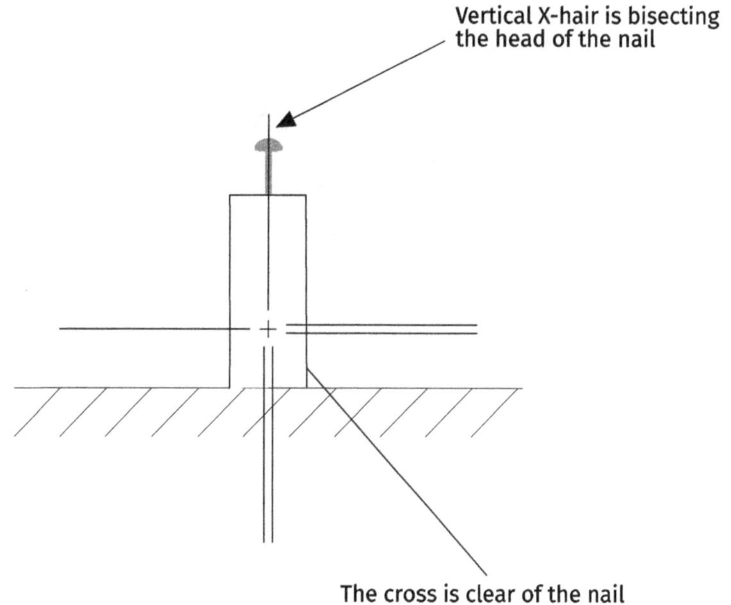

Fig 11-10: Lining up on nail on a pencil for horizontal measurement

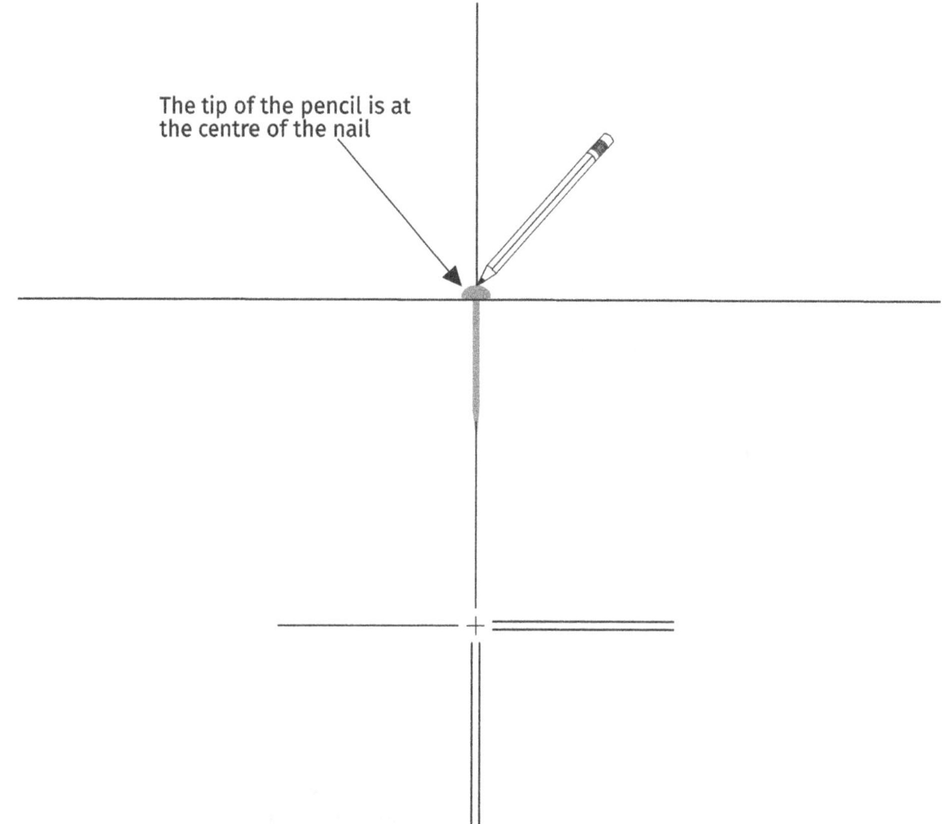

11.15   WHAT IS THE PROCEDURE FOR THE LINING UP THE CROSS-HAIRS FOR MEASURING OR SETTING OUT A LEVEL?

To line up the cross-hairs when measuring or setting out a level:

1. Use the vertical tangent screw to line the horizontal cross-hair through the centre of the arrows (Fig. 11-11), target arrows (Fig. 11-12) or the centre of the cross (Fig. 11-13).

2. Use the horizontal tangent screw to line the vertical cross-hair through the centre of the target arrow or tip of pole. Do NOT adjust the vertical tangent screw even if the centre cross appears off centre vertically.

3. Press the relevant key to take the measurement.

Fig. 11-11: Lining up on the mini prism for a height measurement

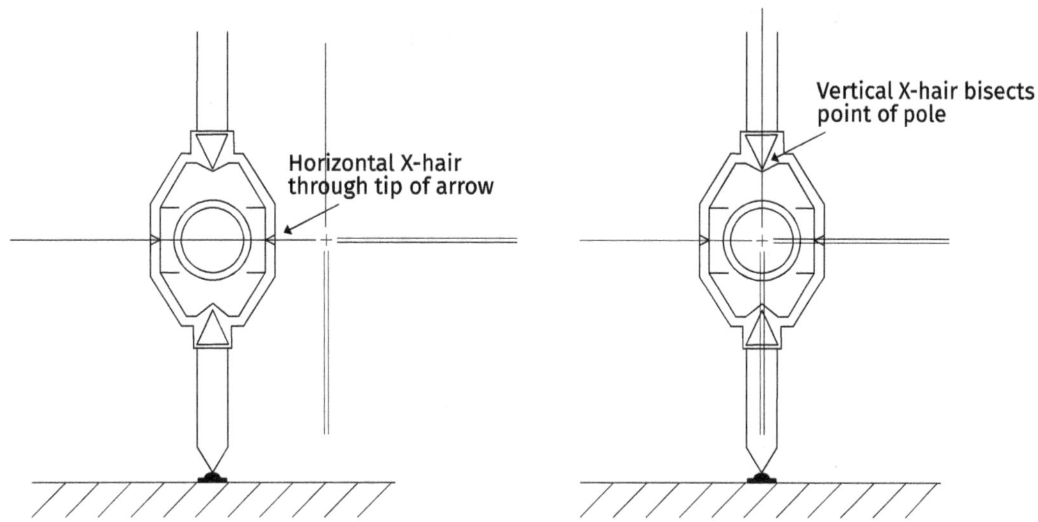

Step 1:
Line up the horizontal x-hair

Step 2:
Scroll sideways until the vertical x-hair is correctly aligned.

Fig. 11-12: Lining up on backsight - offset error

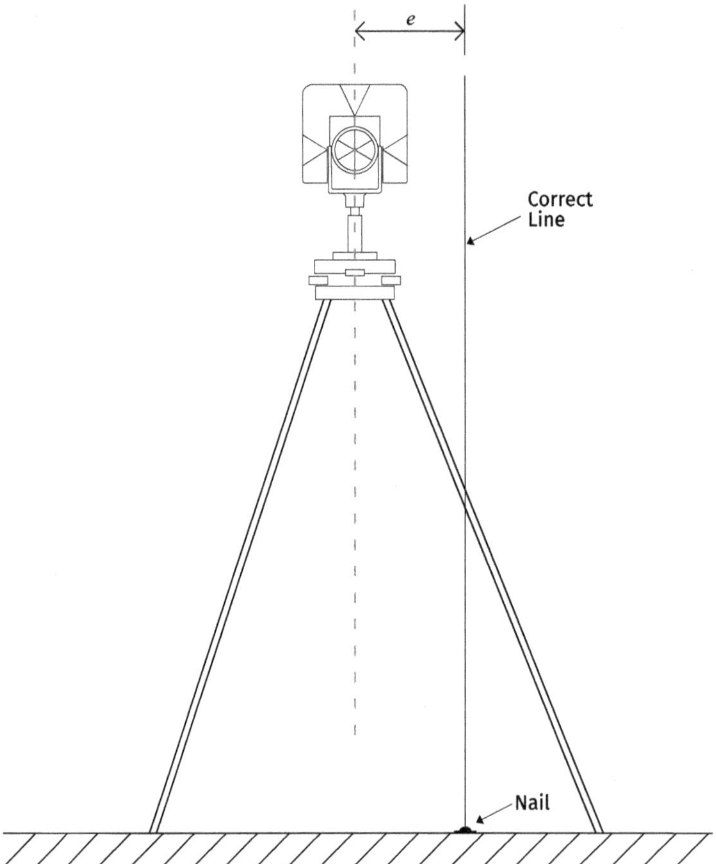

Fig. 11-13: Lining up on a retro for a height measurement

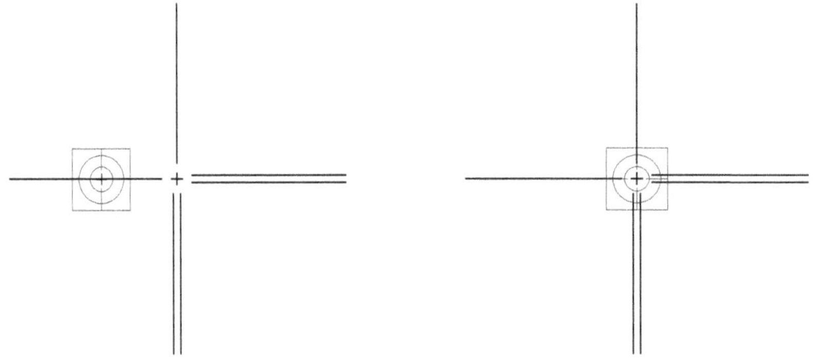

11.16 WHAT IS A BACKSIGHT WHEN USING THE TOTAL STATION?

A backsight is a measurement to a reference point. In levelling, the backsight is the staff reading to a TBM of known reduced level, which allows the height of collimation to be determined. When using the total station, the backsight is to a marker (usually either a PGM or retro target) of known coordinates and/or reduced level.

11.17 WHY SHOULD BACKSIGHTS BE AS FAR AWAY AS POSSIBLE WHEN USING A TOTAL STATION?

When sighting to your backsight, there will be offset errors caused by:

- → lining up the cross-hair on the target
- → the centring of the target over the backsight PGM
- → the instrument set-up (either the resection results or centring over the known set-up point)

This offset error then translates into an angular error. This angular error in the backsight then translates into an angular error in the foresight, i.e. in the setting out or measurement. This angular error then translates into an offset error which increases as the distance from the instrument increases. Here is a numerical example to illustrate the point:

- → An offset error of 3mm over a backsight distance of 10m would translate into an offset error in the foresight of 30mm at a distance of 100m.
- → If the situation was reversed and a long backsight and short foresight were used, an offset error of 3mm over a 100m backsight would translate into an offset error of 0.3mm over 10m in the foresight.

11.18 WHAT ARE THE DIFFERENT TYPES OF TARGET?

There are several different target types for the total station:

- → retro target – reflective sticker with horizontal and vertical lines
- → prisms – glass prism (including round prism (Fig 11-14) and mini prism. (Fig. 11-15)
- → any surface

Fig. 11-14: Round prism (Image courtesy of Leica Geosystems)

Fig. 11-15 Mini prism (Image courtesy of Leica Geosystems)

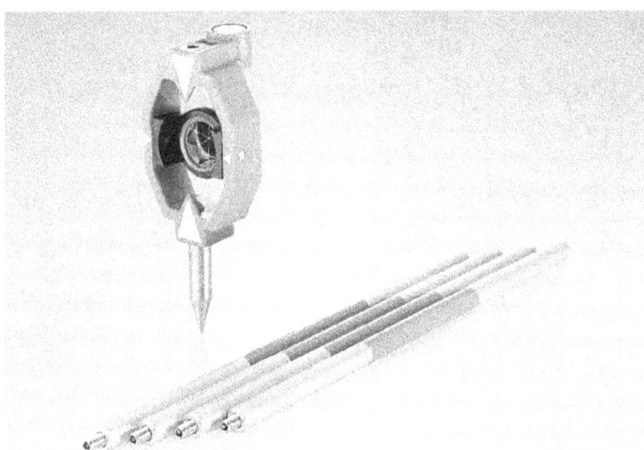

11.19   WHAT IS THE PRISM CONSTANT?

Prisms come in a range of sizes and types. They can be fixed on a moveable pole, set up on a tripod or special post or attached to a permanent holder on a stable surface by a spigot.

The prism constant is defined as the horizontal offset distance from the vertical axis of the prism holder to the theoretical convergence point of the focal point of the prism (Fig. 11-16), taking account of refraction of light as it enters the prism and the time it takes to enter and leave the prism. A negative prism constant indicates that the focal point is in front of the axis, and a positive prism constant indicates that the focal point is behind it.

Prisms which pivot about a point are known as 'nodal prisms'. They give the correct measurement to the vertical axis of the prism holder, even when they are slightly tilted or twisted away from the direct line of sight of the total station.

Fig. 11-16: Prism constant, k

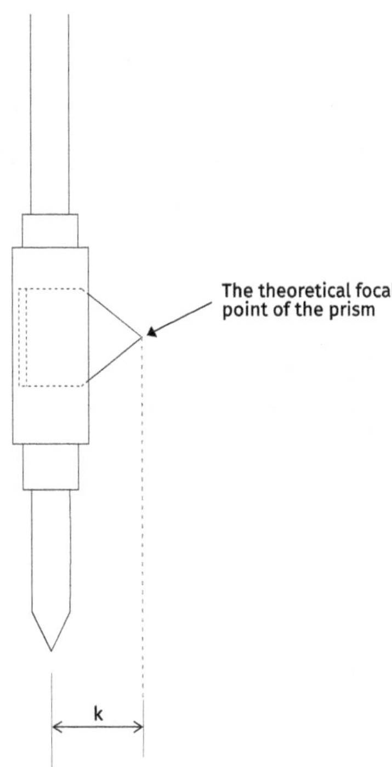

11.20 WHAT WILL HAPPEN IF I USE THE WRONG PRISM CONSTANT?

If you mistakenly use the wrong prism constant during a resection set-up, the total station will only alert you if the accuracy position result is outside of the limits you have set. If you have not manually set the limits, the default limits may be anything up to 25mm, which would be unacceptable for work requiring a higher level of accuracy than this. Ensure that you always check the accuracy position as part of your setting-up process, rather than relying on the total station to alert you to any unacceptable errors.

If you mistakenly use the wrong prism constant during an 'known point' set-up, the total station will not alert you to this. In fact, even if you have sighted to completely the wrong point, the instrument will assume that you are on the correct line. It is essential that you always measure the coordinates of another known point and verify that they are within the required limits.

If you set up your instrument using the correct constant for sighting to retro targets, and then forget to change to the prism constant for the prism you are using for setting out (or surveying), all your setting out (or measurements) will be incorrect by the difference between the actual and selected prism constant. No matter how many times you check the points from this set-up position, they will appear to be exactly right because you are incorporating the same error into the checking process.

Once you have completed a survey or setting out task, it is essential to set up the instrument in a different position and ideally using different control points and a different function, and check the points you have set out. Using a tape measure to check between as-set-out points eastablished from the different set-up positions would also be likely to detect any prism constant errors.

11.21 HOW DO I SELECT THE CORRECT PRISM CONSTANT?

Each time you take a distance measurement, the total station will compensate for the prism constant based on which target type it is set to. Each time you change from one target type to another, you must select the new type. You only need to change the settings when you change from one type to another. When you switch the instrument on, remember to check which target type is set to as some instruments default to the last type used, however some may default to a certain type.

Each instrument has a different process for selecting the target type. The instructions for the most common total stations available are given in the appendices of this book.

11.22 WHY DO LEICA HAVE A DIFFERENT PRISM CONSTANT SYSTEM TO OTHER MANUFACTURERS?

For most total station manufacturers including Topcon, Nikon, Sokkia and Trimble, the prism constant is defined as the distance between the vertical axis of the prism holder and the theoretical focal point of the prism. When the two are coincident, the absolute constant is 0mm. With Leica however, the prism constant is defined as the distance between the vertical axis of the prism holder and a point 34.4mm behind the theoretical focal point. This means that when the vertical axis and turning point of the light are coincident, i.e. for an absolute constant of 0mm, the Leica prism constant is -34.4mm.

This constant originates from Leica's original nodal prisms, which were designed with a 34.4.mm offset to provide optimal accuracy when the prism is tilted up or down. When selecting the prism type within a Leica, both the absolute and Leica constants are displayed. The constant shown on the side of the Leica prism is the Leica constant. To calculate the absolute constant, add on 34.4mm.

11.23 WHAT IS A 360-DEGREE PRISM?

A 360-degree prism is a prism which can be viewed from all sides and is used with robotic total stations. See Fig. 11-17.

Fig. 11-17: A 360-degree prism (Image courtesy of Leica Geosystems)

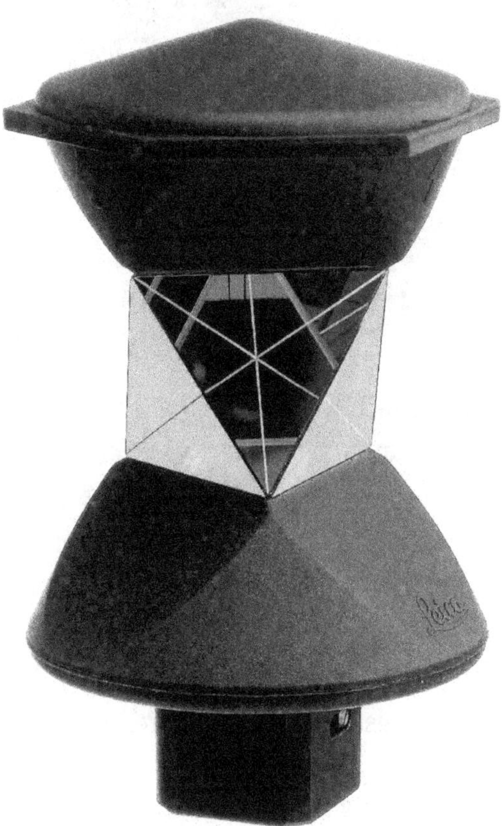

11.24 WHAT IS A MULTI-TRACK PRISM?

A multi-track prism is an active target which has multiple channels for sending out radio signals. The channel is selected on both the target and the total station. This allows multiple robotic total stations to be used on the same site, without the risk of the total station locking onto the wrong target. See Fig. 11-18

Fig. 11-18: Multi-track prism (Image courtesy of Korek Group)

11.25 WHAT'S THE DIFFERENCE BETWEEN AN ACTIVE AND PASSIVE TARGET?

A passive target is a prism which relies on its reflective surfaces to provide location information to the robotic total station.

An active target sends out a radio signal which enables the robotic total station to lock onto the prism with the correct ID. They are used on sites where there are multiple prisms in use, to prevent the total station locking onto the wrong target.

11.26 WHAT ARE RETRO TARGETS?

Retro targets are small squares of reflective material with adhesive backing which can be positioned on vertical surfaces. These sacrificial targets are either silver or neon yellow and are typically available in sizes of 10mm, 25mm, 30mm or 40mm square. See Fig. 11-18.

Fig. 11-18: Retro target

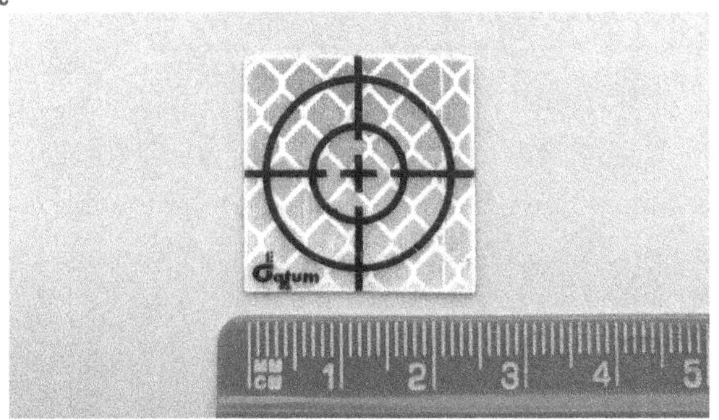

11.27 WHAT ARE THE BENEFITS OF RETRO TARGETS?

Retro targets can be positioned on vertical surfaces, liberally and in convenient locations. There are four main benefits to using retro targets:

→ when sighting to a mini-prism, the accuracy depends on the steadiness of the prism and its position in relation to the centre of the nail. When a backsight prism is set up on a tripod, there is potential for a centring error. Once a retro target is in position however, there is zero movement (assuming there is no movement of the element to which they are attached)

→ there is potential to lose some accuracy when sighting to a prism. The retro target has much clearer horizontal and vertical centrelines than the prism, enabling very accurate lining up

→ you can set up a resection without the need for an assistant

→ retros can be positioned in higher locations where they are less likely to become obstructed or damaged

11.28 WHAT ARE THE LIMITATIONS OF RETRO TARGETS?

Although they have many advantages over prisms, retro targets also have some limitations. They are that:

→ the line of sight needs to be perpendicular to the face of the retro to achieve optimum accuracy

→ if the retro is too far away from the instrument the signal may not be strong enough to produce a measurement.

→ the accuracy with which instruments lock onto retro targets varies from instrument to instrument

→ rain or dirt on the retro can affect the accuracy

→ If the signal is not strong enough when the EDM mode is set to retro, try setting it to reflectorless mode ('Non-Prism)

11.29 WHAT IS REFLECTORLESS MODE?

The reflectorless mode enables measurements to be taken to surfaces other than prisms and retro targets. The advantages of the reflectorless measurement function are that:

→ you don't need an assistant to hold the prism

→ you don't need access to the point you are sighting, so it can be much safer and quicker

→ it can achieve a high level of accuracy for distance measurement (up to 1mm+1ppm depending on the model)

→ it can be used for checking the plumb of walls and columns. See section **24.67** and **24.69** for instructions

→ it can be used for checking the alignment of elements. using the reference line function

11.30 WHAT ARE THE DISADVANTAGES OF REFLECTORLESS MODE?

The disadvantages of the reflectorless function are that:

→ a solid surface is needed, so it is impossible to sight exactly to an edge there for some accuracy can be lost by moving slightly in from the edge to sight to a solid surface

→ some surfaces may not be suitable

→ you don't know where on the surface you are sighting to

→ even though high accuracy can be achieved in distance measurement, there is still the same scope for angular error when lining up

11.31 SIMPLE DISTANCE MEASUREMENT USING THE TOTAL STATION

One of the most basic functions of the total station is to measure the slope distance from the total station to the target. The total station uses the measured slope distance and vertical angle to calculate the horizontal (plan) distance. To measure the distance between two points:

→ Centre and level the total station over one of the points.

→ Sight to a target at the second point.

→ In survey mode, press 'Dist' and the distance will be displayed on the screen.

The slope distance is indicated by a right-angled triangle with a parallel line adjacent to the hypotenuse. The horizontal distance is indicated by a line parallel to the bottom edge of the triangle.

11.32 HOW DO I MEASURE A DISTANCE USING THE TOTAL STATION?

There are two methods for measuring the distance between two points using a total station:

1. set up on point 1 and measure the distance to a prism at point 2 (or to a surface using the reflectorless mode)
2. tie distance/compute inverse function (see section **13.2**)

The method described below is the simplest and involves centring the total station over one of the points and a prism over the other. The steps are as follows:

1. Centre the total station over one of the points.
2. Centre a prism over the other point. The prism may be a mini-prism, a detail pole or a backsight target on a tripod.
3. On the home screen, press the function key for 'dist'.
4. Press the 'Rec' key if you would like to store the measured distance in the memory.

11.33 WHAT'S THE DIFFERENCE BETWEEN THE PLAN DISTANCE AND SLOPE DISTANCE?

When you measure a distance using the total station, the instrument takes a direct measurement between the trunnion axis of the instrument and the point being sighted, as well as measuring the angle on the vertical circle. The slope distance and vertical angle are then converted into the horizontal and vertical distances. The three measurements are displayed on the screen (Fig. 11-19):

→ the horizontal, or plan, distance which is indicated by the letter H or a right-angled triangle with a line adjacent to the bottom face

→ the slope distance which is indicated by the letter S or a right-angled triangle with a line adjacent to the hypotenuse

→ the vertical distance, or level difference, which is indicated by the letter V or a right-angled triangle with a line adjacent to the vertical edge.

Fig. 11-19: Plan distance, slope distance and level change

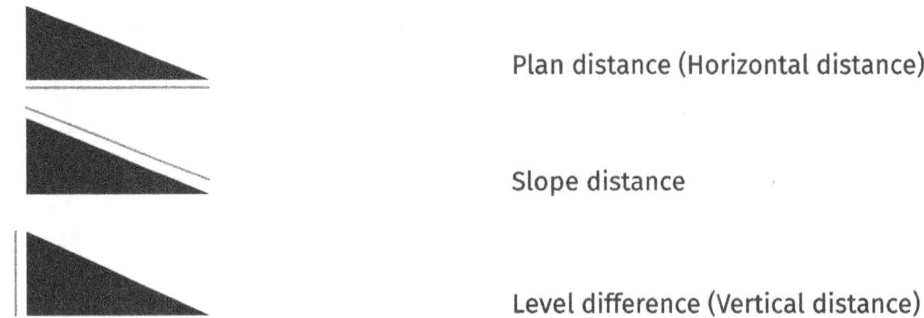

11.34 WHAT IS THE ACCURACY OF THE DISTANCE MEASUREMENT FUNCTION?

The distance measurement accuracy of most modern total stations is 1 – 2 mm (+ 1.5 - 2mm/km). Note that this figure pertains purely to the accuracy of the total station to measure between two points and does not take any account of external human error.

11.35 WHAT ARE THE SOURCES OF ERROR IN DISTANCE MEASUREMENT?

Other than the accuracy of the total station, the two key sources of error in distance measurement are:

→ centring of the total station over the point

→ centring of the target over the point (including verticality and steadiness of the detail pole or mini-prism)

11.36 HOW CAN I MINIMISE DISTANCE MEASUREMENT ERRORS?

If you are using a mini-prism or detail pole as your target, you can improve your accuracy by taking 5 consecutive readings and averaging them.

This is because, if you just take one reading, it is possible that you have taken it at a moment when the bubble wasn't centred. By taking only one reading, you have no way of knowing whether the prism pole was truly vertical or happened to be wobbling about at the exact moment that you pressed the 'Dist' button.

This could result in a small error of a few millimetres or could potentially be much worse; over long distances it's difficult to tell whether your assistant is ready for you to take the reading, or even if the prism is on the correct point.

Recording 5 consecutive readings has 3 purposes:

→ to improve the accuracy of the reading. The greater the number of readings taken, the closer the average will be to the true value.

→ to increase reliability. A single reading could be erroneous and without any evidence to the contrary, your reading is unreliable. Taking 5 consecutive readings provides this evidence. Note that this does not rule out the possibility of other gross errors such as being on the wrong nail or having the EDM set to the wrong prism constant; separate checks should be carried out to eliminate these possibilities.

→ to check precision, i.e. to gauge the steadiness of your assistant's hand.

11.37 WHAT IF THE PRISM IS TWISTED/ROTATED?

For simple distance measurement, even if the prism seems to be slightly twisted away from the total station, the measurement returned will be correct.

11.38 WHAT IF THE CROSS-HAIR IS NOT EXACTLY AT THE VERTICAL CENTRE OF THE PRISM?

If the centre cross is anywhere on the face of the prism, the distance reading will be correct. Try this for yourself by lining up the centre cross at the centre of the prism and taking a reading. Then take readings at random points on the face of the prism, including close to the edge. You will notice that the distance reading is the same each time.

11.39 HOW DO I SET UP THE TOTAL STATION?

When you set up the instrument using one of the on-board functions, the total station uses the inputted station and backsight data to determine the position and orientation. The type of instrument set-up required depends on the task. There are several ways to set up the total station:

→ **known point set-up** - sets the position and orientation of the total station by setting up over a point of known coordinates and backsighting to another point of known coordinates

→ **resection** - sets the position and orientation of the total station by setting up at an unknown (random) point and backsighting to at least two points of known coordinates, enabling the total station to mathematically resolve its coordinates and orientation

→ **level the instrument** – when the position and orientation of the total station are not needed, the instrument can simply be levelled.

11.40 WHAT IS THE PROCESS FOR SETTING UP ON A KNOWN POINT? 📄

The position and orientation of the instrument can be determined by setting up the instrument over a point of known coordinates and backsighting to another point of known coordinates.

To set up on a known point, the steps are as follows:

→ Select the correct function

→ When prompted, enter (or select from the memory) the coordinates of the point you are set up over.

→ When prompted, enter (or select from the memory) the coordinates of your backsight point

→ Sight to the backsight and press the measure key.

→ You can now choose to compute the position and orientation based on the measurements taken so far or to take additional measurements to improve the accuracy of your set-up

11.41 WHY WOULD I NEED TO SET UP OVER A POINT?

For setting out activities which require a high level of accuracy, such as installing control points or setting out bolt positions for steel columns, it is necessary to set up the instrument over a point of known coordinates and backsight to another point of known coordinates.

11.42 WHAT IS THE PROCEDURE FOR SETTING UP THE TOTAL STATION (WITH AN OPTICAL PLUMMET AND LONG BUBBLE) OVER A POINT?

The method described below is one of several quick and systematic procedures. Even if it seems long-winded when you first attempt it, you will soon become much quicker with practise.

To centre and coarsely level the total station (or backsight target) over a nail:

1. Stand several paces away from the nail you are setting up over.

2. Extend the tripod legs equally. To do this, keep the legs of tripod closed together. Release the clamps and screws on all legs and bring the tripod head up to about chest height. If your tripod has both clip clamps and screw clamps, use only the clip clamps and leave the screw clamps undone for the remainder of the set-up process.

3. Space out the legs in an equilateral triangle. You should still be standing several paces away from the nail.

4. By eye, check the tripod head is roughly level. It may help to place your palm on top to get a feel for it.

5. Lift up the whole tripod, walk forwards towards the nail and place the tripod down with the centre of it over the nail. The tripod hinges should be stiff enough to hold their position. If the hinges are slack, you may need to hold them in position as best you can to prevent them falling inwards.

6. Stand back 1 or 2m and check the alignment of the centre of the tripod head in relation to the nail. Repeat on each of the three sides. Move the tripod laterally as necessary until the centre of the tripod head is lined up with the nail from all three sides.

7. Hold the tripod screw fixing vertically and at the centre of the tripod head. Look down from directly above to make sure it is centred over the nail. If the nail is not in view, adjust the position of the tripod head until it is.

8. Check again that the tripod head looks and feels horizontal. Now, your tripod should be set up directly over the nail and the tripod head should look level. If these two conditions are not met, do not continue. Adjust the position of the tripod until it is centred over the nail by moving the feet.

9. Stamp the feet of the tripod firmly into the ground by putting all your weight on the protruding foot platform. Look down through the screw fixing to check that the tripod is still centred over the nail and check that the tripod head is still level. If it has moved significantly off centre, lift the tripod, move it laterally as necessary and repeat step 8.

10. Take the total station (or backsight prism) out of the box and attach it to the tripod head centrally.

11. Adjust each foot screw so that the top of it is touching the line which indicates it is half way up its travel. If there is no line, estimate its halfway position.

12. Look through the optical plummet. If it is more than, say, 200mm, it may be worth putting the total station back in the box and re-centring the tripod over the nail.

13. Looking through the optical plummet, adjust the foot screws until the black circle is centred over the nail. If the nail is not in view, whilst looking through the optical plummet, move your foot around in the view until you can see the tip of your toe. Leaving your foot in that position, move back from the eye piece and look directly at your foot to see where it is in relation to the nail. This should give an indication of which way you need to turn the foot screws to sight the nail.

14. Adjust the length of two of the tripod legs, one at a time, until the pond bubble is centred. Place your foot on the foot platform of the leg you are adjusting to ensure that it doesn't lift out of the ground or change position laterally. It is important that you only adjust the length of the leg, but do not move the position of the foot. The bubble will move towards a leg if you raise it, and away from the leg if you lower it. Adjust any two tripod legs, but not all three.

15. Repeat steps 12 to 14 as many times as necessary until the pond bubble is centred and the black circle is within, say, 30mm of the nail.

16. Whilst looking through the optical plummet, loosen the screw fixing from the base of the tripod so that the total station can be slid around the tripod head, leaving the screw fixing still lightly attached (if you detach it completely, your view through the optical plummet will be obstructed). Slide the base of the total station around the tripod head until the nail is within the black circle. Tighten the screw to fix the total station in position. The total station should now be centred over the nail, and the circular bubble is centred. If not, centre the circular bubble using the foot screws and repeat step 14.

To accurately level the total station (or backsight prism):

NB Before continuing to step 17. the optical plummet should be centred over the nail and the circular bubble centred. If these two conditions are not met, do not continue. Repeat the process of centring and coarsely levelling the total station (or backsight station) until they are met.

17. Rotate the instrument until the long bubble is parallel to two of the foot screws (Fig. 11-20). Call them A and B. Foot screw C is the remaining foot screw.

18. Rotate foot screws A and B in opposite (counter) directions until the bubble is centred. Turn the foot screws at exactly the same rate as each other.

19. Rotate the instrument horizontally through 90 degrees. The line of the long bubble should now be perpendicular to the axis line through A and B (Fig. 11-21).

20. Use foot screw C on its own to centralise the long bubble.

21. Rotate the instrument horizontally through 90 degrees. The line of the long bubble should now be parallel to the axis line through A and B, but on the opposite side of the instrument to your original position (Fig. 11-22)

22. Rotate A and B in opposite (counter) directions and at the same rate as each other until the bubble is centred.

23. Rotate the instrument horizontally through 90 degrees. The line of the long bubble should now be perpendicular to the axis line through A and B, but on the opposite side of the instrument (Fig. 11-23).

24. Use foot screw C on its own to centralise the long bubble.

25. Turn through 90 degrees, back to the starting position (Fig 11-20). The long bubble should be centred. If not, repeat from step 2.

26. Look through the optical plummet to check that you are still centred over the nail. If the nail has moved out of the black circle, loosen the screw fixing, reposition the instrument over the nail by sliding it across the tripod head and then repeat steps 1 – 9. The instrument should be centred over the nail, and the long bubble centred which way it is turned. If these two conditions are not met, see the troubleshooting checklist. (Section **11.50**)

Fig. 11-20 Bubble position 1

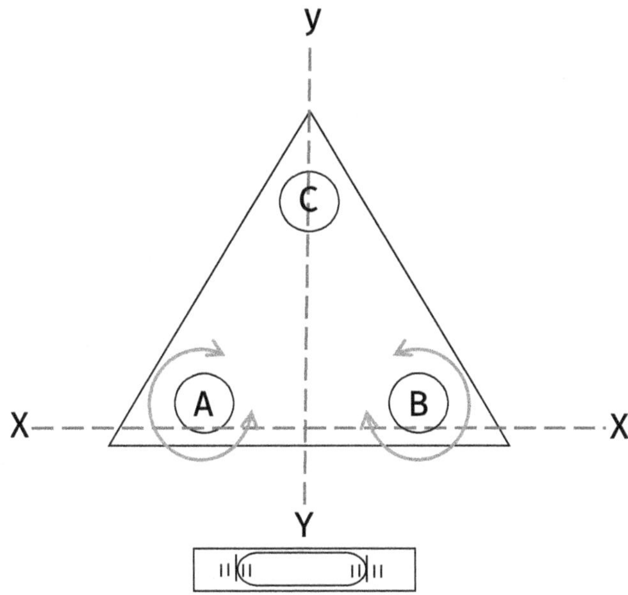

Fig. 11-21 Bubble position 2

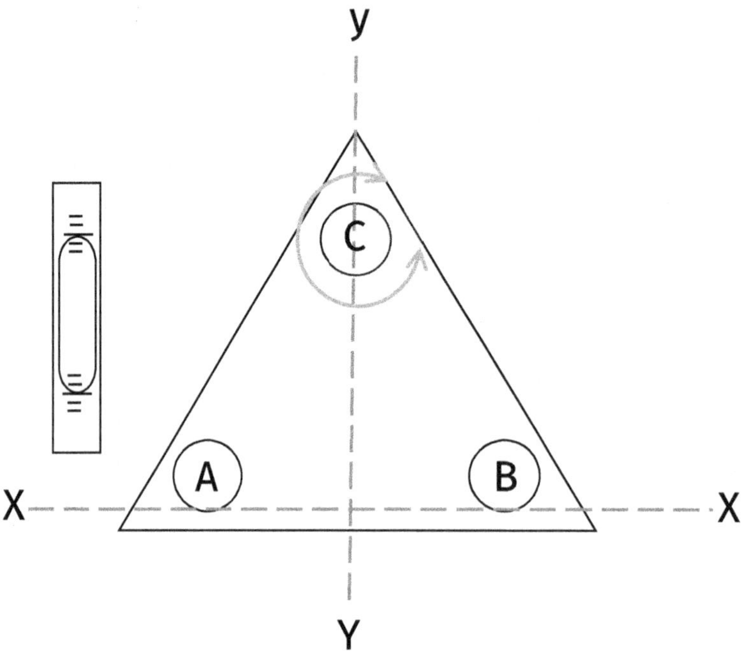

Fig. 11-22: Bubble position 3

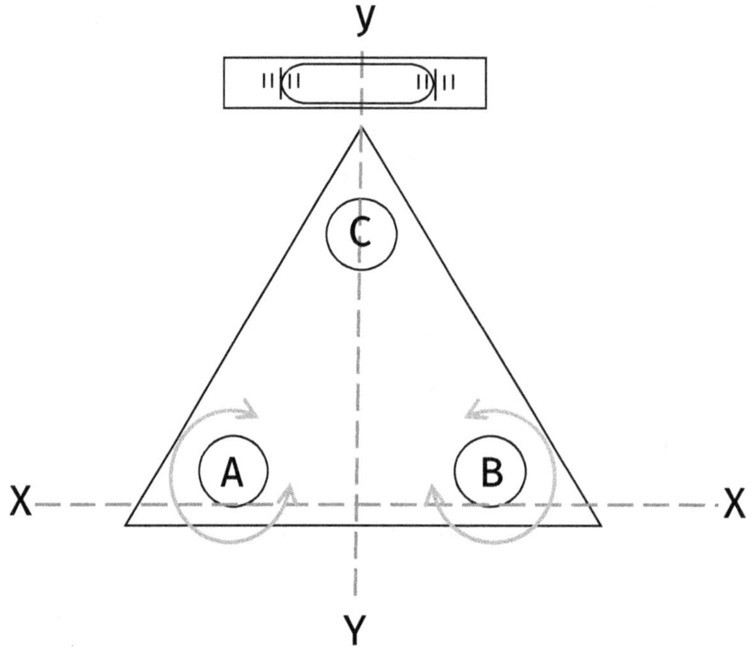

Fig. 11-23: Bubble position 4

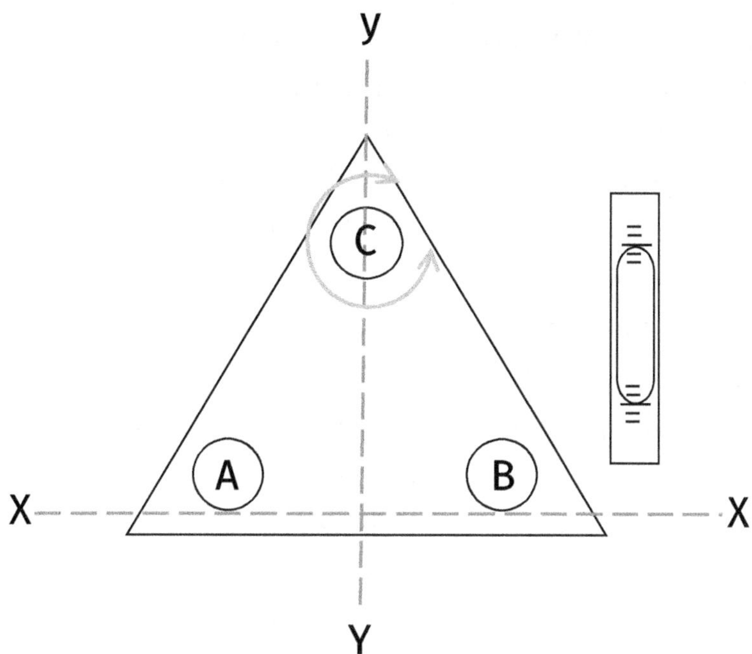

11.43 WHAT IS THE PROCEDURE FOR SETTING UP THE TOTAL STATION (WITH A LASER PLUMMET AND DIGITAL BUBBLE) OVER A POINT?

Steps 1 – 11 are exactly the same as for levelling a total station with an optical plummet.

12. Switch the total station on. A red laser dot will appear on the ground. If it is not visible, place the palm of your hand underneath the screw fixing so that you can see the laser dot. Move your hand gradually downwards to locate the position of the dot on the ground. If the ground is grassy, place your foot under the laser dot so that the dot is visible. On most total stations, the laser dot appears automatically when the instrument is switched on. If it does not automatically appear, you will need to select the correct function which may be called 'level/plummet', 'user set up' or similar. If the dot is more than, say, 200mm away from the nail, it may be worth putting the total station back in the box and re-centring the tripod over the nail before moving onto the next step.

13. Adjust the foot screws until the laser dot is centred over the nail.

14. Adjust the length of two of the tripod legs, one at a time, until the pond bubble is centred. Place your foot on the foot platform of the leg you are adjusting to ensure that it doesn't lift out of the ground or change position laterally. It is important that you only adjust the length of the leg, but do not move the position of the foot. The bubble will move towards a leg if you raise it, and away from the leg if you lower it. Adjust any two tripod legs, but not all three. The laser dot will move off the nail, but don't worry, this is just part of the process.

15. Repeat steps 13 and 14 as many times as necessary until the circular bubble is centred and the laser dot is within, say, 30mm of the nail.

16. Loosen the screw fixing from the base of the tripod so that the total station can be slid around the tripod head, leaving the screw fixing still lightly attached (if you detach it completely, the laser dot will be obstructed). Slide the base of the total station around the tripod head until laser dot is centred over the nail. Tighten the screw to fix the total station in position. The total station should now be centred over the nail, and the pond bubble is centred. If not, centre the pond bubble using the foot screws and repeat step 14.

17. The diagram on the screen displays the digital bubble. Rotate the footscrews as directed on the screen until a tick appears at each footscrew.

11.44 CAN I CENTRE OVER THE NAIL BY ATTACHING THE TOTAL STATION FIRST AND THEN POSITIONING THE TRIPOD HEAD?

There is a method of setting up over a nail which involves attaching the total station to the tripod, pressing one of the tripod feet into the ground, lifting the two remaining legs and moving them about until the optical or laser plummet is pointing at the nail.

The problem with this method is that as you move the total station about, it is not necessarily vertical. Even if the plummet is pointing at the nail, you are not directly over it and the tripod head is not necessarily level. This means that the next stage can take several attempts.

Some people do find this method successful, however, be aware there is a definite knack required. If you find that this method works well for you, there is no reason not to use it. However, if you are new to using a total station, I would recommend starting with the method described in section **11.43** and experiment with adapting it or speeding it up, once you are confident.

11.45 WHY IS THE CIRCULAR BUBBLE GOING WRONG WHEN I ADJUST THE LONG BUBBLE?

While following the steps for levelling the long bubble, ignore the circular bubble completely. During the process, the circular bubble will become uncentred, however if you continue until the end of the process, it will end up back in the centre. If it doesn't, it may be that the circular bubble is out of adjustment.

The circular bubble is only used for the rough levelling in the initial stage and is superseded by the long bubble in the accurate levelling stage. If the outcome is that the long bubble is centred and the instrument is directly over the nail, don't worry if the circular bubble is slightly out. You may wish to have this attended to later, but it shouldn't prevent you working with the instrument.

11.46 HOW CAN I SET UP OVER A POINT MORE QUICKLY?

It may seem like a long-winded process at first, however with repetition, you should be able to set up over a point in a minute or less. Why not try timing yourself and trying to beat the clock?!

The two most important things to remember when aiming to reduce the time taken to set up over a point are:

→ the initial step of getting the tripod over the nail and the tripod head level will have the most impact on the overall process. If you get this bit right, the rest will be very quick.

→ a bit like life, if what you are doing isn't working, try something different. Don't keep repeating the same mistake over and over and wonder why you are not getting anywhere. If you are getting tangled up, start again, follow the steps, use the troubleshooting checklist in section **11.50** if necessary and above all, practise, practise, practise!

11.47 HOW DO I SET UP ON SOFT OR UNEVEN GROUND?

It is much easier to set up on a hard, flat surface, however it's not always possible. Here's a suggested sequence for setting up on soft uneven ground:

1. Start by making the lengths of the tripod legs even.

2. Stand the tripod over the nail, without stamping the feet in.

3. Position the tripod to get the tripod head over the nail, but don't worry if it is not level.

4. Adjust the length of the legs to get the tripod head level.

5. Check that the tripod head is still over the nail. If not, lift the whole thing up and move it over the nail.

6. Adjust the length of the legs to get the tripod head level if necessary.

7. Check that it is still over the nail.

8. Stamp all three feet into the ground as far as they will go.

9. Adjust the length of the legs until the tripod is level.

10. Once the tripod head is level and over the nail, attach the total station and continue as normal.

11.48 DOES IT REALLY MATTER HOW I SET UP OVER A POINT?

There are various methodical procedures for setting up the total station over a point. No one method is either right or wrong. In fact, if the outcome is that the total station ends up centred over the point and is accurately levelled, the quality of the set-up will be the same, no matter how you got there.

However, approaching the task in a random and haphazard way can be time-consuming and frustrating. It makes much more sense to develop your own iterative, methodical procedure and practise it until it becomes second nature.

11.49 HOW LONG SHOULD IT TAKE TO SET UP THE TOTAL STATION OVER A POINT?

Your first attempt to centre and accurately level a total station may well take half an hour or more, but don't worry. After several attempts of setting up using the same system, you will soon reduce that to 5 minutes or less. Experienced engineers may take less than two minutes to set up the instrument over a point.

11.50 I CAN'T GET THE INSTRUMENT OVER THE NAIL AND LEVEL! WHAT AM I DOING WRONG?

The most likely explanation is that you have not followed the steps exactly. You may have missed something out or misinterpreted an instruction. The troubleshooting checklist below is designed to help you isolate exactly what is going wrong so that you can fix the problem. This list is relevant for instruments with optical plummets and laser plummets.

→ **The tripod is poorly positioned** - your initial tripod position may not have been good enough. It is important to make sure the head of the tripod is directly over the nail and looks level before attaching the total station. To increase the chance of being able to slide the total station over the nail, use the heavyweight tripod which has a larger area for the total station to move about on.

→ **You are raising or lowering the legs in the wrong direction** – remember if you want the bubble to move towards a leg, raise it, or to move in the opposite direction from a leg, lower it.

→ **You are getting the steps mixed up** – one of the most common mistakes is using the foot screws to centre on the nail, then using the foot screws again to level the pond bubble. If you do this you will be going in circles endlessly, as you are doing work and then undoing it repeatedly. The correct procedure is to use the foot screws to centre on the nail whilst looking through the optical plummet, then adjusting the length of the legs to centre the pond bubble. Repeat these two steps until you are close enough to loosen the screw fixing and slide the total station over the nail.

→ **The tripod feet are moving position** - when increasing and reducing the length of the legs to centre the pond bubble, it is important that the positions of the feet do not move at all, otherwise the process will not work. If you are on hard ground or a shiny surface, make sure that you are pressing your foot down on the foot plate while you are extending the legs.

→ **The tripod legs are not working properly** – the clamps on your tripod legs may be loose. To check that this is not the case, remove the total station from the tripod, make sure the feet cannot slip outwards and press down with as much force as possible on the head of the tripod. Press down over each individual leg. The clamps should not slip at all. If they do, the bolt may need to be tightened with a spanner.

→ **Poor control of the leg adjustment** – if you are finding it difficult to centre the circular bubble by adjusting the length of the legs, it could be that you are not gripping the legs in the optimum way. If your tripod has screw clamps as well as clips, check that they are loose. Stand facing the leg you are adjusting, with the palm of your right hand facing the leg, little finger at the top and thumb at the bottom. Grip the right-hand pole so the knuckle of your forefinger is wedged between the flat surface at the top of the lower section, and the pole. Use your left hand to release the clip. The leg should not move, because you are holding it steady with your right hand. Adjust the bubble until it is either centred, or if this is not achievable, so that it is opposite one of the other two legs. When you are happy with its position, your left hand is free to tighten the clip.

→ **Back lash in the leg clips** – when you centre the circular bubble by adjusting the length of the legs, as you tighten the clip, the bubble moves out of the circle. To resolve this, identify which direction the bubble moves and compensate by setting the bubble slightly in the opposite direction before tightening the clip.

→ **Not enough travel on the foot screws** – if your foot screws reach the end of the run and will not move any further, do not force them as this can cause damage. You have two options. The first is to adjust the length of the tripod legs to overshoot the bubble and then reverse the foot screws to bring the bubble to the centre. The second is to set all the foot screws to the mid-point of their run, remove the total station from the tripod, re-position the tripod over the nail and re-level the head, and start again.

→ **The pond bubble is out of adjustment** - if all other possibilities have been exhausted, it is possible that the circular bubble is out of adjustment. On the total station there are two circular bubbles, one on the total station itself and one on the tribrach. You can compare these bubbles to each other. Another way to check the adjustment of the circular bubble is given in section **10.25**

11.51   I CAN'T GET THE LONG BUBBLE CENTRED! WHAT IS GOING WRONG? 🚚

This section is relevant for backsight prisms and total stations which have a long bubble rather than a digital bubble.

If you've followed the steps exactly, turned through 360 degrees in 4 stages, and returned to the starting position and the long bubble is not centred, there are several possible reasons for this:

→ **You turned the foot screws at different rates** – to isolate the direction of the bubble, when the bubble is parallel to two of the foot screws, both the foot screws must be rotated in opposite directions at exactly the same rate as each other. If you only turn one, or you turn one faster than the other, you are tilting the bubble in the perpendicular direction, but it is not registering. This means you are undoing what you did in the previous stage.

→ **You didn't rotate the bubble through exactly 90 degrees** – if the bubble is not exactly parallel to the two foot-screws you are adjusting, or if it is not perpendicular to the axis of the single foot screw you are adjusting, you will not have fully isolated the direction of tilt. This means that you will be tilting the bubble in two directions but it will only be registering in one.

→ **You turned the wrong foot screws** – when the bubble is parallel to two foot screws, you should adjust those two foot screws. When the bubble is perpendicular to two foot screws, you should adjust the third foot screw. If you get this the wrong way round, you will be tilting the bubble in a direction perpendicular to the line of the bubble and the changes will not show up.

→ **The long bubble is out of adjustment** – if you have eliminated all other sources of error and the long bubble is still not centred when you return to the starting position, it could be that the long bubble is out of adjustment.

→ **The ground is moving** – the long bubble is extremely sensitive and even minute ground movements can affect it. If you have no choice but to set up on very soft or wet ground, avoid placing your weight anywhere near the tripod feet even as you move from one position to another. Ensure that the feet are stamped into the ground as far as they can go otherwise they could be gradually sinking in under the self-weight of the tripod and total station.

→ **The clips on the tripod legs are not working properly** – see section **9.18** for how to check if the clips are working properly.

11.52 WHY CAN'T I SEE THE NAIL THROUGH THE OPTICAL PLUMMET?

If you don't have a good view of the nail there are a few possible reasons for this:

→ **The plumb bob suspension hook is in the way** - there is a small hook suspended from the centre of the screw fixing. This may be obstructing the view through the optical plummet to the nail. Try holding it out of the way or clipping it to the edge of the screw-fixing so it is not dangling down.

→ **The nail is not in view** – if the nail is not actually within the field of vision, it is hard to tell if the optical plummet is in focus. To move the nail into the field of vision, move your foot so your toe is touching the black circle and focus up your foot. Then take your eye away from the optical plummet and look down at the ground to see where your foot is in relation to the nail. This should help you adjust the foot screws to get closer to the nail. If the nail cannot be reached by turning the foot screws, the best thing to do is remove the total station, reposition your tripod and start again.

→ **The optical plummet is damaged** – although this is the least likely reason for not being able to see the nail clearly, it is a possibility. To check that the optical plummet is working properly, follow the steps in section **11.53** on how to focus the optical plummet.

→ **The light is not good enough** – this may be a problem if you are indoors or if you are working in poor daylight. It can help to first of all, focus the optical plummet as in section **11.53**. It can also help to lay a pencil or other contrasting object on the floor, with the point touching the nail.

11.53 HOW DO I FOCUS THE OPTICAL PLUMMET ON THE NAIL?

1. Move your foot around on the ground until it is at the centre of the field of vision.

2. Bring your foot into clear focus using the focus screw closest to the instrument.

3. Bring the circle into focus using the focus screw closest to your eye.

4. You may need to repeat steps 2 and 3 until both your foot and the circle are in clear focus and in clear sharp contrast to each other.

5. Move your foot out of the way.

11.54 WHAT IS A RESECTION?

A resection is a method of determining the position and orientation of your total station by backsighting to two or more points which you know the coordinates of. The total station uses the distance measurements to the known points and the angles between them, to mathematically resolve its position and orientation using the sine rule. See Fig. 11-24. All modern total stations have this function.

Fig. 11-24: Resection geometry

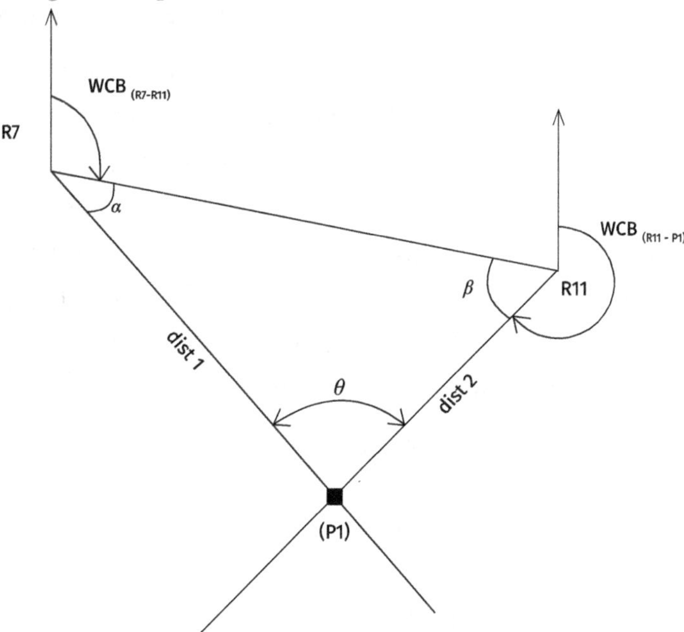

11.55 WHAT ARE THE ADVANTAGES OF A RESECTION?

The two main advantages of the resection function are the flexibility and speed it affords to the user.

The resection allows you to set up your instrument at any position where you have a line of sight to two or more of your control points. Since you are not limited to setting up directly over a control point and backsighting to another control point, you can choose the optimum set-up position for the points you want to survey or set out. This is particularly useful when access to primary control points is not possible.

In addition, retro targets can be placed on vertical surfaces in convenient locations to form control points. This allows you to complete the instrument set-up without setting up a backsight target on a tripod or an assistant to hold the mini-prism on the known point.

Another advantage is that there is no need to centre and level the total station over a nail which is a process that many engineers find frustrating and time consuming.

11.56 WHAT ARE THE POTENTIAL PITFALLS IN USING A RESECTION?

Although there are many advantages to the resection function, caution should be exercised. Due to the simplicity of the process, many engineers use the function without a full awareness of what constitutes good practice. Here are some things to consider:

→ the resection results can give a false illusion of accuracy - even when the accuracy position is displayed as 0mm, this does not necessarily mean that there is a 0mm error in the position of the instrument. It simply means that mathematically, there is 0mm error based on the measurements taken. There may be a hidden error in the position and orientation of the instrument due to either errors in the control points or a cancelling effect of sighting errors. Once the resection set-up is complete, the coordinates of a known point (preferably a primary control point) should be measured and compared to their expected values.

→ resection should only be used for appropriate tasks - some engineers use the resection function when in fact they should be setting up over a point and sighting to a known point. This may be due to lack of awareness of hidden inaccuracies or as a means of avoiding learning how to set up over a point. The resection can disguise a lack of skill and understanding.

→ poor lining up on the target can lead to unacceptable errors in the set-up

→ a suitable level of skill and knowledge is required in lining up on the targets (see section **11.13** for the correct procedure)

11.57 WHAT DOES THE ERROR MEAN IN A RESECTION?

The set-up error, sometimes called the position accuracy or residual error, is calculated using the method of least squares adjustment based on the readings taken to the backsight.

Errors in the resection set-up are caused by:

- → errors in the control points you are sighting to
- → errors in lining up on the points
- → errors in the instrument
- → poor geometry (see Fig 11-26)

Even if the instrument indicates a zero error, this does not necessarily mean that there is 0mm error in the position of the instrument. It simply means that there is 0mm error in the calculations based on the coordinates of the control points and the observations you have taken.

To check the accuracy of your resection:

1. Set up your instrument using the resection function.
2. Measure to several points with a well-defined mark and record the coordinates.
3. Set up the instrument again using the resection function, this time backsighting to different control points.
4. Use the set-out function to mark out the same coordinates you measured in step two.
5. The difference between your original point and the newly marked out points is the error. This may be due to the quality of the set-up or the relative accuracy of your control points.

11.58 HOW DO I IMPROVE THE ACCURACY OF MY RESECTION?

To ensure the best possible results when using the resection function, do the following:

- → **Choose the optimum set-up position in relation to backsight locations** - the optimum angle between backsights is approx. 90 degrees as shown in Fig. 11-25. When you are using more than two backsights, it's not possible to achieve this condition for all of them, but make sure it is met for at least two of them. Fig. 11-26 show's poor set-up positions in relation to the backsight targets.

- → **Use the correct procedure for lining up on the backsight targets** - when setting out or measuring positions, line up the vertical line cross-hair of the telescope on the vertical line of the target, then scroll up so the cross is on the centre of the target. When setting out levels, line up the horizontal cross-hair on the horizontal line first, then scroll sideways until the cross is on the centre of the target. See section **11.13**.

→ **Use a minimum of two backsights** - the more backsights you incorporate into the resection, the better the accuracy of the set-up.

→ **Check to two relevant known points** – once you have completed the resection, select the stakeout or survey mode and measure the coordinates of two other known points. Compare them to their expected value. The relevant known points may be primary control points, as-built or as-set-out points relating to the element you are about to set out

→ **Minimise the number of set-ups** – even if there is an accuracy position error in your resection set-up, all the points you set out from that set-up will be correct in relation to each other. However, each time you set up a resection, you are increasing the risk of a discontinuity error between different set-ups. Therefore, you should carry out as much work from a single resection, e.g. a whole building, or as much of it as possible, from one set-up.

→ **For internal accuracy, use a baseline rather than a resection** – where internal accuracy is critical, for example a steel-framed building, avoid using the resection function altogether. Set up a baseline and use the reference line function for setting out all parallel and perpendicular lines on that structure.

Fig. 11-25: Resection - Good geometry

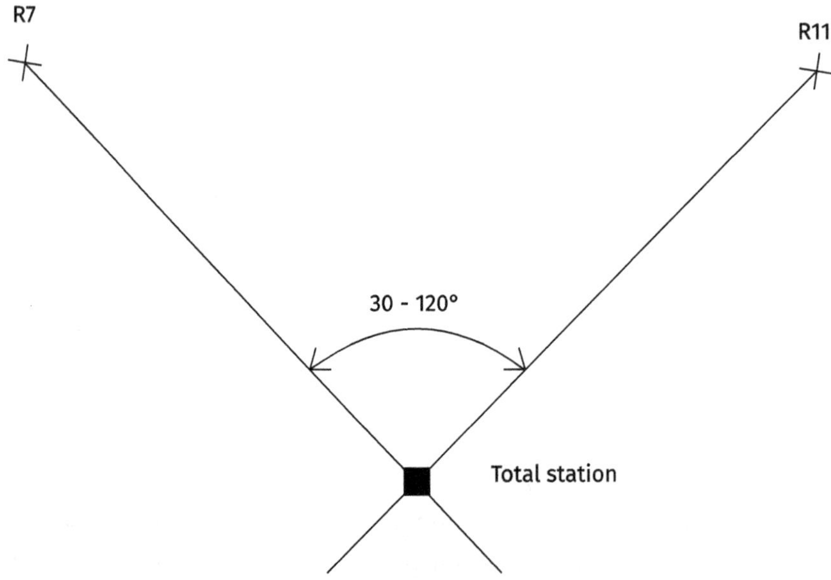

Fig. 11-26: Resection - Poor geometry

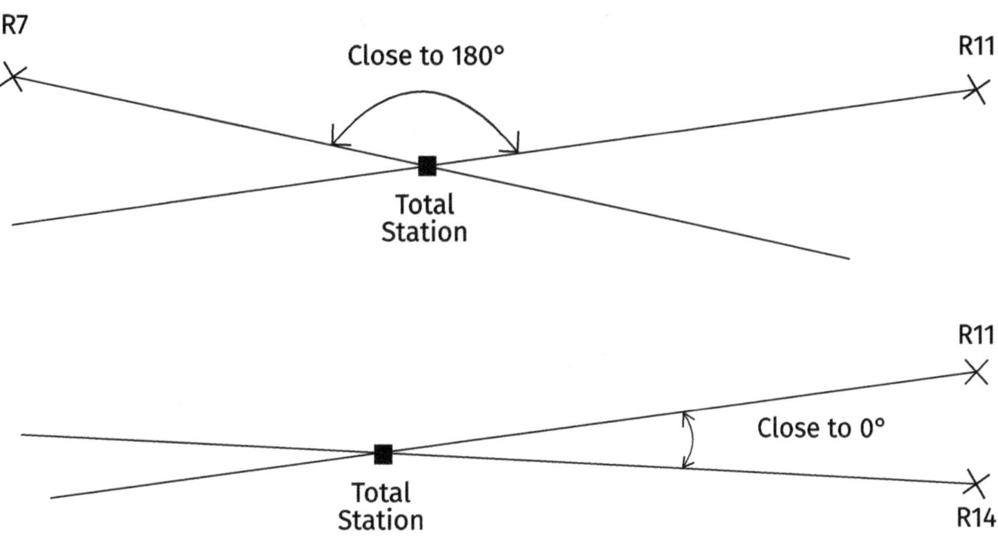

11.59 WHY SHOULD THE ANGLE BETWEEN BACKSIGHTS OF A RESECTION BE 90 DEGREES?

To calculate its coordinates and orientation, the total station uses the sine rule. By sighting to each backsight, you are measuring the distance to each target and the angle between them. The total station uses the measurements to resolve the point where these two lines intersect.

Because the error in lining up on the targets, and in the coordinates of the backsights themselves, cannot be infinitely small, the lines are not infinitely thin. They can be considered to have a thickness. Therefore, rather than intersecting at an infinitely small point, they can be thought of as two overlapping strips. The area of overlap indicates the solution.

When two perpendicular strips cross, the overlapping area forms a square (Fig. 11-27). Mathematically, an intersection angle of 90 degrees creates the minimum possible area of overlap. If the two lines have an intersection angle of, say, 160 degrees, the overlapping area will be much greater (Fig. 11-28), giving a much larger error. If the angle of intersection is 0 or 180 degrees, a solution cannot be found since two parallel lines will never cross. If you attempt to set up a resection and the intersection angle is too close to 0 or 180 degrees, the total station will give an error message.

Fig. 11-27: Resection error (Good geometry)

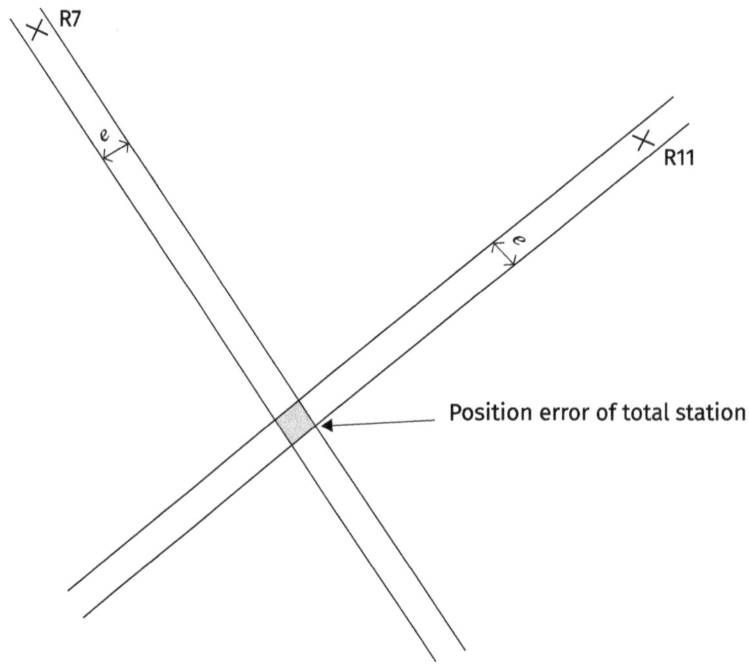

Fig. 11-28: Resection error (Poor geometry)

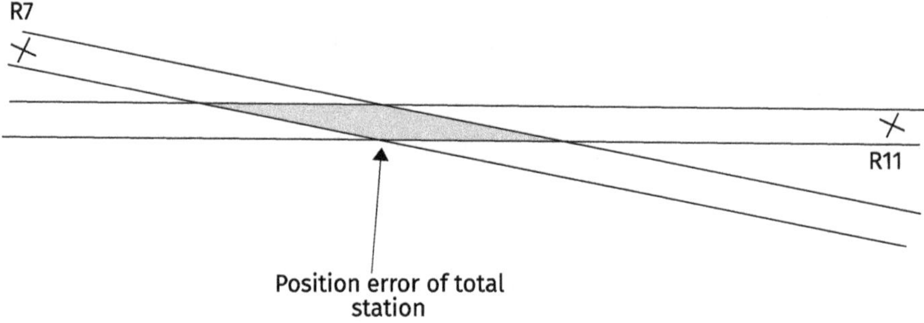

11.60 WHEN IS A RESECTION NOT SUITABLE?

There are many situations when setting up the total station using a resection is not suitable for the task. Examples include:

→ determining the coordinates of secondary control points

→ any setting out task which requires a high degree of external accuracy such as bridge bearings

→ some structural monitoring tasks

11.61 HOW DO I USE THE TOTAL STATION FOR MEASURING AND SETTING OUT REDUCED LEVELS?

There are 3 ways you can use your total station for measuring and setting out levels. They are:

→ **known point set-up** – this method requires you to input the instrument and reflector heights and the reduced levels of your set-up and backsight points. This is the least accurate method and is suitable for large-scale surveys and setting out of low accuracy-elements, such as earthworks

→ **3D resection** – use the resection function to set up at an unknown point and backsight to retro targets of known eastings, northings and reduced level. From this position, you can measure or set out reduced levels. This method is suitable provided the accuracy of your control points is proven and the correct procedures are followed. See section **11.13**

→ **tie distance/compute inverse** – the position and orientation of the instrument do not need to be determined to use this function to measure the difference in height between two points. This method is the most accurate and is the only method suitable for establishing reduced levels of control points. This method is more suitable for measuring than for setting out.

11.62 HOW DO I USE A 3D RESECTION FOR MEASURING AND SETTING OUT REDUCED LEVELS?

A 3D resection determines the position and reduced level of the instrument. This can be a quick and convenient of measuring or setting out levels. Caution and extreme care are required as large errors can develop if the correct procedures are not followed.

To set up your total station using a 3D resection, you will need a set of control points with both coordinates and reduced levels.

The process for setting up a 3D resection is similar to a 2D resection. There are two differences:

→ with the 3D version, you will need to decide whether the position or level will have priority in terms of accuracy. If the position is most critical, you will line up the vertical cross-hair first. If the level is most critical, you will line up the horizontal cross-hair first. See sections **11.13** and **11.15** for tips on lining up the cross-hairs

→ as well as remembering to select the correct target type and prism constant for each reading you take, you will also need to enter the vertical distance from the point being measured to the centre of the target. This is known as the reflector height. For retro targets, this is 0mm. For prisms, it is the distance from the tip of the detail pole to the centre of the prism (Fig. 11-29 and Fig. 11-30). For a backsight target on a tripod, it is the vertical distance from the PGM to the centre of the horizontal arrows (Fig. 11-31).

Once your instrument is set up, as well as checking the coordinates of a known point, you must also check the level of a known TBM.

Fig. 11-29: Mini prism reflector height

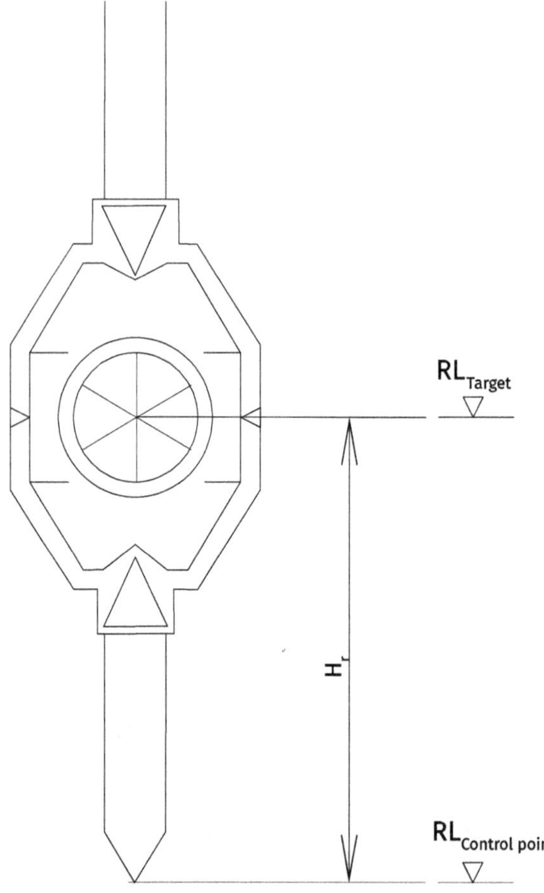

Fig. 11-30: Detail pole reflector height

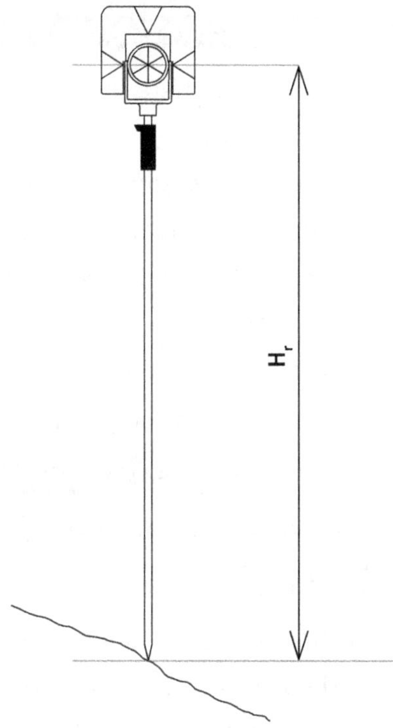

Fig. 11-31: Target on tripod 'reflector height'

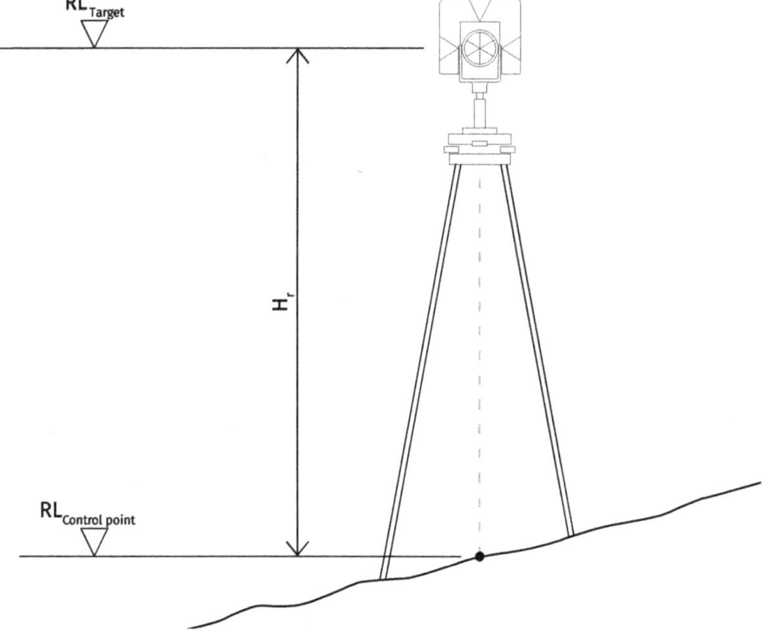

11.63 WHAT CAN GO WRONG WITH USING A 3D RESECTION FOR REDUCED LEVELS?

In addition to the usual total station errors, there are two other things to note when using a 3D resection for setting out reduced levels:

→ the instrument uses the vertical circle reading and the slope distance to calculate reduced levels. If there is a vertical collimation error in the instrument, then this will translate into an error in your setting out. Always carry out a vertical collimation check on your instrument before assuming that it is ok to use for reduced levels.

→ remember to change the reflector height between backsighting to the retro targets for the instrument set-up and sighting to the prism.

11.64 HOW DO I USE THE KNOWN POINT SET-UP METHOD FOR MEASURING AND SETTING OUT REDUCED LEVELS?

This method of measuring and setting out reduced levels is suitable for low accuracy tasks such as topographical surveys and earthworks. To use this method, you will need to know the reduced levels of your set-up point, RL_s. The total station will calculate the RL of the point RL_p by measuring the slope distance and the vertical angle, θ. It will convert this information into a vertical distance, and can then calculate the RL_p using:

$$RL_p = RL_s + H_i - VD - H_r$$

The procedure is as follows (🗎):

1. Centre and level the total station over a known point.

2. Using a tape measure, measure the vertical distance from the known point to the centre of the vertical circle which is marked on the side of the total station. Depending on the physical nature of the known points, you may need to use a spirit level at the point to avoid bending the tape. This value is the instrument height, H_i (Fig. 11-32).

3. Select the correct function for setting up at a known point.

4. Input the coordinates of the point you are set up over (either by selecting from the memory if you have previously stored the information or by manually entering the coordinates).

5. When prompted, enter the height of the instrument, H_i.

6. Input the coordinates of the target you are backsighting to.

7. Enter the reflector height, H_r. This is the vertical distance from the level of the known backsight point to the vertical mid-point of the prism.

8. Select the correct prism type and constant for the target you are backsighting to.

9. Sight to the target point giving priority to whichever is more critical, the horizontal or vertical angle.

10. Press 'Measure'

11. Proceed to measure or 'add' more points.

12. Repeat steps 6 – 10 for at least one more known point.

13. You can now compute the position accuracy. If necessary, to improve the accuracy, measure more points.

14. When you have achieved the required set-up accuracy you can choose to set the instrument position and orientation.

15. Enter survey mode and measure the coordinates of a reliable retro target or PGM remembering to adjust the prism constant and reflector height accordingly.

16. Check the reduced level against the known value of the point. If ok, continue to the survey function or Stakeout function as required.

17. Remember to input the correct prism constant and reflector height for the mini-prism or detail pole you are using for your survey or setting out task.

Fig. 11-32: Instrument height

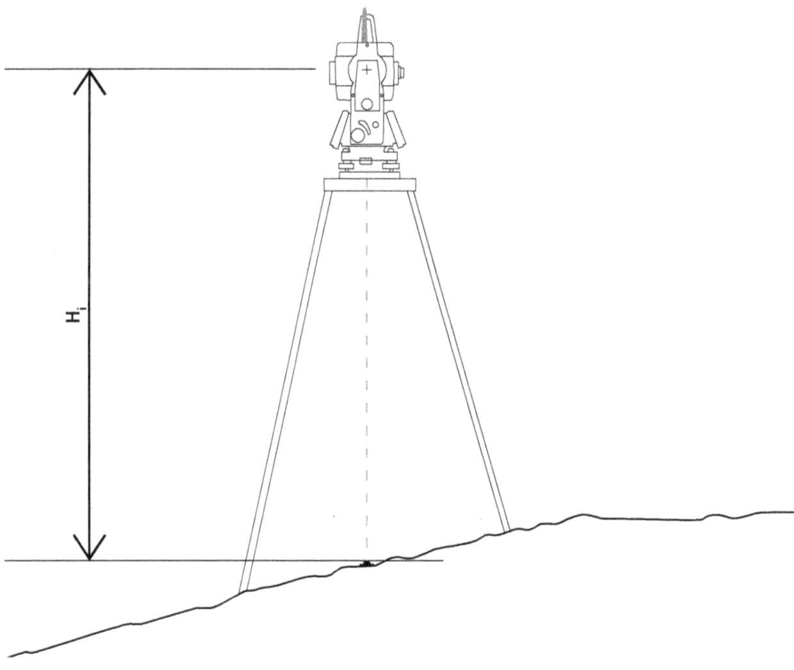

11.65 HOW DO I USE THE TIE DISTANCE/COMPUTE INVERSE FUNCTION FOR MEASURING AND SETTING OUT REDUCED LEVELS?

The tie distance/compute inverse function is better suited to checking levels rather than setting them out as it does not have the tracking mode option.

The procedure is as follows (🖹):

1. Select the tie distance/compute inverse function.

2. Select the correct prism type and constant.

3. When prompted to sight target point 1, sight to the mini-prism on a reliable TBM, lining up the horizontal cross-hair first and then moving the cross sideways until it is on the prism.

4. Press 'Meas'.

5. When prompted, sight to target point 2, the mini-prism at the point you are measuring, lining up the horizontal cross-hair first and then moving the cross sideways until it is on the prism.

6. Press 'Meas'.

7. The level difference between the two points will be displayed on the screen.

8. Use this level difference to calculate the reduced level of the point you are measuring in relation to your TBM.

9. To measure further points, in relation to the TBM, select 'New point 2' and repeat steps 5 – 8.

11.66 WHAT ARE THE DIFFERENT MANUFACTURERS OF TOTAL STATIONS?

There are a range of total stations in common use in the UK. The most common are Leica, Trimble and Topcon. Others include Nikon and Sokkia, Spectra and Geomax. Equipment-specific instructions for Leica, Trimble and Topcon can be found in the appendices of this book.

11.67 WHAT ARE THE DIFFERENCES BETWEEN TOTAL STATIONS OF DIFFERENT MANUFACTURERS?

The main differences between total stations produced by different manufacturers is the user interface and workflow (what buttons you press and in which order). Each manufacturer produces a range of total stations of different capabilities, accuracies and price.

11.68 WHAT IS A DIGITAL THEODOLITE?

A digital theodolite is similar to a total station, but does not have distance measurement capability or software. It has an optical or laser plummet so that it can be set up over a point, can be accurately levelled and displays the horizontal and vertical angle on an LCD screen. They are most commonly used on site by steel erectors for plumbing up steel columns as described in section **24.65**.

11.69 WHAT IS A ROBOTIC TOTAL STATION?

A robotic total station, sometimes known as a one-man total station, has the same capabilities as a standard total station as well as some additional functions such as measuring stockpile volumes by automatically scanning more than 20 points per second. Fig. 11-33 shows a robotic total station, Fig 11-34 shows a controller, and Fig 11-35 the active prism.

Instead of the user having to sight through the telescope to line up on the target, the telescope of a robotic total station can lock onto the prism and the instrument is controlled remotely from a handset. The controller, which communicates with the total station via radio signal or Bluetooth, can be attached onto the prism pole for convenience.

Fig. 11-33: Robotic total station (Image courtesy of KOREC group)

Fig. 11-34: Robotic total station controller (Image courtesy of KOREC group)

Fig. 11-35: Active prism and controller

11.70 WHAT ROBOTIC TOTAL STATION SKILLS SHOULD I HAVE?

You should be able to:
- → set the position and orientation of the total station
- → enter data into the controller manually
- → connect the controller with the total station
- → navigate around the menus and programs
- → navigate the stored information in list format and map view
- → view, edit, add and delete data
- → enter the correct settings
- → select the correct prism constant
- → take a topographical survey and record the results systematically
- → set out points of known co-ordinates using on board functionality
- → set out points in relation to a given line using on board functionality
- → set out points at given chainages and offsets along a radius e.g. road centrelines using on board functionality
- → transfer large amounts of data from the total station to the computer and vice versa
- → extract line data e.g the horizontal distance and level difference between two points
- → measure irregular areas and volumes such as stockpiles
- → import and work with DXF and DTMs
- → create and export DXFs and DTMs
- → import and export data to and from Excel and survey software

11.71 DO I NEED A MECHANICAL TOTAL STATION OR A ONE-MAN ROBOTIC?

The decision to hire or buy a robotic total station rather than a mechanical two-man total station depends on your budget and the tasks you will be using it for. Here are a few things to consider:
- → the robotic total station can be more than three times the price of a standard total station

- → insurance and maintenance costs will be higher

- → there is a higher risk of theft

- → even though you don't need an assistant to hold the prism, you may still need someone to stand next to the instrument in public places to prevent it being stolen

- → the robotic total station is quicker for repetitive tasks such as staking out lots of points, but would not necessarily be more cost-effective if used infrequently

- → the robotic total station has the potential to achieve higher accuracy if the user is properly educated about the limiting factors; the accuracy achievable is often far greater than is necessary for task such as earth works

- → there are additional functions on the total station for specialist tasks such as measuring volumes of stockpiles.

11.72 I HAVE CONNECTED THE TOTAL STATION TO THE CONTROLLER, BUT I CAN'T FIND DATA WHICH I HAVE INPUT. WHY NOT?

Once the controller is connected to the total station, the total station is completely over-ridden and acts only as a seeing eye and measurement device. The controller acts as the 'brain' and any jobs or data in the total station will become inaccessible. All data should be input into the controller.

11.73 IS A ROBOTIC TOTAL STATION MORE ACCURATE THAN A STANDARD TOTAL STATION?

The total station can accurately lock onto the 360-degree prism. This eliminates any errors associated with sighting to the target. It does not, however, reduce positional errors such as centring over the point, detail pole verticality or errors inherent in the control points.

The robotic total station is more accurate than the standard total station for measuring level differences using the compute inverse function, but will not reduce errors in measuring the height of the instrument or setting up a 3D resection.

Be aware that on some 360-degree prisms, there is a level difference between alternate prism faces of up to 6mm. If this is the case, there is usually an arrow or triangle on one of the faces so that you can distinguish between the faces. This type of prism should only be used for low accuracy work such as earthworks. If you wish to use it for measuring accurate levels, you will need to calibrate the prism against two reliable points and use your hands to cover up the faces either side of the correct face.

11.74 THE ROBOTIC TOTAL STATION CANNOT FIND THE PRISM. WHAT MIGHT BE WRONG?

Possible reasons that the total station cannot find the prism are:

→ you have set the wrong target type

→ there are one or more other total stations operating nearby.

11.75 CAN THE ROBOTIC TOTAL STATION LOCK ONTO THE WRONG TARGET?

It is possible for the robotic total station to lock onto either the wrong prism or a static point. In survey mode, this would not necessarily be obvious to the operator at the time of the error. Each time the user presses the 'measure' key, a reading will be registered, but unless the user is checking the co-ordinates of each reading to check that they are roughly what they expect, then there would be no way of knowing there is a problem.

The issue would be noticeable immediately in stake out mode, as the relative position of the target to the required point would not change as anticipated.

The likelihood of this error occurring can be minimised by having an awareness that this can happen, and by consciously checking that the coordinates of the measured points are roughly as expected based on previous measurements and relative coordinates of nearby features. Coordinates can be checked numerically, or visually using the map view. The potential for this error can be eliminated with the use of an active prism which sends a radio signal to ensure that the total station is locked on to the prism with the correct ID. Further details of active prisms can be found in section **11.25**.

11.76 CAN I USE THE ROBOTIC TOTAL STATION AS A TWO-MAN INSTRUMENT?

Most, but not all, robotic total stations can be used as a two-man instrument. Simply leave the controller switched off and use the total station in the same way as a standard one.

11.77 WHAT IS A POINT ID?

Each point entered or measured is given a point ID. Beware that each point ID is not unique. If you measure several times to a point called P34, each time it will be recorded in the memory, so you will have several measured points called P34. If you wish to measure a point and view the point data on the screen without storing it in the memory, select 'DIST'.

Only select 'Meas' if you would like to keep a record, for example once you have set a point out and you wish to record the as-set-out position or if you are doing a survey.

11.73 WHAT IS A POINT CODE?

Points can be classed into categories indicating the type of feature they relate to, for example:

- → kerb (K)
- → lamppost (LP)
- → spot height (SH)
- → centre line (CL)
- → crown (CR)
- → gulley (G)

You can use:

- → your own unique system
- → a site-wide system
- → a company-wide system
- → a pre-determined list within the total station

11.74 WHAT IS A CODE LIST?

A code list is a list of codes to categorise the points you are surveying. For example, kerb, channel, bollard, lamp post, tree, road marking, spot level etc. When taking surveys, codes can be attributed to points so that they can be grouped or filtered when processing the information. Each code has an associated property which can be a:

- → Point
- → Line
- → Area

By assigning a code with a 'line' property to a point, a 'string' can be created. A string is multiple points joined together in sequence to create a continuous line. Most good total stations will allow you to switch easily between different strings of the same code so you can create multiple lines at once. For example, to survey a road you could set up the following codes:

→ K1 – kerbline 1 (line)

→ C1 – Channel 1 (line)

→ CL – Centre line (line)

→ C2 – Channel 2 (line)

→ K2 – Kerbline 2 (line)

→ MH – manhole (point)

In this way, you could zig-zag across the road and back, selecting the corresponding code for each measurement and simultaneously creating five different strings. Codes can also be assigned to a layer for example, kerb lines, channel lines and centre lines could be on a layer called 'Roads' and manholes, gullies and inspection chambers could be on a layer called 'Ironwork'. The use of codes and layers allows the information to be easily filtered. In table format, the information can be ordered by code or layer. In the map view within the total station or once the information has been exported to survey software, layers can be turned on and off.

The benefit of using codes in this way is that once the survey is complete, the data is already coded, joined and labelled and needs little or no extra work in order to make it useful and presentable to the end user.

CHAPTER 12

INSTRUMENT CHECKS ON THE TOTAL STATION

12.1 HOW OFTEN SHOULD I CARRY OUT TOTAL STATION FIELD CHECKS?

Field checks on your total station should be carried out at regular intervals between yearly services and calibrations. Not only will this habit minimise the risk of errors caused by faulty equipment, but it will demonstrate that you are taking proper care of your equipment. At a minimum, you should carry out the total station field checks:

→ when the equipment is first delivered to your site

→ when you have lent your equipment to another engineer

→ every two weeks

→ at the frequency specified in the relevant company policy

→ at the frequency specified in the contract documents of the project you are working on

→ if you have reason to suspect that there is a fault with your instrument

→ when measurements are not tying in as expected, to rule out the possibility that there is something wrong with your instrument.

Fig. 12-1: Total station checks - booking example

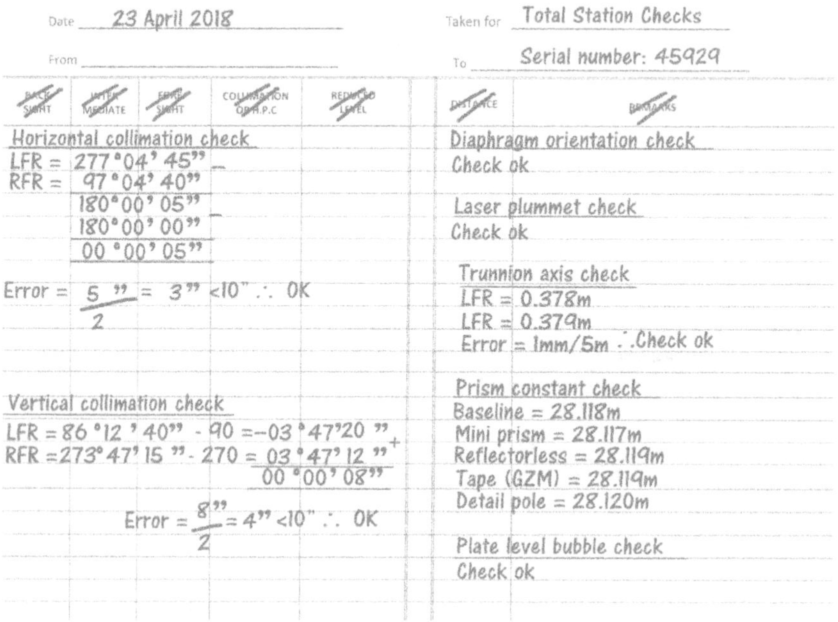

Date: 23 April 2018   Taken for: Total Station Checks

From: To: Serial number: 45929

BACK SIGHT	INTER MEDIATE	FORE SIGHT	COLLIMATION OR H.P.C	REDUCED LEVEL		DISTANCE	REMARKS

Horizontal collimation check
LFR = 277°04' 45"
RFR = 97°04' 40"
    180°00' 05"
    180°00' 00"
    00°00' 05"

Error = $\dfrac{5"}{2}$ = 3" <10" ∴ OK

Vertical collimation check
LFR = 86°12' 40" - 90 = −03°47'20"
RFR = 273°47' 15" - 270 = 03°47' 12"
    00°00' 08"

Error = $\dfrac{8"}{2}$ = 4" <10" ∴ OK

Diaphragm orientation check
Check ok

Laser plummet check
Check ok

Trunnion axis check
LFR = 0.378m
LFR = 0.379m
Error = 1mm/5m ∴ Check ok

Prism constant check
Baseline = 28.118m
Mini prism = 28.117m
Reflectorless = 28.119m
Tape (GZM) = 28.119m
Detail pole = 28.120m

Plate level bubble check
Check ok

12.2 HOW SHOULD I RECORD THE TOTAL STATION CHECKS?

Your company may require you to record the total station checks on purpose-designed forms and if this is the case, it is important that you comply with your company's procedures. I recommend that your total station checks are recorded in your field book as shown in Fig. 12-1.

12.3 HORIZONTAL COLLIMATION CHECK

The check for horizontal collimation error involves taking a reading to the same point from both faces. If the collimation error is zero, then there will be a difference of 180 degrees. Any discrepancy between the measured difference and the theoretical difference of 180 degrees, constitutes twice the error.

This is based on several assumptions:

→ the point you sight to is fixed and does not move

→ all other sources of errors are eliminated, which isolates the residual error in the instrument itself.

It is essential to sight to a sharp point (Fig. 12-2) rather than a vertical line such as the edge of a building, lamppost or feature (Fig. 12-3). There are two reasons for this:

→ it is not safe to assume that the line is truly vertical. If you were to sight to an apparently vertical line, after changing faces, you wouldn't be able to identify the exact same point

→ lining up on a line is not as accurate as lining up on a point.

To achieve optimum accuracy, ensure that:

→ the single vertical line, rather than the double line or the small cross at the centre of the cross-hairs, is lined up on the point

→ you have good visibility

→ your total station is set up on firm ground

→ the clips of your tripod legs are working properly

→ you take time to observe the point as accurately as possible

The procedure for checking the horizontal collimation error is as follows:

1. On Face Left, line up the single vertical cross-hair (not the horizontal line!) on a sharp point which protrudes horizontally onto a contrasting background (Fig. 12-2). If possible, find something that has sky behind it so that there is a sharp contrast between the point and the sky behind it. The further away the point, the more accurate this exercise will be.

2. Using the fine tuner (the tangent screw), move the cross-hair clear of the point and back onto it again several times until you are certain that you have stopped at the point where the single line first meets the point. To minimise any error caused by backlash in the mechanism, it is necessary to move the cross-hair smoothly towards the point and stop on the point rather than lining it up on the point and making tiny adjustments.

3. Record the horizontal angle, Hz (not the vertical angle!)

4. Change faces and repeat steps 1 – 3, observing to the exact same point.

5. Subtract the right face reading from the left face reading.

6. Subtract the answer from 180 degrees.

7. The answer is twice the error, so divide this by two to find the collimation error.

Fig. 12-2 Horizontal collimation check - correct lining up

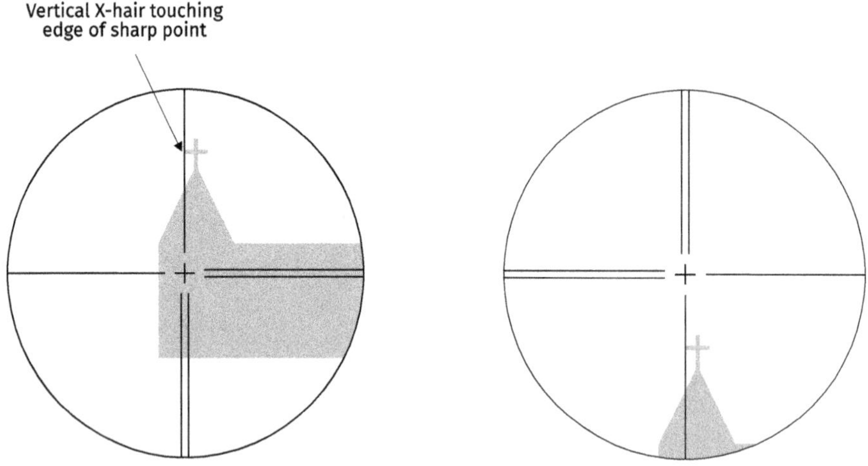

Fig. 12-3 Horizontal collimation check - incorrect lining up

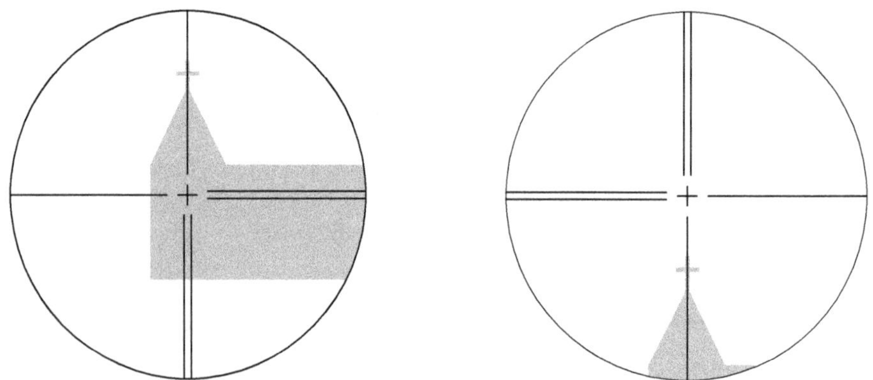

12.3 VERTICAL COLLIMATION CHECK

The process for checking the vertical collimation error is very similar to that of the check for the horizontal collimation, in that it involves taking a reading to the same point from both faces. In this case though, the angle measured on each face is the vertical angle of the telescope relative to its horizonal position.

The procedure for checking the vertical collimation error is as follows:

1. On Face Left, line up the single horizontal line (not the vertical line!) on a sharp point which protrudes vertically upwards onto a contrasting background (Fig. 12-4). If possible, find something that has sky behind it so that there is a sharp contrast between the point and the sky behind it. The further away the point the more accurate this exercise will be (see section **19.10** for the theory behind this). Ensure you use a sharp point rather than a flat surface (Fig. 12-5)

2. Using the fine tuner (the tangent screw), move the cross-hair clear of the point and back onto it again several times until you are certain that you have stopped at the point where the single line first meets the point. To minimise any error caused by backlash in the mechanism, it is necessary to move the cross-hair smoothly towards the point and stop on the point rather than lining it up on the point and making tiny adjustments.

3. Record the vertical angle, V (not the horizontal angle!)

4. Change faces and repeat steps 1 – 3, observing to the exact same point.

5. One of the vertical angle readings will be close to 90 degrees. Subtract this angle from 90 degrees.

6. One of the vertical angle readings will be close to 270 degrees. Subtract this angle from 270 degrees.

7. Add the two angles together.

8. The answer is twice the error, so divide this by two to find the collimation error.

Fig. 12-4: Vertical collimation check

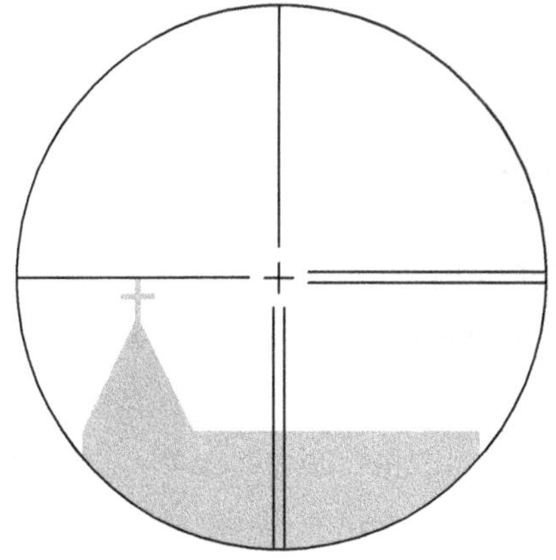
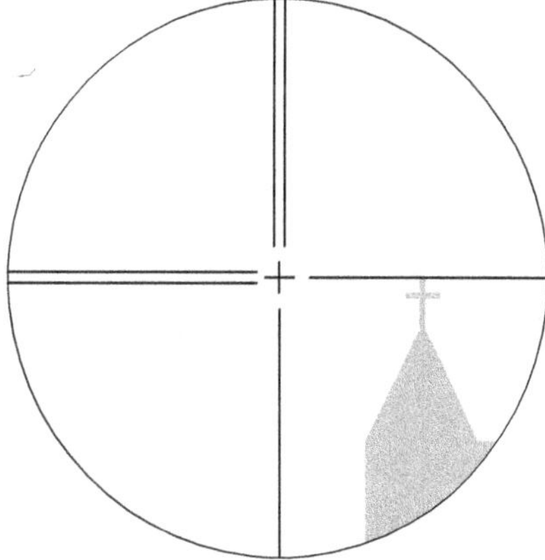

Fig 12-5: Vertical collimation check - wrong

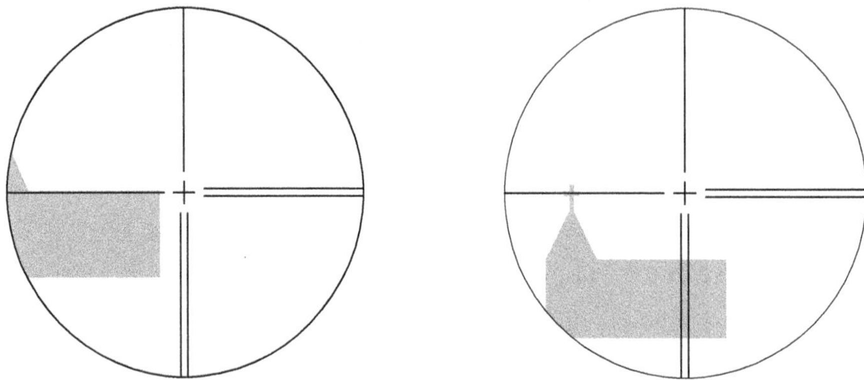

12.4 WHAT IS AN ACCEPTABLE ERROR FOR THE HORIZONTAL OR VERTICAL COLLIMATION CHECK?

In the absence of any other written information, use an acceptance criterion of 5" error. However, as with all instrument checks and construction tolerances any figures provided in the documents of the specific project you are working on or your company's policies take precedence over the guidance given in this book.

12.5 DIAPHRAGM ORIENTATION CHECK

The cross-hairs are on a diaphragm which can potentially rotate within the telescope. To verify that the cross-hairs are in correct adjustment, the steps are as follows (Fig. 12-6):

1. Line up the lowest point of the single vertical cross-hair on a sharp point in the distance which protrudes horizontally onto a contrasting background. If possible, find something that has sky behind it so that there is a sharp contrast between the point and the sky behind it.

2. Using the horizontal fine tuner to move the cross-hair well clear of the point and back onto it again several times until you are certain that you have stopped at the point where the single line first meets the point. To minimise any error caused by backlash in the mechanism, it is necessary to move the cross-hair smoothly towards the point and stop on the point rather than lining it up on the point and making tiny adjustments.

3. Use the vertical fine tuner to scroll up to the highest point of the single vertical cross hair. If the line stays in contact with the point, then the cross-hairs must be truly vertical. If the cross-hairs are not correctly aligned, the line will move away from the point or cross over it as you scroll to the extents of the cross hair.

This test will also check that the instrument is levelled up correctly.

Fig. 12-6: Diaphragm orientation check

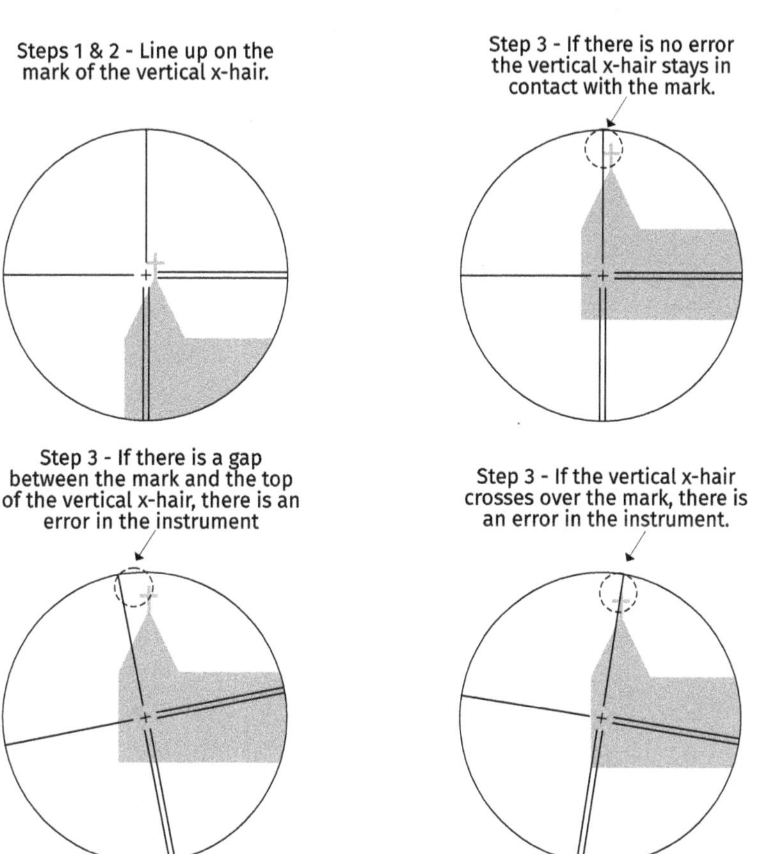

12.6 TRUNNION AXIS CHECK

To check that the transverse axis of the instrument is perpendicular to the vertical axis:

→ Line up the single vertical cross hair on a well-defined point, with the telescope at an elevation of at least 45 degrees to the horizontal.

→ Lay a levelling staff or tape measure transversely on the ground with the scale facing you, approx. 5m from the total station.

→ Scroll down and take a reading on levelling staff or tape measure (Fig. 12-7).

→ Repeat on the opposite face.

→ The error is the difference between the two readings.

If an error is found, setting out should be carried out using both faces. The correct point is the mid-point between the face left and face right mark.

Fig. 12-7: Trunnion axis LFR

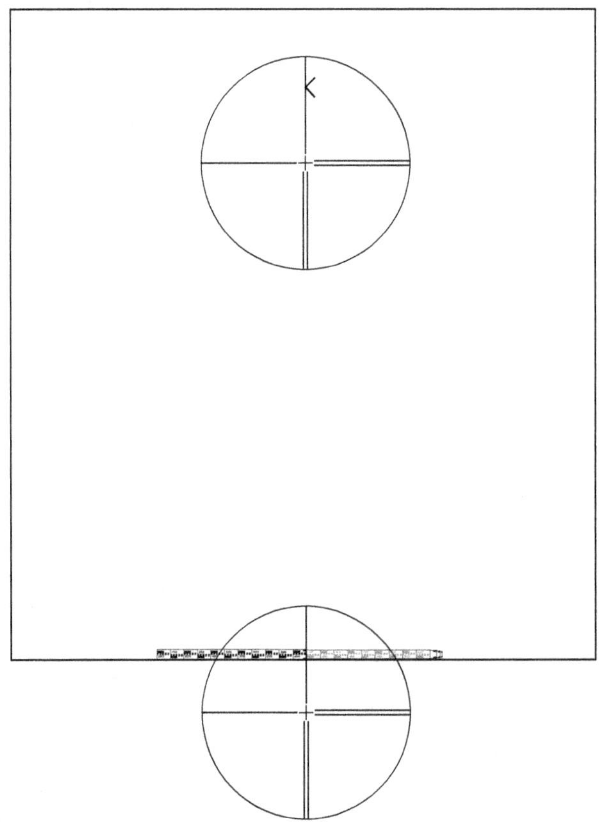

EXERCISE:

3. What is the trunnion axis error if the total station is set up approx. 5m away from the point being sighted to and LFR and RFR to a tape measure on the ground are 0.668m and 0.667m respectively?

12.7 OPTICAL PLUMMET CHECK

To check the optical plummet of a backsight target or a total station with an optical plummet:

1. Centre and level the instrument over a point.

2. While sighting through the optical plummet, rotate the total station through at least 180 degrees.

3. If the circle (or cross) stays on the point, then the optical plummet is truly vertical. If it moves off the point, then there is an error and should be returned to the workshop for repair.

12.8 LASER PLUMMET CHECK

To check the laser plummet of your total station, the procedure is the same as for the optical plummet, but observing the red laser dot as you rotate the instrument.

12.9 PRISM CONSTANT CHECK

To check the prism constant, the distance measurement between the total station and the prism should be calibrated against a reliable baseline. Even if the prism constant is marked on the prism and appears to match exactly the prism type listed in the total station, you should still carry out the check. Check all the different prism types as well as retro targets and the reflectorless function (by holding a flat surface on the centreline of the point). The steps are as follows:

1. Using a tape measure or a total station with verified prism, establish two points a known horizontal distance apart. Alternatively, locate two points of known coordinates and calculate the distance between them.

2. Centre and level the total station over one of the points.

3. Position your chosen prism over the other point.

4. Select 'EDM', and select the prism type.

5. Press 'Dist'.

6. Compare the horizontal distance displayed on the screen to the known distance.

7. Repeat steps 3 – 6 for each different type of prism and target.

CHAPTER 13

TOTAL STATION FUNCTIONS

13.1 WHAT IS THE DIFFERENCE BETWEEN THE 'MEAS' AND THE 'DIST' BUTTONS?

To take a measurement with the total station, there are three options you can choose from. The terminology used varies from instrument to instrument.

→ Measure and store – 'Meas', 'Set' or tape measure with disk icon ('Quick measurement')

→ Measure only (to measure and display the results without storing) – 'Dist', 'Start' or tape measure icon ('Precise measurement')

→ Store (to store the results for the measurement you have just taken) – Store, Set or disk icon

Choose 'measure and store' when you want the instrument to take that measurement and do something with it e.g. a calculate a result or store it in the memory. Here are some examples:

→ when you are setting up the instrument, when prompted, sight to the backsight and press 'measure and store'. The instrument will take the angle and distance measurements and use it in the calculations to determine the instrument's position and orientation.

→ when you have marked out a point in the correct position using the stake-out function, pressing 'measure and store' will store the position as 'measured data'. You will have evidence that the 'as-set-out position' is within tolerance.

→ when using the tie distance/compute inverse function, use the 'measure and store' when measuring to the beginning and end point of the line.

Choose 'measure only' when you want to display the information on the screen but you do not want the instrument to store that data or use it to progress to the next stage of the operation. Here are some examples:

→ when sighting to a known point to check your instrument set-up is correct, press 'measure only'. This will display the coordinates on the screen, so you can manually compare them to the expected value.

→ when setting out a point, sight to the prism and press 'measure only'. This will display on the screen the relative position of the correct point. Repeat until you have located the correct position within the required tolerance. If you pressed 'measure and store' each time, you would have dozens of results for the same point cluttering up the instrument, which increases the risk of confusion.

Press 'Store' when you have taken a measurement using 'measure only' and you would like to store it in the memory. Examples of when to use the 'Store' function include:

→ when you have set out a point and have pressed 'measure only' to check the position is correct, press 'Store' to save the point in the memory, even if the prism has been removed from the point.

→ when you have surveyed a point and have pressed 'measure only' to check that you are happy with the results, press 'Store' to save the point in the memory, even if the prism has been removed from the point.

13.2 TIE DISTANCE/COMPUTE INVERSE FUNCTION (ALSO CALLED 'COMPUTE INVERSE' OR 'MISSING LINE MEASUREMENT')

The tie distance or compute inverse function is used to measure accurately between two points, from a third position which is neither of the points (Fig. 13-1).

It is a simple, yet extremely useful function which allows great flexibility as it is not necessary to set the position and orientation of the total station in relation to control points or to set up over a fixed point. In other words, you don't need any control points whatsoever for this function to work.

Two points are sighted from a random set-up position and the total station calculates an accurate plan distance, slope distance and level difference between the points. The information is displayed on the screen and can be recorded in the memory if required. The reflectorless mode can be used when the points are not accessible.

Any number of consecutive measurements can be taken, relating either to the start or end point of the original line measured. When consecutive end-to-end distances are measured it is good practice to measure the total distance in relation to the start point to avoid a build- up of errors (Fig. 13-2).

Fig. 13-1: Tie-Distance (Horizontal distance)

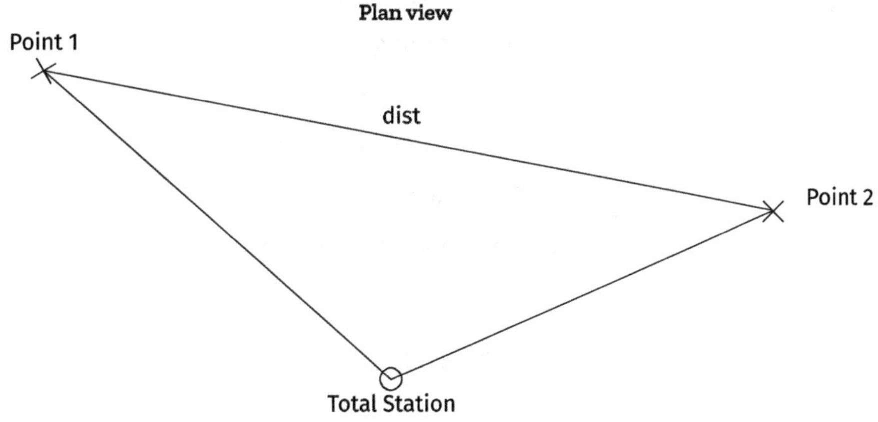

Fig 13-2: Tie-distance/compute inverse
(Multiple horizontal distances in relation to original point)

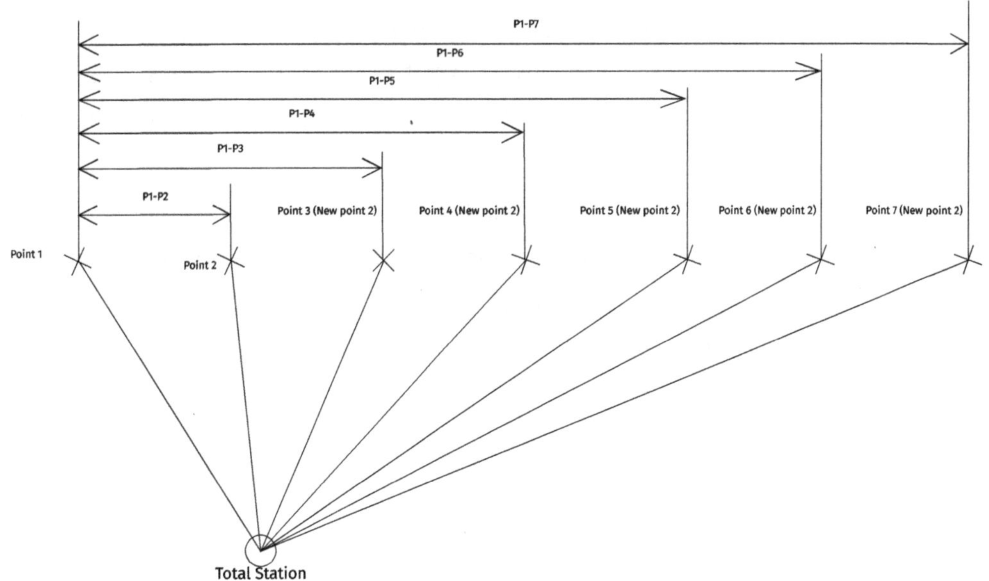

13.3 WHAT IS THE TIE DISTANCE/COMPUTE INVERSE FUNCTION USEFUL FOR?

The tie distance compute/inverse function is useful when:

→ there is a visual obstruction directly between the two points

→ it is not possible to set the total station or target prism up over one of the points (i.e. when they are on a wall or inaccessible ground)

→ you have set out points using the set-out function or a tape measure and you need to check the accuracy of your setting out using a different method to the one you used to set the points out in the first place

→ it is quicker than setting up the instrument over a point

→ the distance to be measured is greater than the length of your tape measure

→ control points are not visible, or none exist.

13.4 HOW DO I USE THE TIE DISTANCE/COMPUTE INVERSE FUNCTION TO MEASURE A LEVEL DIFFERENCE?

The method for measuring the vertical distance between two points (Fig. 13-2) is the same as for measuring horizontal distances, except for the lining-up process in steps 5 and 6. When measuring levels, you should give priority to the vertical angle (horizontal cross-hair) as described in section **11.15**

The screen will display the horizontal distance, slope distance and level difference between the two points. Note that in this case the accuracy of the vertical distance has been given priority over that of the horizontal distance, so the horizontal distance displayed here should not be relied upon for high accuracy measurements.

When measuring a series of levels, each new level should be measured in relation to the original point to avoid a cumulation of errors.

Fig. 13-3: Measuring level differences with the total station

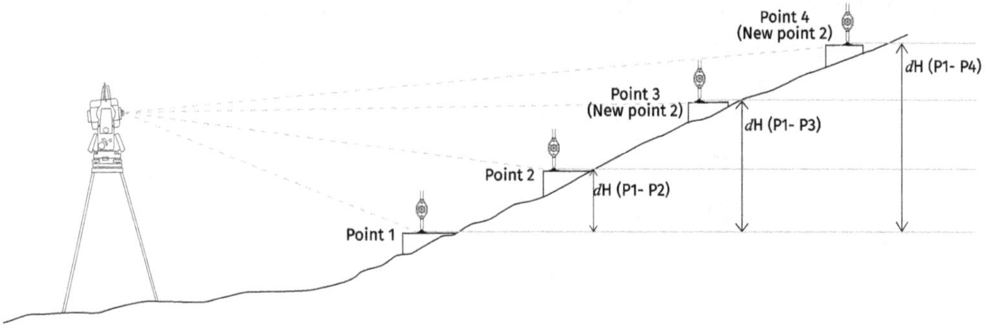

13.5 HOW DO I MEASURE THE HORIZONTAL AND VERTICAL DISTANCE AT THE SAME TIME?

Every time you take a measurement using the tie distance/compute inverse function, the horizontal, slope and vertical distances will be displayed, however the accuracy of one will be greater than the other depending on how you lined up the cross-hairs on the target.

13.6 WHAT IS THE ACCURACY OF THE TIE DISTANCE/COMPUTE INVERSE FUNCTION?

The tie distance/compute inverse function can be accurate within 2mm, however this level of accuracy requires extremely careful lining up of the cross-hairs. To improve your accuracy, carry out separate measurement for horizontal distances and level distances, lining up the cross-hairs as described in section **11.13** and **11.15**.

13.7 HOW CAN I CHECK THE ACCURACY OF THE TIE DISTANCE/COMPUTE INVERSE FUNCTION?

If you require an accuracy of less than 10mm, the simple distance measurement function is the best method. If this is not possible, and you have no option but to use the tie distance/compute inverse function, you should first of all check the accuracy of your own tie distance/compute inverse measurement using this simple exercise:

→ Measure the distance between two points by setting up over one point and taking a simple distance reading to the other. Record the distance.

→ Measure the distance between the same two points by setting up the total station at a random point and using the tie distance/compute inverse function.

The two points you use for this test should be a similar distance apart as the points you will be measuring. If you are measuring a range of distances, carry out separate checks for the shortest distance and longest distance you are likely to be measuring.

13.8 HOW DO I MEASURE BETWEEN TWO POINTS I HAVE PREVIOUSLY MEASURED OR INPUT USING THE TIE DISTANCE/COMPUTE INVERSE FUNCTION?

Rather than measuring to the start and end points of the line, you can select them from the memory. In this way you can check the distance and level difference from any two Fixpoints or measured points.

13.9 STAKEOUT FUNCTION (ALSO CALLED 'SET OUT' OR 'LAYOUT')

The stake-out function is one of the most common functions used in setting out. It enables points of known coordinates to be set out in relation to control points. When the target is sighted and measured, the screen indicates how far to the left or right and how far towards or away from the instrument the target needs to be moved in order to achieve the correct position. The target is repositioned according to the directions on the screen and then and remeasured. The process is repeated until the target is in the correct position and a mark is made.

There are several options for setting out the point. Which one you use will depend on your personal preference. You can:

→ set the horizontal angle to the correct position and direct your assistant into the line of the vertical cross-hair, take a distance reading and then direct your assistant forwards or backwards by the correct amount

→ sight to the prism and follow the arrows on the screen which indicate how far left or right and how far backwards or forwards the prism needs to be moved in relation to the total station

→ display the eastings and northings of the point on the screen and have the prism moved until the coordinates are within tolerance

→ display the difference in eastings and northings and have the prism moved until the difference between the measured and the required is within tolerance

13.10 HOW DO I OPTIMISE ACCURACY WHEN USING THE STAKE OUT MODE?

Where there is a tolerance of less than 5mm, take the following measures:

1. set the position and orientation of the total station using the 'known point' set up rather than a resection.

2. once you have marked the points, repeat the setting out on the opposite face as described in section **11.10.**

3. where internal accuracy is critical, set out as many points as possible from a single set-up position. If it is not possible to set out all the points in a particular section of works from a single set-up, each time you move instrument position, sight to at least two of the points you set out from your previous set-up position to check that they tie in.

13.11 WHY HAVE THE ARROWS DISAPPEARED FROM THE SCREEN?

In most total stations once you are within a certain distance of the required point (usually set to 0.5m) the large arrows will disappear and instead you will see a diagram of the prism in relation to the required point. There will still be smaller arrows indicated the required direction of movement and the required distances will be also be displayed if you prefer to use this method.

13.12 HOW DO I USE TRACKING MODE?

Once the prism is within say 20mm of its required position you can use tracking mode to speed up the final stage. In tracking mode, the total station will automatically repeatedly take a distance reading. This means that the distance will be displayed on the screen and you can use hand signals to direct your assistant forwards or backwards. You will need to keep checking through the telescope to ensure that the tip of the prism pole is in line with the vertical cross-hair. If it moves out of line, direct your assistant left or right accordingly.

13.13 HOW DO I USE THE RED LASER DOT?

You can switch on the red laser dot which will indicate roughly where the point will be. A reflectorless reading can be taken to the ground. This method is sufficiently accurate for tasks such as setting out piles, but for more accurate work such as column centrelines or corners of shutters, the red laser dot should be used only as a guide with accurate measurements being taken to the mini-prism.

13.14 WHY IS THERE VERY FAST BEEPING?

You can set the total station to alert you when you are within a certain distance of the required position. If you find the beeping distracting you can switch it off.

13.15 WHY IS THE POINT NOT WHERE I AM EXPECTING IT TO BE?

There are a few reasons why the point you are trying to set out is not where you are expecting it to be:

→ you have misinterpreted your information and the point is indeed in the correct place

→ you have entered the wrong point data, either by misreading the source data or by making a typing error during the input stage

→ when you have sighted to the prism, you have pressed 'measure and store' instead of measure only. When you press 'measure and store' you are telling the instrument that you have finished setting out that particular point and are ready to move on to the next one. The direction the instrument is giving you is now, is for the next point in the list rather than the one you think you are setting out. Select the point ID and use the arrow keys to scroll left to go back to your previous point

→ there is a gross error in your instrument set-up. Measure to several known coordinates to verify that they are correct. If in doubt, set up your instrument from scratch and start again.

13.16 WHY IS MY SETTING OUT NOT TYING IN WITH OTHER POINTS?

There are several reasons why your setting out may not be tying in with existing features, as-built or as-set-out points as you would expect it to. Here are some possible reasons:

→ the quality of your current or your previous instrument set-ups

→ the quality of the control points you are using – were your primary and secondary control points installed using the procedures described in chapter **19**? If you didn't personally install them, have you verified them?

→ the quality of your previous setting out – did you set out the previous points or did someone else? Could it be that the previous setting out contained errors?

→ are you using the wrong prism constant? If your target type is not set correctly, or was not set correctly during the station set-up process, this could lead to errors of either 18 or 34mm in all your setting out.

→ have the control points been disturbed or have the values been reassigned and you are using the old values?

13.17 SURVEY FUNCTION

The survey mode is used for measuring and recording coordinates and reduced levels of points. The most common uses of this function are:

→ as-built surveys

→ topographical surveys

13.18 WHAT SHOULD BE THE ACCURACY OF A SURVEY?

When deciding the accuracy for your survey, use the information you have available to you combined with your engineering judgment. Consider the following:

→ Who has asked you to do the survey? Have they specified an accuracy?

→ What is the purpose of the survey? For a topographical survey, generally +/- 30mm is sufficiently accurate, however for something like bridge bearings, you may require less than +/-1mm in internal accuracy in which case you may need to use the total station in combination with other equipment such as a precise level.

→ Is an overall profile needed or do localised undulations need to be mapped in detail?

13.19 REFERENCE LINE FUNCTION (ALSO CALLED 'REFLINE' OR 'POINT TO LINE')

The Reference Line function can be used in 3 ways:

→ to set out points in relation to a defined line

→ to set out a grid in relation to a defined line (chainage and offset method)

→ to indicate the perpendicular distance from a defined line at any point along the line

It is very useful for:

- → road centrelines
- → structures or buildings set out on gridlines
- → checking the straightness and plumb of walls and columns (see sections **24.68** and **24.69**).

13.20 WHAT ARE THE ADVANTAGES OF THE REFERENCE LINE FUNCTION?

The advantages of the reference line function are:

- → it can be used without having to backsight to any control points
- → you only need to know dimensions and not coordinates
- → it can be used as a check for points you have set out using the stake out function
- → it is good for internal accuracy, as the line can be defined as 'as-built' or as 'as-set-out' points
- → you can check the perpendicular offset at any point along the line without having to consider the chainage.
- → once you have set out a point, you can then record the final position in the memory
- → the line is projected to infinity.

13.21 WHAT IS THE REFERENCE ARC FUNCTION?

The reference arc function defines a curve. Points can be set out at any point along the chainage of the defined curve or the tangential offset from it. The radius is automatically increased and decreased according to the offset.

The reference arc function is useful for road centrelines, kerb radii and curved buildings. An arc can be defined as:

- → the coordinates of the start and end point and the radius of the arc (Fig. 13-4)
- → three points on the arc (Fig. 13-5)

Fig. 13-4: Reference arc function - 2 points

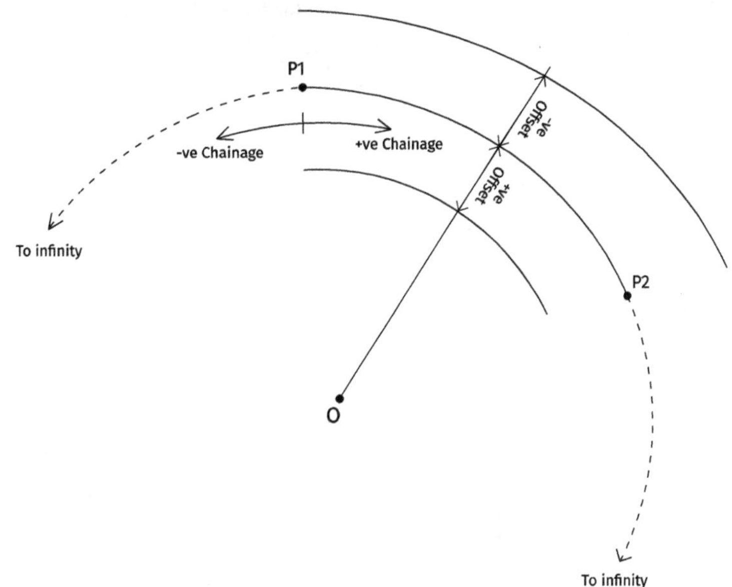

Fig. 13-5: Reference arc function - 3 points

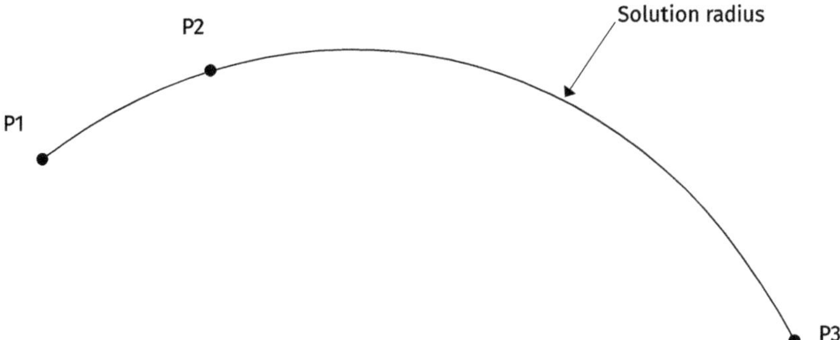

13.22   REMOTE HEIGHT FUNCTION

The remote height function is used for setting out on vertical surfaces where only one of the points can be accessed with a target (Fig. 13-6). The benefit of the remote height function over the reflectorless measurement function is that the vertical distance from the target is displayed on the screen as you move the telescope up and down. This means the fine tuner can be adjusted until the required distance is displayed. The laser dot can be used to indicate the desired position.

Fig. 13-6: Remote height

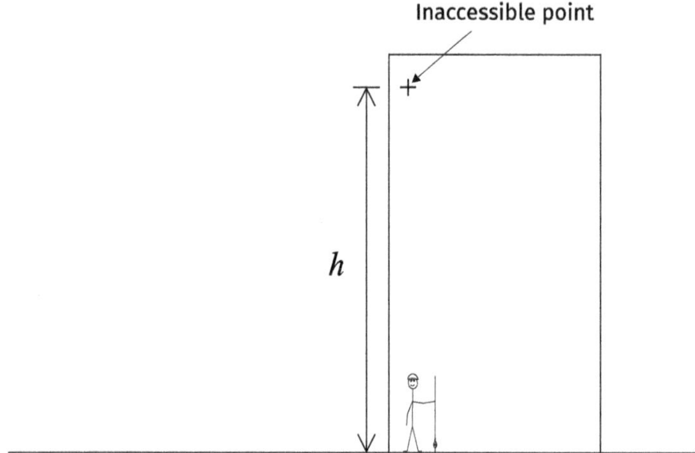

13.23 AREA MEASUREMENT FUNCTION

The area measurement function is used to measure irregular areas (Fig. 13-7). Simply sight to points around the perimeter of the area you would like to measure, and the total station will compute the area. Do not return to your original point as the total station will automatically join the first and last points for you. Any cross over points will either cause the total station to give you an error message or will lead to an incorrect answer.

Fig. 13-7: Area function

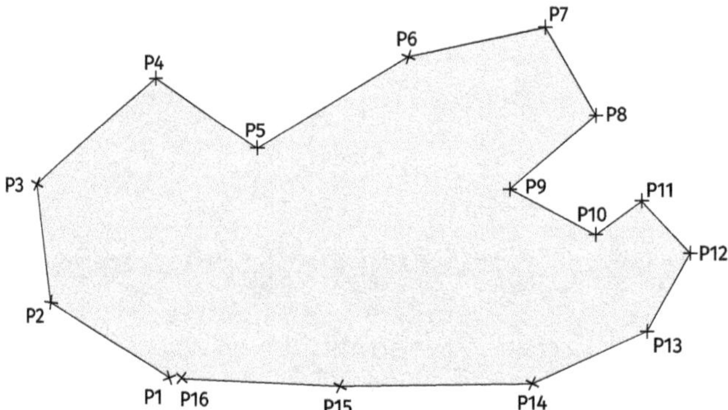

13.24 VOLUME MEASUREMENT FUNCTION

The volume measurement function is used to measure irregular volumes such as stockpiles or excavated voids (Fig. 13-8). There are two components to measuring a volume: the boundary and the surface. The boundary is the external perimeter and the surface is the outer face of the element. For volume calculation purposes, the boundary points will be used to create a surface. In some total stations the boundary must be defined by the uses. In others, the total station can define the boundary by automatically selecting the outermost points as if a rubber band has been placed around a plan of the points.

Fig. 13-8: Volume function - All points visible

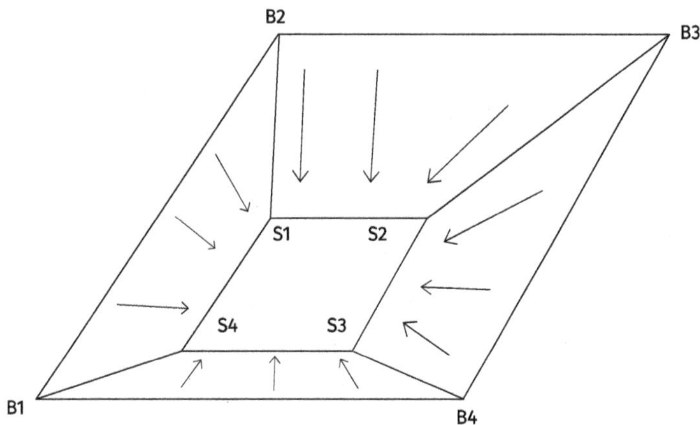

13.25 HOW DO I MEASURE THE VOLUME OF SOMETHING IF I CAN'T SEE ALL FACES OF IT FROM A SINGLE SET-UP?

For large stockpiles, even with a fully extended reflector height, it may not be possible to view all surfaces from a single set-up position. In this case, the element should be measured in two or more parts. The element should be spilt up into sections defined by boundaries. You may wish to physically mark the boundaries using spray paint. To measure a stockpile in two sections, two separate parts should be created. Each part is defined by the longitudinal crown line and the outer perimeter at ground level. Measure the two parts separately and add the volumes together to calculate the final volume.

13.26 HOW CLOSE TOGETHER SHOULD THE POINTS BE FOR VOLUME MEASUREMENT?

The interval of the points depends on the required accuracy of your volume calculation. If you have flat vertical or horizontal surfaces, you need only measure the extents.

CHAPTER 14

GNSS EQUIPMENT

14.1 GNSS

This chapter is intended as an overview of Global Navigation Satellite System (GNSS) equipment, to provide terminology and to highlight the key information needed for setting out. A good source of in-depth information on GNSS is the book Surveying for Engineers by Uren and Price. The user manual for the specific equipment you are using will also be helpful.

14.2 WHAT IS GNSS?

Global Navigation Satellite System (GNSS) is a technique used in surveying and setting out which uses satellites to pinpoint the location of a receiver anywhere in the world.

GNSS has similar applications and capabilities to a total station, but with the added benefit that it can measure positions in relation to the any coordinate system, whether global, national or local, without the need for control points.

GNSS equipment can be used to:

- → measure distances between two points
- → measure or set out coordinates and levels in relation to a local site datum or to global coordinate systems
- → set out offsets from lines
- → set out arcs and offsets from arcs
- → measure areas and volumes

14.3 WHAT GNSS SKILLS SHOULD I HAVE?

You should be able to:

- → use GNSS terminology
- → explain the basic principles of GNSS technology
- → list situations where GNSS is appropriate and situations where it is not
- → carry out a coordinate transformation (also known as grid transformation, localisation or site calibration) to create a local coordinate system
- → list the potential sources of error in GNSS
- → list the factors affecting accuracy
- → check and monitor the level of accuracy achieved
- → select the correct scale factor for the task
- → enter data into the controller manually
- → connect the controller with the GNSS rover
- → navigate around the menus and programs
- → navigate the stored information in list format and map view
- → view, edit, add and delete data
- → enter the correct settings
- → carry out a topographical survey and record the results systematically
- → set out points of known co-ordinates using the onboard functionality
- → set out points in relation to a baseline using the onboard functionality
- → set out points at given chainages and offsets along a radius e.g. road centrelines using onboard functionality
- → transfer large amounts of data from the controller to the computer and vice versa
- → extract line data e.g. the horizontal distance and level difference between two points
- → measure irregular areas and volumes such as stockpiles
- → import and work with DXF and DTMs
- → create and export DXFs and DTMs
- → import and export data to and from Excel or IFC files.

14.4 GNSS TERMINOLOGY

Antenna - the part of the instrument which receives and amplifies radio signals from GNSS satellites. The antenna can take many different forms for example, a circular beacon about 100mm thick an 300mm diameter.

Atmospheric errors – errors caused by variation in atmospheric pressure and humidity

Base and rover – GNSS equipment which includes a base station and rover

Base station (also referred to as a reference station) - the receiver with an antenna which re-broadcasts the GNSS signal and allows greatly improved precision and accuracy. The base station may set up on a tripod or for long-duration works, may be set up permanently. The use of a base station means that RTK subscription fees are not needed.

Constellation – a group of satellites which work together. Examples of constellations are GPS, GLONASS, Galileo, BeiDou and QZSS.

Controller – the controller is the handset which is connected to the rover. All operations are carried out using the controller. The controller is connected to the rover by Bluetooth or cable.

Ellipsoid – an ellipse rotated about an axis. The ellipsoid is the surface formed by this revolution about an axis. The Ellipsoid is a geometric model of the shape of the Earth and approximates the geoid.

Ephemeris errors – errors in the orbital data transmitted by the satellites.

Epoch – the measurement interval of a GNSS receiver.

ETRS89 - the European coordinate system which coincides with WGS84.

Geoid – the hypothetical shape of the earth's surface oceans if the whole world was at mean sea level (MSL) and there were no waves or tides. The geoid is an irregular shape and provides the reference system for altitudes. The geoid, or a regional geoid model, is used to compute orthometric elevations.

Grid north – the direction, north, along the gridlines of a map. Grid north is distinct from true north which is the direction to the north pole and magnetic north which is the direction to which a compass needle points and is slightly different from true north. Grid north for Great Britain Ordnance Survey maps varies slightly from true north. The variation depends on the location within Great Britain Where a local grid is used, grid north is completely unrelated to OS grid north, true north or magnetic north.

Grid transformation – the action of mathematically converting a set of coordinates from one grid system to another. This may be from global (e.g. WGS84) to an (assumed) local grid. Functions of the transformation can include shift, rotation and scale factor.

Initialisation – this is the process of starting up the receiver and connecting to the geostationary PPP satellites, local UHF radio link or internet . Initialisation should be done in the same conditions that it will be used in. Don't do initialisation in a car since initialisation is the process by which a rover receiver is computing its position relative to a base, by resolving ambiguities in the data received. It requires good-quality input data to initialise.

Latitude – latitude references the north-south component of a coordinate on the earth's surface. It is expressed as an angle, zero degrees at the equator and 90 at the north pole. Lines of constant latitude run east–west and form parallel bands around the earth.

Longitude – longitude is the east-west component of a coordinate on the earth's surface. Longitude lines are also known as meridian lines and are notional vertical lines that run around the earth vertically and pass through the north and south poles. There is one longitude line for each degree of rotation. Zero degrees is physically marked by a brass strip in the ground at Greenwich, UK.

Magnetic north – the direction in which the north end of a compass needle or freely suspended magnet will point. The direction varies slightly depending on the location of the compass.

Mean sea level – the average level of the earth's surface oceans over a period of time, taking account of changing tide levels.

Multipath errors – errors caused by unwanted reflections as received by the GNSS antenna. A reflected signal is longer than a direct signal which is the cause of the error.

OSGB – Ordnance Survey Great Britain.

OSGM15 – Ordnance Survey Geoid Model 2015 – The geoid model for transforming ellipsoid heights to orthometric height across Great Britain.

OSTN15 – Ordnance Survey Transformation 2015 – A look up table for transforming latitude and longitude to OSGB national grid Eastings and Northings to relate to OSGB36 coordinate system.

PPK – Post-processed kinematic - Mainly used in UAV mapping

PPP – Precise Point Positioning – A global correction service delivered by L band from a constellation of geostationary satellites where local networks e.g. radio or mobile internet are not available or are interrupted. The Leica Geosystems' service is known as SmartLink.

Reference station – See 'Base station'

Redundancy (signals) – when there is more signal available than the minimum needed. Redundancy increases the chance of connectivity.

Redundancy (measurements) – when there are repeated measurements of the same point or vectors in from multiple reference stations, whether RTK, PPK or static.

Repeater station – the repeater station amplifies the signal from the base station to increase the range of UHF radio links, or acts as an intermediate point to get around obstructions which block inter-visibilty. UHF radio links are limited to line-of-sight and output power rating limitation in the UK by licensing by OFCOM.

Rover – the mobile 289 receiver, either hand-held or attached to a detail pole or vehicle.

RTCM – Radio Technical Commission for Maritime – a standard format for transmitting differential corrections in GPS and GNSS systems.

RTK – Real-Time Kinematic – this is a system which computes the baseline in real time to give an almost instantaneous answer with cm level accuracy. RTK is the name of the algorithm or process.

Satellite – the hardware which orbits the Earth and sends precise details of its position in space back to the ground component on Earth. The receiver then uses the information received to identify its exact position in relation to the satellites and on Earth.

SmartNet - the first network RTK service in Great Britain and is operated by Leica Geosystems. SmartNet comprises networks globally with more than 4000 reference stations.

Survey style (Trimble) or **Working style** (Leica) – defines in the controller whether the survey is being carried out using the total station or GNSS.

TDL signal – TDL is the model name of a Trimble radio. Pacific Crest and Satel are other widely used radio modems for transmission and reception of differential corrections.

Tilt-compensation – the ability of a GNSS rover to tilt the pole and calculate the coordinate for the tip of the point at the end of the pole. Two approaches exist; magnetometer and compass, require frequent calibration and magnetic free environments and are limited to 15 degrees tilt. Alternatively, IMU technique is calibration-free, immune to magnetic fields and the tilt angle is unlimited. Permits survey and stakeout faster without need to level the liquid bubble and enables measurements to be taken at previously inaccessible points e.g. building corners. As well as productivity and accuracy benefits, there are health and safety benefits of not stretching or leaning to view the liquid bubble.

Trilateration and ranging – the mathematical process used by GNSS software to determine the position of the receiver in relation to the satellites. The radial distance is measured to four or more satellites, using the time taken for the signal to reach the receiver. Satellites contain an extremely accurate clock which generates a unique coded signal for each satellite. The receiver on the ground generates the same coded signal and compares the received code generated by the satellite. These time delays are then converted into distances which are used to compute the global position of the receiver.

True north – points to the North Pole from any point on the earth's surface. This is different to magnetic north which is slightly different, and grid north which is local to a map or site.

VRS – Virtual Reference Station. MAX and iMAX are alternative approaches to network RTK that offer traceability to real reference points as opposed to virtual ones.

VRS Now (also known as VRS) – VRS Now is a Trimble Network RTK service (including hardware and software) which increases accuracy of GNSS equipment from 1m to 1cm. There are approx. 120 ground receivers distributed across Great Britain, which are owned and maintained by Ordnance Survey. The signal from these receivers is augmented by Trimble. A subscription is needed for each controller. The subscription can be paid monthly or yearly. Works with only high grade (cm level capable) GNSS equipment. The Leica Geosystems equivalent of VRS Now is 'SmartNet'.

WGS84 (World Geodetic System 84) – the standard global coordinate system used in the UK for GNSS. It is the reference coordinate and altitude system whose coordinate origin is at the earth's centre of mass.

14.5 WHAT'S THE DIFFERENCE BETWEEN GPS AND GNSS?

The terms GPS and GNSS are often used interchangeably, however there is a difference in the meaning. GNSS is the umbrella term for all global positioning systems, whereas GPS (Global Positioning System) is the name of one particular range of constellations of satellites.

14.6 WHAT ARE THE PROS AND CONS OF GNSS?

In recent years there have been significant advances in the technology which enables GNSS surveying applications. The equipment costs have plummeted in relation to the achievable accuracy while user-friendliness and mobile internet coverage have made the equipment more widely accessible.

When deciding whether GNSS is appropriate for a task or project, there are both pros and cons to think about.

The pros are:

- → **speed** – minimal set-up required and very quick to set out each point
- → **simplicity** – user-friendly software and work flows
- → **accuracy** - sufficiently accurate if used properly and for the right jobs
- → **range** – suitable for large projects stretching over many miles
- → **flexibility** – a line of sight to control points is not needed
- → **productivity** - one-man operation

Cons:

- → **lack of accuracy** – with RTK, use of GNSS is limited to activities which require relatively low accuracy such as earthworks, topographical surveys or highways construction. Accuracy can be enhanced in near-realtime using post-processing software, to better than +/- 10mm for activities such as structural monitoring.
- → **cost** – more expensive to hire or purchase, maintain, insure and run than a total station.
- → **user errors** – like any equipment, there is an increased risk of inaccuracy or gross error if the user is not properly trained. Because GNSS appears to be very straightforward, it is often used without the necessary training.
- → **limitations** – there are a number of environmental factors which limit the conditions in which GNSS can be used.
- → **signal limitation** – when a mobile phone signal is not available a base station is required. A line of sight is needed from the base station to the rover, so repeater stations are needed where there are large obstructions such as hills or mountains. This is becoming less of a problem due to ever-extending and intensifying mobile phone coverage.

14.7 HOW DOES GNSS WORK?

There are three segments to GNSS:

- → **the space segment** – satellites orbit the earth, continuously sending out a signal identifying their location.

→ **the control segment** - the control or ground segment is the infrastructure, systems and network of global monitoring stations that are used to control the hardware in space belonging to the network operator. In the case of GPS, this is the US Department of Defence.

→ **the user segment** – the receiver scans for and then locks onto the radio signal broadcast by the satellites. It uses the principle of trilateration to pinpoint its position in relation to the satellites. The positional accuracy is improved when a receiver is operated in conjunction with one or more reference receivers placed at known position/s.

14.8 WHAT ARE THE LIMITATIONS OF GNSS?

There are some limitations of GNSS to be aware of:

→ it can't be used indoors.

→ a good view of the sky is needed so it can't be used in wooded areas or near tall buildings.

→ sources of radio interference can prevent a good GNSS radio signal reaching the user's receiver. This is quite rare, but most likely to be encountered around military bases, where high-powered radio and radar will be in operation. The user's receiver will not gain initialisation or will obtain poor positional precision when receiving poor signals.

→ buildings with a reflective façade may cause an issue, but the system may still work.

14.9 HOW DOES THE ROVER GET A SIGNAL?

The rover can obtain its position information by three means of connection:

1. **SIM card** - the receiver connects to RTK via an internet connection. Correction services such as 'VRS Now' (Trimble) and 'SmartNet' (Leica) enhance the speed and accuracy of measurements. If the rover SIM card cannot get a phone signal, but your mobile phone can, you can connect to your phone hotspot. Multi-network SIM cards can be used.

2. **Base and rover** – the rover receives a radio signal from the base station. Due to extensive mobile phone signal availability, this method not always required nowadays for general construction site conditions.

3. **PPP global corrections** – satellite delivered corrections from a constellation of geostationary satellites for use in remote areas or to maintain RTK when mobile internet or radio signals are interrupted.

14.10 HOW DOES A BASE AND ROVER WORK?

An RTK system comprises two receivers working together. One is placed at a known position and data is broadcast from this receiver (base) to the user's receiver (rover). The information from the base is processed together with the data received by the rover to compute a vector, thus positioning the rover precisely, relative to the base. Fig 14-1 shows a base station. Fig 14-2 shows rover and controller on a pole.

Conventional RTK uses a real receiver at the base. Data is usually broadcast from the receiver to one or multiple rovers by radio or mobile phone network for extended ranges beyond radio signal range. In the case of Network RTK, Trimble replaces a real base with a virtual base and the data transfer is by the internet (mobile data signal). Large construction sites operate one or more real base receivers, usually mounted on site cabins. Leica SmartNet uses the iMAX and MAX concept which relates measurements to a real (and therefore traceable) reference station.

It depends on the manufacturer's equipment as to how the rover connects to the internet. It is normally by a modem in the receiver or controller.

Fig 14–1. Base station on a tripod. Image courtesy of KOREC Group

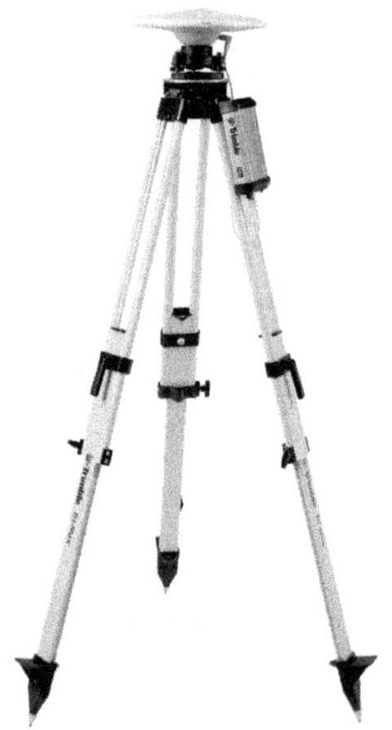

Fig 14–2. Rover and controller on a pole. Image courtesy of Leica Geosystems

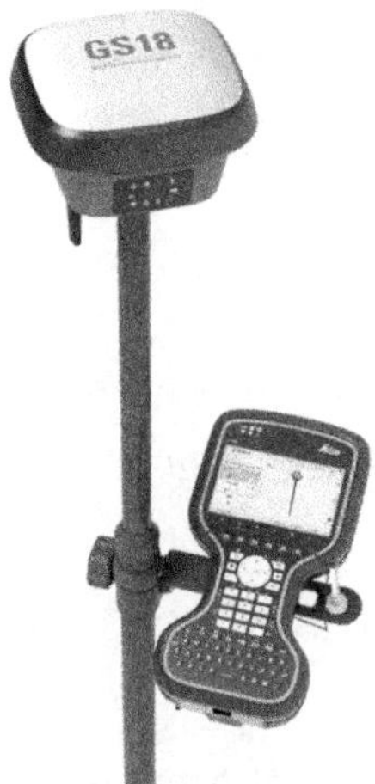

14.11 WHAT AFFECTS THE ACCURACY OF GNSS?

Factors affecting the accuracy of GNSS measurements can be separated into two broad categories: errors affected by the quality of the signal, and user errors.

Errors affected by the quality of the signal can be caused by anything that degrades the quality of the radio signals being received by the receivers in the RTK system including:

→ atmosphere

→ multipath errors

→ poor sky view (thus poor satellite geometry)

User errors can occur when users are not properly trained, are not carrying out the necessary checks and are not aware of the potential pitfalls. Errors caused by lack of user knowledge and awareness include:

→ entering the wrong scale factor

→ failing to notice a high error or PDOP

→ using poor quality control points for creating the local grid (also known as site calibration or localisation)

→ entering the wrong setting out data

→ poor verticality of the pole for equipment which does not take account of pole angle

Factors which do not affect accuracy include:

→ cloud cover

→ other mobile phones

→ overhead powerlines

14.12 WHAT IS PDOP?

PDOP quantifies the quality of the satellite geometry. A minimum of 5 satellites need to be visible at all times for the RTK to work. If the receiver is only picking up 5 or fewer satellites, then the sky view must be obstructed, which in turn will mean the geometry will be poor and thus PDOP will go up.

The PDOP indicates the accuracy of the measurement. The lower the PDOP value, the better the accuracy.

→ **DOP** – dilution of precision – an indicator of the quality of the GNSS position. DOP takes account of each satellite's position in relation to other ones in the constellation.

→ **PDOP** – position (E, N, H)

→ **HDOP** – horizontal (E, N)

→ **VDOP** – vertical (H)

→ **TDOP** – (time)

14.13 WHAT ACCURACY CAN BE ACHIEVED WITH GNSS?

When connected to RTK, the positional accuracy is 10-20mm and the height accuracy is 15-30mm. The accuracy can be reduced to 10mm when base stations are employed.

14.14 WHAT IS A TRANSFORMATION TO LOCAL COORDINATE SYSTEM?

This process converts WGS84 coordinates into a local coordinate system by calibrating them against local control points. The optimum accuracy can be achieved when a grid transformation is performed. Different terms are used in different types of equipment, including 'Site calibration', 'Localisation', 'QuickGrid' and 'grid transformation'.

RTK is used for this process. If using a real base, then a 3 minute measurement is taken. If using a correction service such as Network RTK then 4 x 3 minute measurements should be taken at each control point.

To carry out a site calibration:

1. Select the site calibration programme in your controller.
2. Input the coordinates of your local control points.
3. Select the measurement type (RTK or static).
4. Select your required tolerance.
5. Select the number of epochs or the time for the measurement (the longer the measurement, the greater the accuracy to a point).
6. Place your rover over the first control point.
7. Start the measurement and keep the rover level for the duration.
8. If the residuals are within your required tolerance, accept the measurement.
9. Repeat steps 6 – 8 for all other control points.
10. Select 'report' to view the accuracy results of your site calibration.
11. De-select any measurements which are not accurate enough.

14.15 HOW CAN I IMPROVE THE QUALITY OF A SITE CALIBRATION?

The quality of your local control points affects the quality of the calibration. RTK with a real base works at +/- 20mm and VRS at +/- 30mm. These are on top of whatever precision the control was originally put in with.

To ensure the optimum accuracy of your site calibration:

→ your local control points should have the highest possible order of accuracy. See section **19.2** on primary control points.

→ ensure 'good geometry' of your control points. They should be spaced out evenly around the extents of the site and not clustered together in one location.

14.16 HOW CAN I CHECK MY SET-UP IS CORRECT?

It is good practice to check to a reliable known point at the start and end of your task. This is known as 'checking in' and 'checking out'. This process will identify gross errors from incorrect antenna type, height, coordinate system or other blunders.

14.17 SCALE FACTORS

For sites under 10km, a scale factor of 1 can be used. For over 10km, OSTN15 should be used. A site calibration is like a locally computed coordinate system, specific to each work site. So, if OSTN15 was accidentally selected instead of the local scale factor, it would mean the user would not be working in the correct grid for the site which would result in an error.

14.18 MEASUREMENT OPTIONS

There are 3 measurement options:

→ **'Instantaneous'** (Leica) or **'Rapid point'** (Trimble) – for measuring soft details or 2D points e.g. railway sleeper or peg. This is the quickest but least accurate type of measurement as it interpolates the correction since the signal was last received (up to one second ago)

→ **'Topo point'** – for measuring hard details or 3D points e.g. the corner of a slab. The measurement takes approx. 3 seconds and takes the mean of 3 measurements. The measurements are more precise but slower than the rapid point. This can be configured to set the number of epochs.

→ **'Observed control point'** – for measurement of control points. The measurement can take up to 3 minutes and takes approx. 180 measurements. It is good practice to reoccupy and measure control point more than 45 minutes later to allow for change in GNSS constellation geometry for increased robustness.

14.19 WHAT IF I LOSE INTERNET CONNECTION?

With certain equipment features it is possible to continue working for up to ten minutes after satellites and internet are lost. The functions 'X Infill' (Trimble) and 'SmartLink Fill' (Leica) let you work temporarily after satellites and internet are lost.

14.20 WHAT CHECKS SHOULD I DO ON MY GNSS EQUIPMENT?

Other than checking to a known point, there are no field checks necessary for GNSS equipment.

14.21 WHAT CAN GO WRONG WITH GNSS?

Not performing a site calibration correctly can lead to large horizontal and vertical errors. Site calibration should only be done by someone who understands the principle of what is being performed and how to conduct it. Typically, issues occur when not enough local control points are used, and they aren't distributed around the whole work site.

CHAPTER 15

TAPE MEASURES AND STRING LINES

15.1 THE TAPE MEASURE

Taping is a quick, simple, cheap and accurate method of measuring distances. It is the most basic form of linear measurement and involves stretching a graduated tape between two points to measure the relative distance between them. When carried out under the correct conditions and with the appropriate level of skill, it can achieve an accuracy of +/-1mm over 10m.

It is often overlooked and even considered by some people as redundant, however, it is an essential skill for engineers to develop.

It is suitable for a wide range of low- and high-accuracy setting out tasks including:

- → profile boards and batter rails
- → excavations
- → corners of shutters
- → gridline offsets
- → relative positions of rebar, starter bars, kickers, stop-ends
- → concrete level in relation to the shutter level
- → cast-in pipes
- → box-outs
- → piles
- → cladding fixings
- → bridge bearings
- → structural steel
- → road cross-sections

15.2 WHAT TAPE MEASURE SKILLS SHOULD I HAVE? ⚷

You should be able to:

- → list the sources of error in taping
- → incorporate checks when setting out using a tape measure
- → set out on sloping ground
- → set out right angles and rectangles

15.2 WHAT ARE THE DIFFERENT TYPES OF TAPE MEASURE?

Tape measures come in a variety of lengths, typically 5m, 8m, 30m, 50m and 100m.

5m and 8m tapes are typically steel, however longer tapes may be fibreglass, nylon or steel.

15.3 HOW CAN I IMPROVE THE ACCURACY OF MY TAPING?

The tape measure is used to measure the relative difference between two points on a linear graduated scale. In its simplest form, the zero end is held on point 1, and the reading is taken at point 2. There are several ways to improve accuracy and reduce the risk of gross error.

METHOD 1: SWITCHING THE DIRECTION OF THE TAPE

To reduce the risk of gross error due to misreading the tape:

1. Hold the zero end of the tape at point 1 and take the reading at point 2.

2. Reverse the direction of the tape, holding the zero end at point 2 and take the reading at point 1.

3. If they match within the required tolerance, the mean of the readings is the correct answer. If they do not match, repeat steps 1 to 3.

METHOD 2: ADDING ON A CONSTANT

To improve the accuracy of your reading, use of the zero end of the tape should be avoided. The reading at point 1 should be taken at either 0.1m or 1m and the equivalent constant subtracted from the reading taken at point 2. See Fig. 15-1. There are several reasons for this:

→ There is usually a tab, hook or eye at the zero end which makes it difficult to hold it accurately on your point

→ There is 'waggle' room to allow for the thickness of the tab. The distance of the 'waggle' is unreliable and can come out of adjustment over time.

→ It is not always clear exactly which point is zero.

→ It is difficult to get a good grip on the tape to hold it steady when holding the zero-end on the point. When holding the tape further up, you can grip the tape with both hands, one each side of the correct point.

METHOD 3: RANDOM TAPE READINGS

If you set something out using a tape measure and make a misreading, you are likely to make the exact same mistake when you are checking it. The method of random tape readings gives a completely 'clean' way of checking your setting out as it will be free of errors which may have been built into the original setting out. The process is as follows:

→ At point 1, instead of holding the tape at zero or at a fixed constant such as 0.1m, have your assistant hold it at a random number e.g 0.327m. This is reading 1.

→ At point 2, take the reading. This is reading 2.

→ Subtract reading 1 from reading 2, to give the distance between the two points (Fig. 15-2).

→ Repeat steps 1 to 3, making sure that reading 1 is different each time and does not end in zero.

→ If the two readings match up within the required tolerances, take the mean of the readings as your final answer. If they don't match up, repeat steps 1 to 4 until you have a consistent set of results. Discard any erroneous readings and take the mean of the consistent readings.

Fig. 15-1: Taping - Adding on a constant

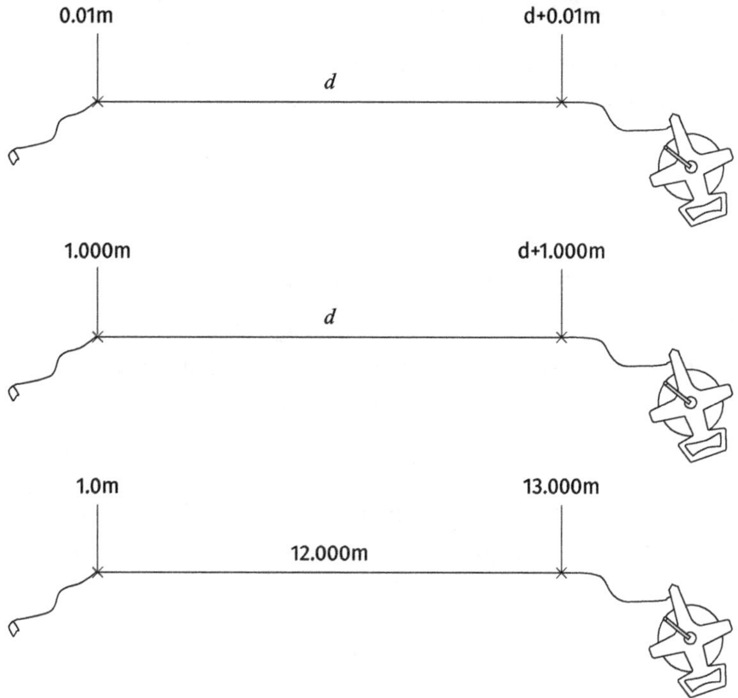

Fig. 15-2: Taping - Random tape readings

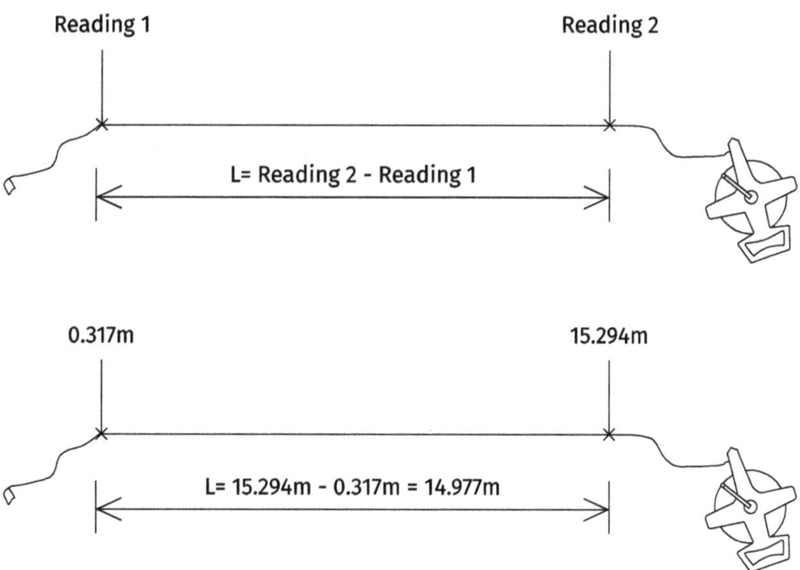

15.4 HOW DO I USE THE TAPE MEASURE FOR SETTING OUT?

When using the tape measure for accurate setting out, always start the tape measure at 1m rather than zero, and add 1m onto the distance being set out, to determine the required setting out reading.

SETTING OUT A RIGHT ANGLE USING A TAPE MEASURE

There is a range of methods for setting out a right angle using a tape measure. The methods below assume that the ground is level. If you are setting out on sloping ground, you will need to calculate the slope distances as described in section **15.5.**

METHOD 1: 3-4-5 TRIANGLE

This method uses the convenient fact that if the two sides of a right-angled triangle are 3m and 4m, then the hypotenuse will be 5m. There are several ways to use the 3-4-5 principle. This is just one of them:

1. Create your first line, L1, by hammering pins into the ground at P1 and P2.

2. If L1 is greater than the length of your tape measure, set up a string line between P1 and P2. If not, pull the tape from P1 to P2 and measure 4m along line from P1 and hammer a pin into the ground at P3.

3. Hook the end of the tape over P1 and then round P3.

4. Have your assistant hold the tape measure at P1 at a reading of 12m.

5. At a tape reading of 9m, pull the tape until all sides of the triangle are taut. Hammer a pin in at this point, P4.

6. Remove the tape and check that the sides of the triangle are 3m, 4m and 5m respectively. If not, adjust accordingly (Fig. 15-3).

7. You now have a right angle P3P1P4. You can extend the line P1-P4 to the required length and install your final pin at P5.

8. Pins P3 and P4 can be removed.

METHOD 2: INTERSECTING ARCS

This method is more accurate than method 1, since it does not involve extrapolating lines.

1. Create your first line by marking at P1 and P2.

2. Calculate the length of the diagonal.

3. From P1, pull the tape taut to the correct length and mark an arc in the expected location of P3. For optimum accuracy, add on a constant as described in section **15.3**. Create the arc by making several accurate marks at intervals.

4. From P2, pull the tape taut to the length of the diagonal and make an accurate mark each side of the arc you created in step 3.

5. Join the sections of both arcs. P3, is the point where the arcs intersect (Fig. 15-4).

6. Repeat steps 3 – 5, to set out P4.

7. Measure between P3 and P4 to check it is within tolerance. If it is not, check your distances from P1 and P2 to P3 and P4 and adjust accordingly.

METHOD 3: METHOD OF TWO TAPES

To use this method for accurate setting out, you will need 3 people; two to hold the tape accurately on P1 and P2 at the chosen constant and one person to set the position of P3.

1. Create your first line by marking at P1 and P2.

2. Calculate the length of the diagonal.

3. Have each assistant hold their tape measure accurately on the point.

4. Pull both tapes taut, and adjust the position where they cross until the length of P1-P3 and the diagonal meet.

5. Make a mark at this point, P3.

6. Repeat steps 3 – 5 to set out P4.

7. Measure between P3 and P4 to check it is within tolerance. If it is not, check your distances from P1 and P2 to P3 and P4 and adjust accordingly.

METHOD 4: TURN 90 DEGREES WITH A THEODOLITE.

This method was traditionally used before the widespread availability of total stations:

1. Create your first line by marking at P1 and P2.

2. Set up your instrument over P2.

3. Sight to P1.

4. Turn 90 degrees in the correct direction.

5. Get your assistant to hold the tape taut from P1, placing their thumb at the required length of P2-P3.

6. Have the tape moved into line with the vertical cross-hair of the theodolite. This is P3.

7. Set up over P3.

8. Sight to P2. Turn 90 degrees in the correct direction.

9. Repeat steps 5 and 6 to set out P4.

10. Measure between P3 and P4 to check that it is within the required tolerance.

11. Calculate the lengths of the diagonals. Check that they are within the required tolerance. If not, work through steps 2 to 10 again to identify the source of the error.

Fig. 15-3: Taping - 345 triangle

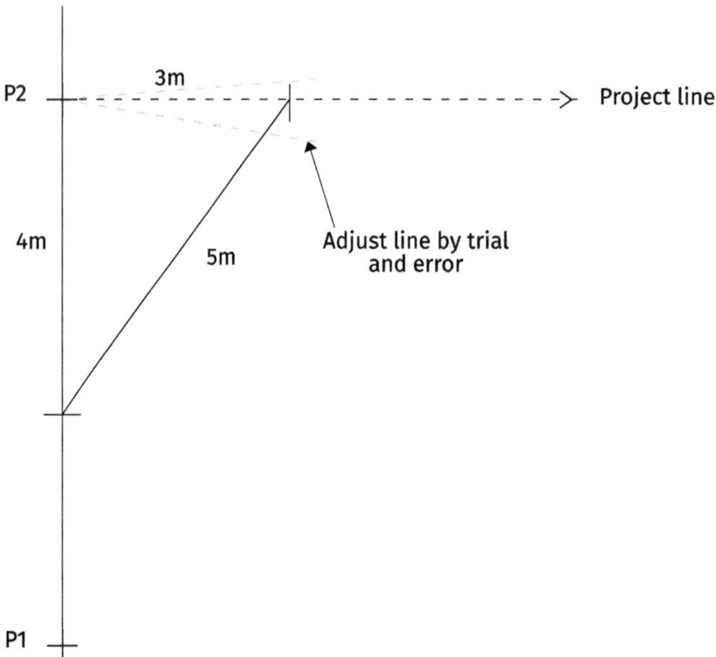

Fig. 15-4: Taping - Intersection of arcs

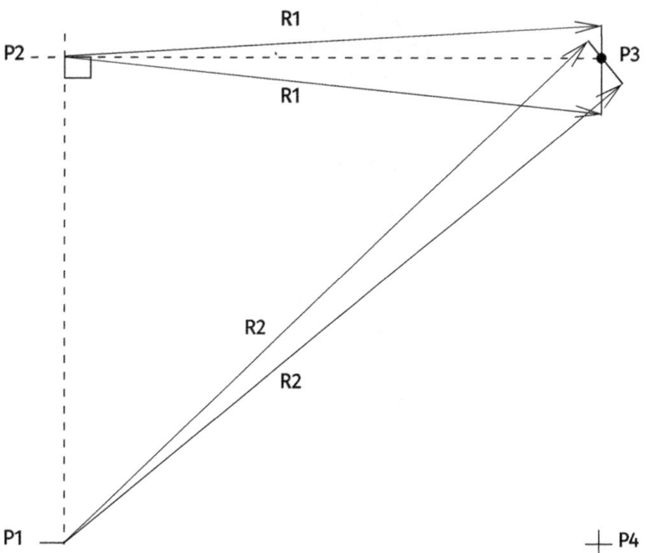

15.5 HOW DO I SET OUT ON SLOPING GROUND?

If there is level change between the points, the reading on the tape (slope distance) will be different to the plan (horizontal distance).

To calculate the slope distance:

1. Determine the level change between the two points. One way to do this is to set up your level and take a staff reading at the start and end of the line.

2. Subtract one staff reading from the other to find the level difference between the two points.

3. Use Pythagoras' theorem to calculate the slope distance using the plan distance and the level change.

4. Set out your point using the slope distance.

15.6 HOW DO I TAKE AN ACCURATE MEASUREMENT ON SLOPING GROUND?

To calculate the plan distance from measured slope distance:

1. Determine the level change between the two points. One way to do this is to set up your level and take a staff reading at the start and end of the line.

2. Subtract one staff reading from the other to find the level difference between the two points.

3. Use Pythagoras' theorem to calculate the plan distance using the slope distance and the level change.

15.7 WHAT ARE THE EFFECTS OF SLOPING GROUND?

The effect of sloping ground depends on the length of your measurement and the level change between the two points. You should calculate the effects for your own specific situation. Here are a couple of illustrations:

Scenario 1:

Slope distance = 28.500m, Level change = 2.3m, Plan distance = 28.407m, difference = 93mm

Scenario 2:

Slope distance = 3.5m, Level change = 0.07m, Plan distance = 3.499m, difference = 1mm

15.8 SOURCES OF ERROR IN TAPING

Sources of error when using a tape measure include:

Slack or obstructed tape measure – if the tape measure is not is a perfectly straight line, the reading on the tape measure will be higher than the true value. This can happen if the tape measure gets caught on grass (Fig. 15-5), is sagging under its own weight (Fig. 15-6) or is blowing in the wind. Ensure the tape is taut and in a perfectly straight line before each reading is taken.

Incorrect zero point – if the point which is assumed to be zero is not zero, an error will be introduced. This can happen if the end of the tape is damaged or if the user is not sure exactly which part of the end of the tape is zero. Always calibrate the end of your tape measure against a ruler and train your assistant accordingly.

Incorrect starting value assumed – when a constant is added rather than using the zero end of the tape, there is potential for confusion. Clearly communicate with your assistant the tape reading which you would like them to hold on the point. If necessary, show them where the point is on the tape measure and mark either side of it with a marker pen or electrical tape. To reduce the potential for confusion, stick with the same constant all the time rather than varying it from task to task.

Misreading the tape – no matter which method you use, there is the potential for misreading the tape measure. The more you practice, the less the risk of error, although it will always exist. To minimise the risk, use the methods described in section **15.3.** It may help to find a tape measure you feel confident with until you have had plenty of practice.

Stretching the tape – if the tape is stretched, the reading on the tape will be lower than the true value. For accurate work, you should use a calibrated steel tape. Avoid using cheap tape measures. Fibreglass and linen tapes are more susceptible to stretching and should only be used for low accuracy tasks such as earthworks.

Cumulative error – when measuring or marking out consecutive stretches, measure each new point in relation to a single origin rather than from the end of the previous one, as the cumulative error will increase for each mark (Fig. 15-7).

Fig. 15-5: Tape measure caught on obstruction

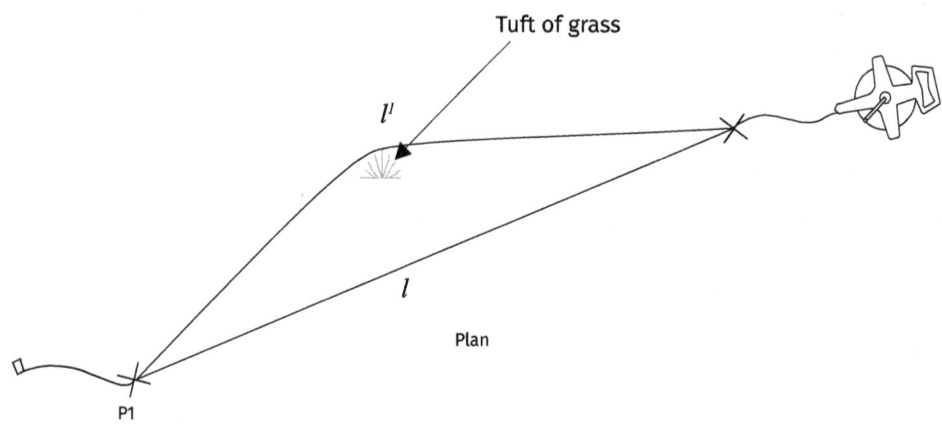

Fig 15-6: Tape sagging

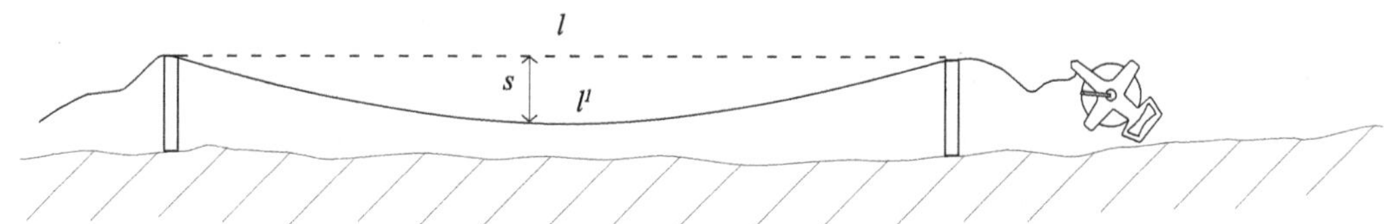

Fig. 15-7: Taping - Cumulative errors

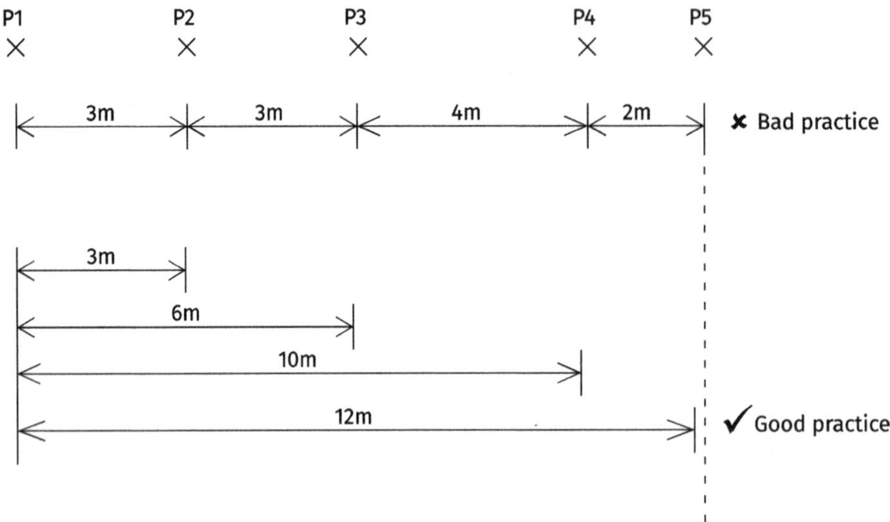

15.9 CARE OF THE STEEL TAPE

To minimise the risk of error, it is important to take good care of your tape measures. Here are some precautions you can take:

→ keep paper towels handy and always clean and dry the tape when you have finished with it

→ avoid catching the tape on sharp edges such as tying wire

→ don't bend or kink the tape

→ avoid getting spray paint on it as this can lead to misreading the measurement if numbers or graduations are obscured.

15.10 HOW DO I CALIBRATE MY PACES?

Calibrating your paces to 1m is a useful skill for engineers and their assistants to have. Stepping out distances using 1m paces can help you visualise layouts, carry out rough checks or roughly locate a hidden feature such as a control point or borehole in relation to nearby features such as fence lines.

You can use paces to measure rough distances, areas and volumes. You may already have a knack for this, but if not, calibrating your paces to 1m only takes 5 minutes to do. The steps are as follows:

1. Walk normally between two points of known distance (at least 30m). If you don't have two points, mark out 30m on a flat surface using your tape measure.

2. If the number of paces is greater than the distance, you need to increase your pace size slightly. Repeat the walk with slightly bigger pace sizes. If the number of paces is fewer than the distance, you need to decrease your pace size slightly. Repeat the walk with slightly bigger pace sizes.

3. Repeat step 2 several times until the number of paces matches the known distance.

15.11 HOW CAN I CALIBRATE MY BOOT SIZE?

Calibrating the length of your boot can be surprisingly accurate. To do this simply measure the bottom of your boot from the heel to the toe. Alternatively, make a mark, place the back of one of your heels on it, place your other heel against your toe and repeat until you have ten footsteps in a line and make a mark at your toe. Measure the distance between the two marks with a tape measure and divide the answer by ten to find the length of your boot.

15.12 STRING LINES

String lines are used for creating straight lines between two points. They may be secured to a nail at each end of the line or held firmly in place.

They are used for:

→ Setting out intermediate points between the ends of a straight line

→ checking the straightness of the edge of a shutter

→ creating a fixed line between two levels, so that vertical measurements (dips) can be taken

→ creating a vertical Reference Line, by suspending a weight at the free end

15.13 WHAT IS A LASER MEASURE?

A laser measure, often referred to as a laser disto, is a hand-held electronic device used for measuring to an object or surface. Laser measures can be used for ranges of up to about 200m and typically have an accuracy of +/- 3mm for distances up to 100m.

In certain situations they can be quicker and more convenient than using a tape measure. They are useful when one of the points, such as a high ceiling, cannot be accessed. Up to 30 measurements can be stored in the memory. Some laser measures contain basic area and volume measurement functions.

To use a laser measure, the bottom of it is placed on the start point of the line being measured. There must be a perpendicular surface or object at end point of the line for the laser to reflect off. When the button is pressed, the distance is displayed on the screen.

CHAPTER 16

PIPE LASERS AND MACHINE CONTROL

16.1 THE PIPE LASER

A pipe laser (Fig. 16-1) is an instrument which is used to set underground pipes to a constant rate of fall during the laying process. The instrument is cylindrical in shape and can be set up either inside the pipe, supported on the invert of the pipe, or set up outside of the pipe using a tripod or trivet. It projects a single laser line at the chosen elevation.

To set up the pipe laser:

1. Calculate the grade of the pipe as a percentage (see section **28.33**)

2. Set the pipe laser grade to the correct value, typically by using 'up' and 'down' arrows

3. The target and the feet of the pipe laser are adjustable. To check that they are both in alignment and that the laser hits the centre of the target, place the target close to the laser.

4. Place the pipe laser on the invert of the pipe (usually at the lowest end)

5. Place the Perspex target at the other end of the pipe

6. The most recently laid end of the pipe is at the correct level when the laser lines up with the centre of the target.

Fig. 16-1: Pipe laser (Image courtesy of Leica Geosystems)

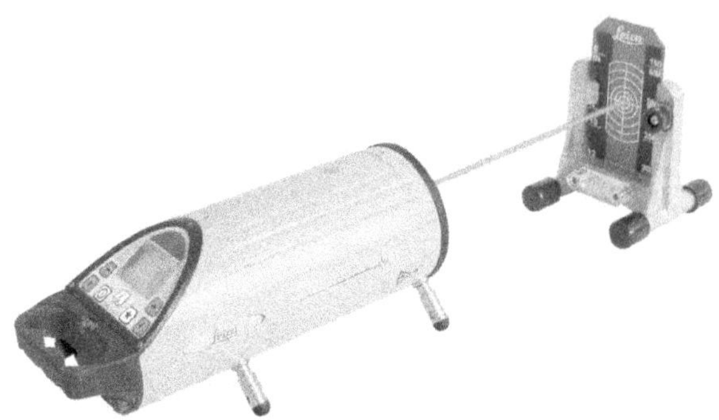

16.2 MACHINE CONTROL

'Machine control' refers to the equipment and software used to accurately position earthworks machinery. It utilises 3Ddesign models and enables levels and positions to be formed relative to GNSS or local grid systems. Construction plant is controlled with GNSS technology and where greater accuracy is required, robotic total station receivers. The technology aids machine operators, and in some cases is effectively 'self-guided'

In the UK, automated machine control is rapidly becoming much more common and is likely to become a vital part of a Site Engineer's role on most future earthworks projects.

Fig 16-2 shows the display unit found in the machine driver's cab. The display screen indicates the position of the bucket or blade in relation to the design levels. In this way, physical markers do not need to be placed in the ground to indicate finished positions and levels.

Fig 16-2 Machine control display unit. Image courtesy of Topcon Positioning Systems'

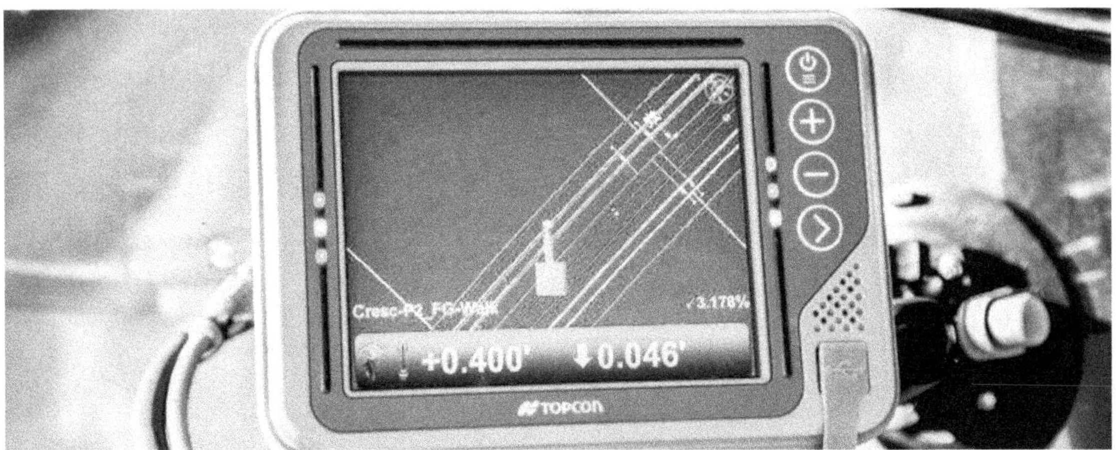

CHAPTER 17

SURVEY SOFTWARE - COMPUTER AIDED DESIGN (CAD)

17.1   WHAT CAD SKILLS DO I NEED AS A SITE ENGINEER?

CAD skills can mean anything from being able to export survey data from the total station into a simple drawing file, to carrying out advanced modelling and data manipulation exercises.

HERE ARE SOME USEFUL SKILLS FOR SITE ENGINEERS:

→]Import images

→ Select points from an electronic file and export them to a setting out schedule

→ Create a contoured plan

→ Compare volumes between subsequent surveys (e.g. to determine how much a stockpile has reduced by)

→ Retrospectively insert or edit break lines

→ Carry out visual checks on 3D models

→ Render slab levels and create a 'heat map' of as-built Vs design levels

→ Select different layers to isolate the information you are interested in

→ Check and convert units

→ Import data into the total station including strings, polylines and design surfaces

→ Create drawings and deliverables and present them correctly

17.2 WHAT IS THE DIFFERENCE BETWEEN CAD AND AUTOCAD?

AutoCAD is a product name that has become synonymous with CAD. A bit like Hoover is to vacuum cleaner.

AutoCAD (including Civil 3D) is a multi-purpose design tool which is not specifically designed for setting out and surveying purposes. Although AutoCAD can be used for a range of tasks including preparing setting out information and converting survey data into a drawing or 3D model, there are specialist survey software packages which are much more intuitive, quicker and much more flexible, not to mention cheaper to run.

In most cases it would make no difference what software you are using, if you are coming up with the right results and providing information in a format that other parties can work with.

17.3 DO ALL SETTING OUT ENGINEERS NEED CAD SKILLS?

Many setting out engineers do not have or need CAD skills for the projects they are working on. Tasks such as setting out reinforced concrete slabs and structures, steel- framed buildings, drainage, houses or small roads do not require it. It's only in the last 20 years that CAD has become commonplace on construction sites. Prior to that, all setting out was done from paper drawings and the rest of the world managed to get built ok! On many projects today, paper drawings are still the primary or only means of communicating setting out information. However, there is no doubt that having CAD skills will give you more opportunities, more flexibility and allow you to be more productive.

17.4 DOES HAVING CAD SKILLS MAKES YOU A BETTER SETTING OUT ENGINEER?

To engineers who aren't conversant with CAD, it can seem like engineers who do have CAD skills are in a different league. Engineers, supervisors and recruiters may wrongly assume that engineers with CAD skills are better at setting out. This isn't necessarily the case as there is no direct relationship between possessing CAD skills and using good practice in the field. In other words, it is possible to be a whizz with the software, but be terrible at the practical and record keeping side of things.

A drawback to CAD is that it can cover up an engineer's shortcomings. It can allow an engineer to 'get by' without knowing the first principles and maths involved, whilst giving a false impression that they have a full understanding. This means that they are less likely to pick up on any errors or mistakes before it is too late.

However, there is no doubt that when used properly and to its full potential, CAD can help engineers to be more productive, enable them to carry out a wider range of tasks, reduce the risk of error and much more.

17.5 WHAT ARE THE BENEFITS OF LEARNING CAD SKILLS?

Although it is possible to work as a setting out engineer even if you don't have CAD skills, there is no doubt that having these skills can open up a whole new world of opportunities. Skilled site engineers are in high demand at the moment. By equipping yourself with CAD skills, you can get more done in less time, store and process your data more effectively, create drawings and 3d models and put yourself forward for a much wider range of projects. You can carry out tasks that would be extremely time consuming to do manually, such as accurate volume calculations for large irregular shapes, sections and elevations or visual comparisons of as-built versus design levels.

Not only can CAD make your own life easier and allow you to take on bigger and more challenging projects, it can make you much more valuable to your employer.

17.6 HOW CAN I GET STARTED WITH CAD?

Although some CAD skills are taught in college and university, the quality of the teaching and the relevance of the content can vary dramatically. As with the practical side of surveying and setting out, some graduates may feel well prepared when they start working, and others may not feel they have the confidence they need.

Whether you are new to the role of site engineer or have been working in the role for many years, there are a range of ways to learn CAD skills. You can learn from other people on the job you are working on, teach yourself by trial and error or attend a training course.

It doesn't matter which survey software you use to develop your skills in the first place. Just choose one and get started. CAD is similar to the practical side of setting out, in that it doesn't matter which type of instrument you learn to set out with; once you have learnt first principles and know the capabilities, you can apply those skills to any instrument. Obviously, the work flow and user interface will be different, so it may take a bit of time to work your way around a new software, but you won't be learning from scratch. When you know the outcome you want, you can work out how to achieve it on the particular software you have. Many engineers and surveyors have a few different software packages and use different ones for different tasks, even within the same job. It can come down to a matter of personal preference, just like which type of smartphone you use.

17.7 WHAT ARE THE DIFFERENT TYPES OF SURVEY SOFTWARE AVAILABLE?

There are many different options including n4CE, LSS, NRG, LISCAD, TBC (Trimble Business Centre) and MAGNET (Topcon's own software).

17.8 WHAT ARE THE BENEFITS TO EMPLOYERS OF TRAINING THEIR ENGINEERS IN CAD?

If you are an employer considering the potential return on investment (ROI) involved in training your engineers in CAD, it's worth thinking about not just how it will help your engineers to perform better on the jobs they are already doing, but how it can increase their flexibility to transfer to other projects. For smaller companies, adding these skills to your workforce could even allow you to bid for bigger and more challenging projects and expand into new areas.

17.9 WHAT ARE SOME COMMON ERRORS WHEN WORKING WITH CAD?

Here are some common sources of error to watch out for:

→ **Wrong units** - Using or assuming the wrong units (e.g. metres instead of millimetres)

→ **Layer overload** – The design consultants sometimes have hundreds of layers which need to be pruned to eliminate unnecessary data.

→ **Lost coordinate system** – The coordinate system is destroyed/rotated to fit nicely on a sheet of paper and needs to be restored.

→ **Wrong scale factors** - Differing scale factors between the measurements taken in the field and the coordinate system in the model.

→ **Mismatch between drawings and labels** - The dimensions on the plans do not actually match what has been drawn. This can happen when the dimensions are manually labelled rather than generated using the correct tool.

→ **Different versions of information within the same drawing** – Multiple copies of the same data in the same drawing, but at different stages of revision.

→ **Using the wrong coordinate system** - When an arbitrary user coordinate system (UCS) has been setup to make the numbers of the coordinate system more human friendly.

→ **Extracting data from CAD without written agreement** – In many contracts, paper copy controlled drawings are the only contractually binding design information. When hard copy drawings or PDFs are used, if there is any question over any of the information, written confirmation should be sought from the designer via the appropriate procedural mechanism, usually a technical query (TQ). When electronic data is provided, the Contractor should have written agreement to extract data from the CAD model otherwise the Contractor could be liable even if the information within the model is incorrect. It is the Contractor's responsibility to check that the electronic information and the controlled drawings match and raise a TQ if they do not.

→ **Just a few levels are dotted over the design** - The Setting out Engineer is then liable for any incorrect interpretation of these levels elsewhere in the design. See the previous point regarding the contractually binding source of information.

17.10 WHAT IS RAW SURVEY DATA?

When you take a measurement with the total station, it measures the horizontal and vertical circle readings (angles) and the distance reading. This is the raw data which it mathematically converts in to coordinates of Eastings, Northings and Height (ENH). The survey software uses the raw data to convert the information into coordinates rather than relying on the coordinates created by the total station. This raw information can be used to check the quality of the survey.

17.11 WHAT IS A BREAKLINE?

A breakline is a way of separating different components of a survey. For example, if two adjacent stockpiles have been surveyed, some of the points may have inadvertently been joined, falsely raising the overall surface between the two features. Breaklines are a way of trimming the incorrect lines back to the correct feature so that the 3D model reflects the true conditions.

CHAPTER 18

STAGES OF SETTING OUT

18.1 WHAT ARE THE DIFFERENT STAGES OF SETTING OUT?

Setting out can be broken down into three stages: primary, secondary and tertiary.

Primary setting out is the installation and measurement of primary control points to the highest appropriate level of accuracy.

Secondary setting out is the installation and measurement of secondary control points, which are one step removed from the primary control points and therefore will inherently be less accurate.

Tertiary setting out is the final stage of setting out which is the final marker placed in the field which will determine the position and level of the construction. The level of accuracy required here will vary from task to task.

18.2 WHAT IS PRIMARY STAGE SETTING OUT?

Primary stage setting out involves the installation of the original set of control points from which all other secondary and tertiary setting out will be referenced. The primary control network consists of strategically placed physical markers whose values are all verified in relation to each other.

A separate set of markers may be established for horizontal (positional) control points and vertical (level) control points or they may be combined into a single set of control points which have both coordinates and reduced levels attributed to them.

The process for establishing the coordinates of horizontal primary control points is called a 'traverse' and is described in detail in section **19.2**.

To establish the reduced levels of vertical control points, the process is called 'transferring a TBM' and is described in section **20.4**

18.3 WHAT IS SECONDARY STAGE SETTING OUT?

Secondary stage setting out involves the installation of a set of secondary control points from which tertiary setting out will be referenced. The secondary control network consists of strategically placed markers whose positions and/or levels have been determined and verified in relation to the primary control points.

As with the primary control points, a separate set of markers may be established for horizontal (positional) control points and vertical (level) control points, or they may be combined into a single set of control points which have both coordinates and reduced levels attributed to them.

The process for establishing the coordinates of secondary horizontal control points is described in detail in section **19.28**

Even when levels are attributed to retro targets, it is advisable to have an additional set of vertical control points set on horizontal surfaces on which a levelling staff can be held.

To establish the reduced levels of secondary vertical control points, as with primary control points, the process is transferring a TBM and is described in section **20.4**

18.4 WHAT IS TERTIARY STAGE SETTING OUT?

Tertiary stage setting out is the mark you make which indicates the position and/or level of the element you are setting out.

18.5 WHAT IS THE PRINCIPLE OF WORKING FROM THE WHOLE TO THE PART?

The principle of working from the whole to the part means that the highest level of accuracy appropriate is used to establish a network of control points covering the whole area. This is the 'whole'. Then these control points are used as a frame of reference for the final markers or measurements. This is the 'part'.

This method of working ensures that errors do not accumulate to unacceptable levels and that a conscious decision can be made about the accuracy required for each specific task.

18.6 WHAT ARE CONTROL POINTS?

Control points are physical points which are used as reference points from which other points are measured or set out in relation to.

18.7 WHAT IS THE DIFFERENCE BETWEEN HORIZONTAL CONTROL AND VERTICAL CONTROL?

Horizontal control points are for positions (coordinates). Vertical control points are for levels (heights).

18.8 WHAT SHOULD PRIMARY CONTROL POINTS BE MADE OF?

Primary control point markers normally take the form of permanent ground markers (PGMs) and must be sufficiently robust to endure the entire duration of the construction project without any damage, movement (including ground movement), or wear and tear. Examples include:

→ a ground nail with a recess in the top for a nail or the centre of the mini-prism pole to rest (Fig. 18-1)

→ a nail in a timber peg set in concrete, protected by timber barriers if necessary (Fig. 18-2)

→ a waist-height vertical pipe set into the ground, filled with concrete and a tribrach set into the top which a backsight target or total station can be clipped into (Fig. 18-3).

Fig. 18-1: Setting out nail

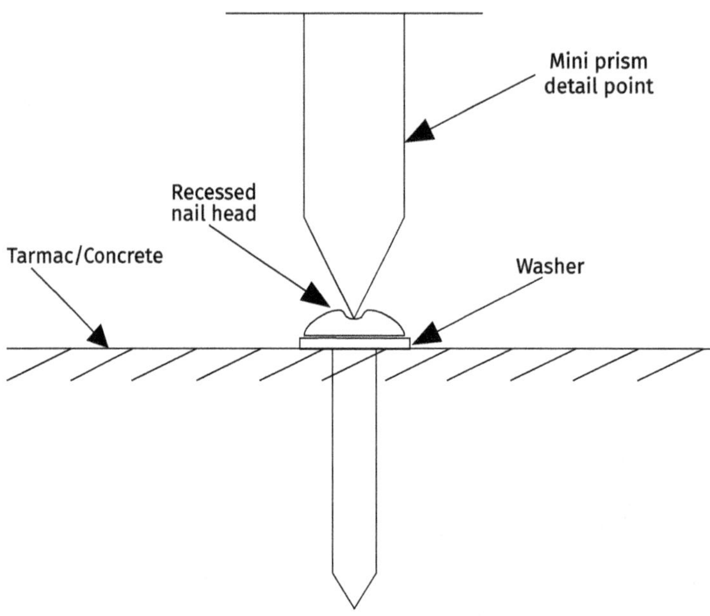

Fig. 18-2: Timber peg with nail, set in concrete

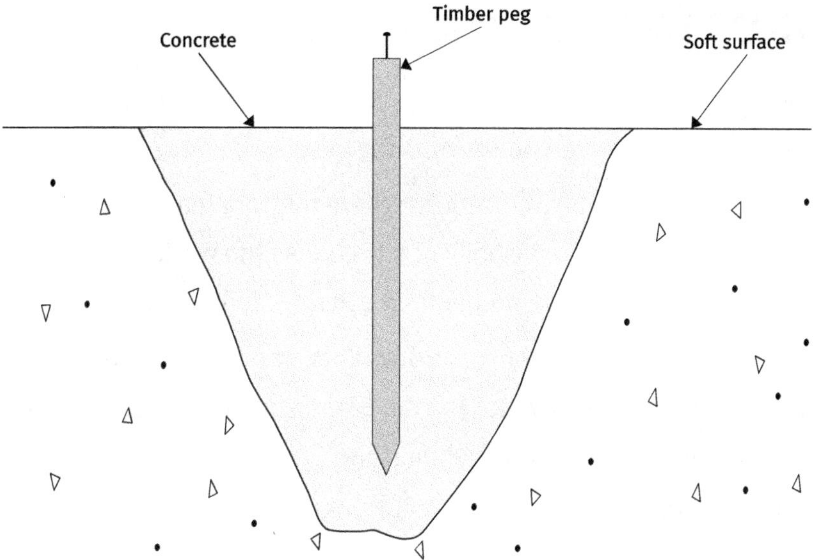

Note: Peg 40mm wide by 600mm deep, with 100mm above ground

Fig. 18-3: Pipe filled with concrete, with spigot for tribbach (Backsight)

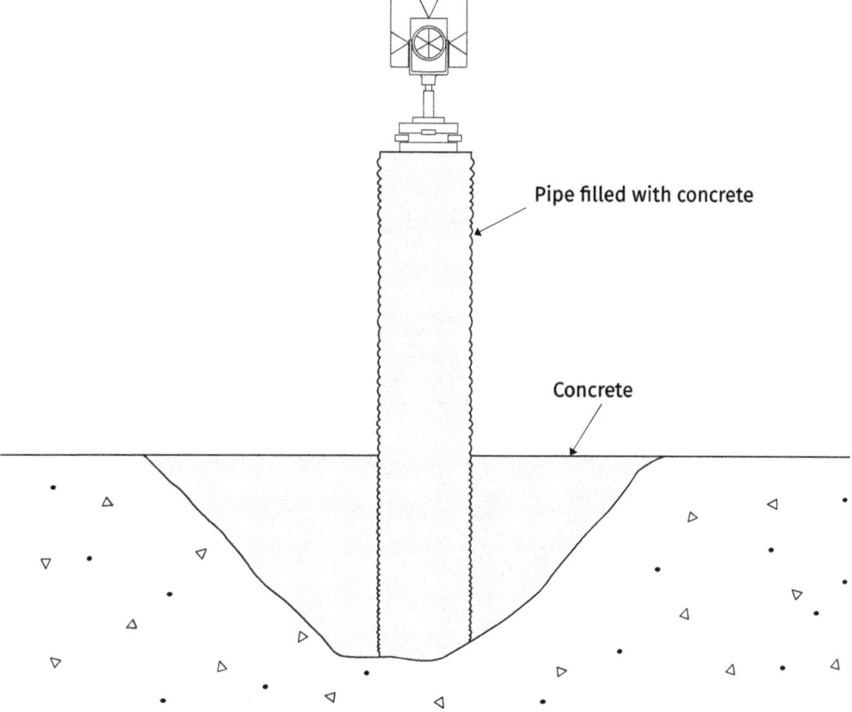

18.9 WHERE SHOULD PRIMARY CONTROL POINTS BE LOCATED?

Primary control points should be strategically located as close to the extents of the site boundaries as possible. There are two reasons for this:

→ the control points should be in locations where they are least likely to be damaged by construction machinery or visually obstructed as the works progress. They must be located where there will be no planned construction activity

→ the greater the distance to the control points, the greater the accuracy. Avoid positioning control points any closer than 50m from each other. See section **11.17** for an explanation of the reasons behind this.

When planning the location of control points consider:

→ stockpiles

→ site huts

→ cranage

→ traffic routes

→ the staging of the construction

→ locations of secondary control points

It is unlikely that it would be possible or convenient to set out the full works directly from the primary control points, so consider how primary and secondary control points will work together as a single network.

18.10 HOW MANY PRIMARY CONTROL POINTS SHOULD THERE BE?

The number of primary control points depends on the size, nature and complexity of the works as well as the likelihood of damage or obstruction.

Ensure there are sufficient points that there is a degree of redundancy. Assume that at least half of your control points will be damaged or obstructed at some stage during the project. Plan the number and locations so that the worst-case scenario is that you will never have any less than three inter-visible primary control points available to you.

18.11 CAN PRIMARY CONTROL POINTS BE LOCATED OUTSIDE OF THE SITE BOUNDARIES?

Primary control points can be located outside of the site boundaries. Here are some things to consider:

→ ensure you have full agreement from the land owner

→ if the land is public land, be conscious that the markers are not protected from the public so there is a risk that they may be vandalised or inadvertently damaged. Position them discreetly, e.g. a nail in tarmac

→ in a public place, you may not be able to leave a tripod with a backsight on unattended. This means they are not ideal for tasks where you need to set up your total station for several hours and periodically check back to your backsight. You will need to send your assistant to the point with a mini-prism which may take more time

→ ensure that they could not cause a trip hazard.

18.12 HOW DO I ESTABLISH THE COORDINATES OF PRIMARY CONTROL POINTS?

There are various procedures for establishing the coordinates of primary control points including least squares adjustment and a 'traverse'. In this book we will focus on the traverse method as it can be carried out using even the most basic total station and does not require any special software.

The principle of the traverse is that we physically measure the angles and distances between points and then use that information to calculate the coordinates of the points in relation to our starting point. For details of how to carry out a traverse, see section **19.2**.

18.13 HOW DO I ESTABLISH REDUCED LEVELS OF PRIMARY CONTROL POINTS?

Reduced levels for primary control points can be established using the procedure for transferring a TBM as described in section **20.4** and/or by using the tie distance/compute inverse or survey function in your total station as described in section **13.4**.

18.14 WHY DO I NEED SECONDARY CONTROL POINTS?

There are several situations in which primary control points may not be suitable to act as the only control points for the duration of the works. When this is the case, it is necessary to set up secondary control points.

Possible reasons include:

→ primary control points being located where it is not appropriate to leave a target set up unattended, and where it is too time consuming to send your assistant to hold a prism each time you need to set up your instrument or check your set-up position

→ the works area not being visible from a primary control point

→ primary control points not being visible from your desired instrument set-up point.

The purpose of secondary control points is to improve the flexibility of your set-up location making tertiary stage tasks easier, quicker and more convenient.

18.15 WHAT IS THE RELATIONSHIP BETWEEN PRIMARY AND SECONDARY CONTROL POINTS?

Secondary control points are derived from primary control points. Primary control points are of the highest order of accuracy.

By nature, secondary control points have a lower order of accuracy than primary control points as they are one step removed from the original source. It would be mathematically impossible to create secondary control points which were of a higher order of accuracy than the primary control points.

18.16 HOW MANY SECONDARY CONTROL POINTS DO I NEED?

The number of secondary control points required depends on the specific situation. Mathematically, the minimum number of retro targets required to set up your total station using a resection is two, but it is good practice to use a minimum of three. Depending on the size of the site, you may need dozens in total, or even more, to achieve visibility of three retro targets from any given point on the site.

Secondary control points can be added as and when they are needed so you don't need to decide on the required number at the outset. It is good practice to add them in batches rather than one at a time. They need to be installed in relation to primary control points, so it is important that there are sufficient primary control points for this purpose.

18.17 WHERE SHOULD SECONDARY CONTROL POINTS BE LOCATED?

Where possible, secondary control points should be positioned on features that will not be obstructed, removed, vibrate or blow in the wind.

They should be located as close to the extents of the site as possible. From your set-up position, the distance to your secondary control points should be greater than the distance to the works you are setting out.

18.18 WHAT SHOULD SECONDARY CONTROL POINT MARKERS BE MADE OF?

Secondary control points can either be a PGM on a horizontal surface or a retro target on a vertical surface.

There are several advantages to using retro targets:

→ you can set up a resection without an assistant since you don't need to have a prism held on the point

→ they are fixed in position so there is no centring or levelling error which you might get with a mini-prism or backsight target set up on a tripod

→ they have a well-defined black cross on a bright contrasting background so you can line up your cross-hairs accurately.

There are also some drawbacks:

→ you cannot set up directly over them, so you can only use them for resection set-ups and not for occupied point set-ups

→ there is a risk of forgetting to change the target type to prism once you have set up your resection. To minimise this risk, get into the habit of checking to a known point using the mini-prism before you set out any new points.

18.19 WHY BOTHER WITH PRIMARY CONTROL POINTS AT ALL IF SECONDARY CONTROL POINTS ARE SUFFICIENTLY ACCURATE?

Primary control points come first. You can't have secondary control without primary control. If you miss out the traverse and go straight to setting up retros by measuring their coordinates using the survey function, all you have is a set of primary control points which are more likely to be poor quality. This may cause problems either in the short term or later down the line as the project progresses from earthworks to activities which require higher levels of accuracy, such as construction of reinforced concrete or structural steel. The traverse is a robust process which involves rigorous checking and extreme care to create a network of control points. These primary control points then provide accurate reference points from which to measure your secondary control points.

18.20 SHOULD I REFERENCE PRIMARY OR SECONDARY CONTROL POINTS FOR TERTIARY STAGE SETTING OUT?

When doing your final stage of setting out, you can reference either primary or secondary control points, whichever is most convenient and appropriate. Note that for high accuracy tasks, primary control points should be used.

If you have discovered that some of your setting out is not tying in correctly with existing or as-built features, it is prudent to check it using primary control points. It may be a sign that something you have previously set out was incorrect, or it may indicate that there is an error in your secondary control points. In fact, it is a good habit to cross-reference primary and secondary control points where possible when setting out.

18.21 I HAVE BEEN GIVEN THE VALUES OF THE CONTROL POINTS BY ANOTHER ENGINEER. HOW DO I KNOW IF THEY ARE SUFFICIENTLY ACCURATE?

The only way that you will know for sure that control points are sufficiently accurate is if you start from scratch, using all the correct procedures, and end up with the same values.

It is not always appropriate to invest a large amount of time in this this so here are some things to consider, to help you make a judgment:

→ is there a robust Survey Management File on the project? If not, is that a reflection of poor culture and poor quality setting out? Or is there some other watertight system in place that gives you confidence?

→ who installed the control points? Was it a credible engineer with a good reputation or someone who is muddling their way through? Was it an experienced engineer, a freelance engineer, a trainee engineer or an external company? Is that person still working on the project? Did they have a vested interest in the project?

→ have the values of the control points been adjusted at any time since they were originally installed? If so, why?

→ have the control points been independently verified?

→ have there been any problems so far? What stage is the job at? If it is still at the earthworks stage, then it won't necessarily be apparent yet if there are any errors in the control points. If the job is at the reinforced concrete or M&E stage and everything is running smoothly, this could be an indication that the control points are accurate and reliable.

→ is there any potential for movement of the control points? If so, when were they last checked?

If possible, get sight of original records, for example:

→ traverse observation data and calculations

→ data files if the total station traverse programme was used

→ TBM field data

→ two-peg tests and total station calibration checks that were carried out on the equipment at the time

Based on the above you can make an informed decision as to whether to carry out some basic checks and accept the existing control points or to re-measure them from scratch.

18.22 WHAT IF I ARRIVE ON A SITE AND THERE ARE RETRO TARGETS, BUT NO CONTROL POINTS IN THE GROUND (PGMS)?

If you arrive on a site which is already up and running and you find that retro targets are the only control points in use, there are a few things to consider:

→ is this an indication of poor culture or is there a good reason for the lack of Permanent Ground Markers (PGMs)?

→ were primary control points installed at the beginning, but have since been destroyed?

→ how were values for the retro targets determined? Was a robust checking procedure used?

→ if there are no PGMs, how should new control points be installed? What happens if the existing retros get obstructed or destroyed?

→ how far along is the project? It may or may not be worth installing some PGMs

Even if the control points are sufficiently accurate for the stage of the project you are at, think about the future. Don't wait until the point when you are forced into using bad practice and then say, 'it wasn't my fault that the proper control points weren't in place sooner'. If you don't take a proactive approach at this stage, later down the line you might find yourself in a position where you:

→ are forced to use retro targets for installing further control points

→ are under pressure and need to install PGMs in a hurry

→ are tempted to rely on control points which you don't have confidence in

If you arrive on a site where there are no PGMs, here's how to check the quality of the retro targets in use:

1. set up a resection using the retros available

2. measure the coordinates of two different points on the ground, P1 and P2

3. set up a resection in a different position back-sighting to different retro targets if possible

4. measure the coordinates of P1 and P2 again

5. compare with the results from your first resection

6. repeat this exercise until you have checked all the retros on the site, always measuring to P1 and P2 if possible

The above exercise will give you an indication of the precision of the control points, but will not necessarily reflect their accuracy. You will need to do further work to check their accuracy in relation to external features and interfaces. If this is not critical, however, good precision may be adequate.

If you decide that it is necessary to retrospectively install PGMs, the process is as follows:

1. install at least three PGMs

2. ascertain the coordinates of two of the ground markers either by referring to reliable external features or from a resection using the retros available

3. carry out a traverse exercise on the ground markers to determine their coordinates

4. set up over one of the ground markers using the occupied point set-up

5. backsight to another ground marker

6. measure the coordinates of a different ground marker and check that they are correct

7. measure the coordinates of the retros

8. if there is a discrepancy, you will need to communicate this to any whose work will be affected. If it is your responsibility to make a decision about what to do, see section **18.23** on what to do if you have been using the wrong control points.

18.23 I'VE BEEN USING THE CONTROL POINTS FOR A WHILE BUT HAVE DISCOVERED THEY ARE WRONG. WHAT SHOULD I DO?

If you believe that the control points you have been using are incorrect, it is important to take ownership of the problem. 'Someone else set up the control points' is no defence. You should only ever use control points once you are 100% satisfied that they are reliable. If you are using control points and they turn out to be wrong, then it is your own fault, no matter who installed them and however certain they were about their accuracy. If you find yourself in this situation:

→ stop using the control points until further investigation has been done. Highlight the problem in the Survey Management File accordingly to ensure no one else uses the incorrect values

→ inform your line manager about your concerns

→ inform anyone else using the control points about your concerns

Next you will need to determine the extent of the problem. Use all the resources available to you to answer these questions:

→ what is the magnitude of the error?

→ was it caused by an accumulation of accuracy errors or an isolated gross error?

→ are the control points correct in relation to each other, but all out of the correct position?

→ are all the control points affected or only some of them?

→ has this caused any errors in the setting out and construction? If so, what is the extent and what are the implications of the work done so far?

Then, make a judgment on what to do next, taking the following into consideration:

→ how will it affect future work if you neglect the effects of the error? Will it mean the whole project is out by a constant amount, but in fact it all ties in to itself?

→ will there be programme or cost implications?

→ will remedial work be needed?

→ do you need to consult the Designer?

→ do you need to inform the Client?

→ even if there have been no major problems so far, will there will be problems with interfaces and tie-ins later?

→ will the errors lead to 'lack of fit' when installing reinforced concrete, structural steel work or mechanical and electrical (M&E) elements?

→ can the values be adjusted or do you need to scrap them and start from scratch?

CHAPTER 19

HORIZONTAL CONTROL POINTS

19.1 PRIMARY HORIZONTAL CONTROL POINTS

The procedure for establishing primary horizontal control points is called a 'traverse'. The secondary control points are then determined in relation to the primary control points.

19.2 THE TRAVERSE

A traverse is a survey of a group of control points. The geometrical arrangement of the points is determined by physically measuring the relative horizontal angles and distances between the points. The coordinates of the control points are then calculated relative to a baseline, of which the start and end coordinates are known or assumed. Based on the measured data, a closing error is calculated. The error is then distributed proportionally across the traverse points. Traverse points, known as stations, are a series of inter-visible points distributed around the perimeter of the work area as shown in Fig. 19-1

Traversing is the most common way of setting up primary control points. It achieves a high level of accuracy and incorporates rigorous checks on the quality of observations.

Fig. 19-1: Traverse layout

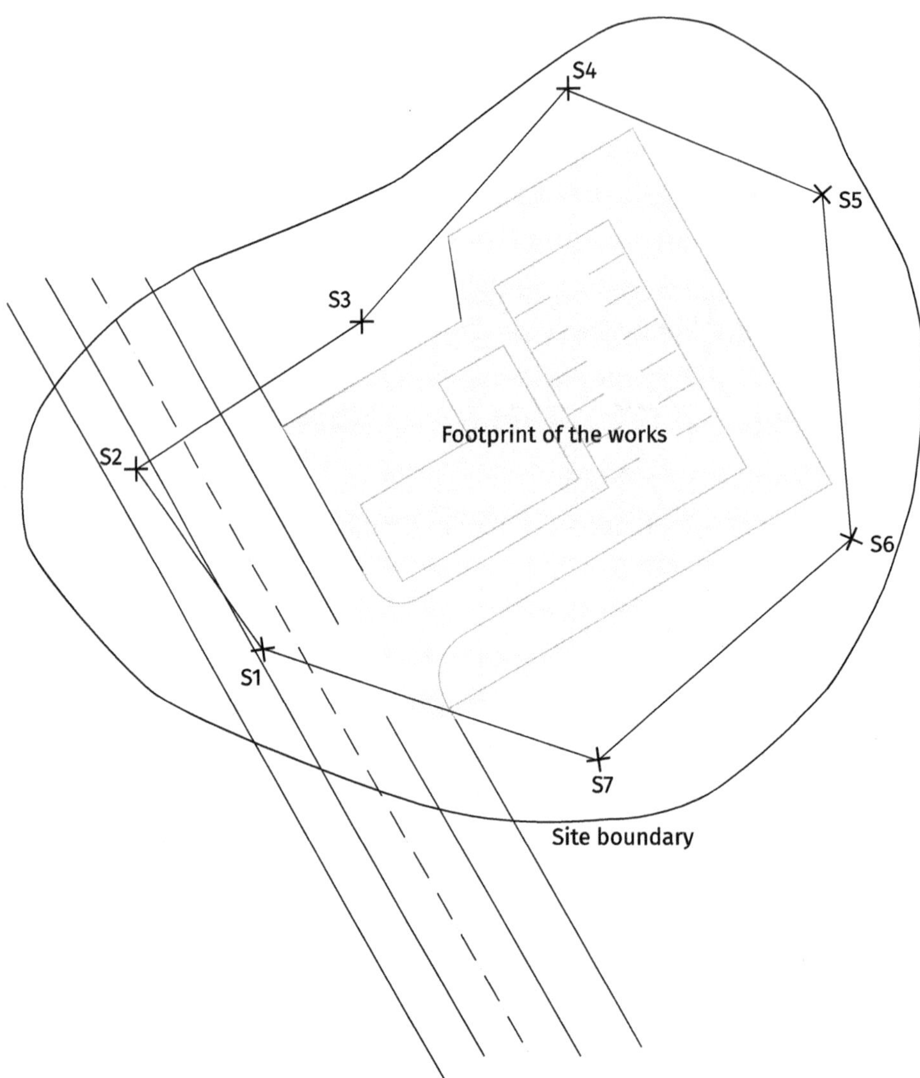

19.3 WHAT ARE THE DIFFERENT TYPES OF TRAVERSE?

There are two types of traverse:

→ an open traverse

→ a closed traverse.

A traverse can form a loop or a long zig-zag line. Whether it is open or closed does not relate to its shape, but to where it finishes. If it finishes at the same point it started on (in loop arrangement) or a separate point of reliable known coordinates, then it is a closed traverse. If it finishes on an unknown point, it is an open traverse. The different traverse arrangements can be seen in Fig. 19-2

In a closed traverse, the misclosure, or closing error, can be calculated, allowing the overall accuracy to be calculated. In an open traverse, there is no way of knowing whether the accuracy of the measured data is within acceptable limits.

In the case of a zig-zag line from a known point to an unknown point, once the final point has been reached, the survey should be repeated in reverse, returning to the original point. This will allow a closing error to be calculated.

It is preferable to use a closed traverse over an open one.

Fig. 19-2: Types of traverse layout

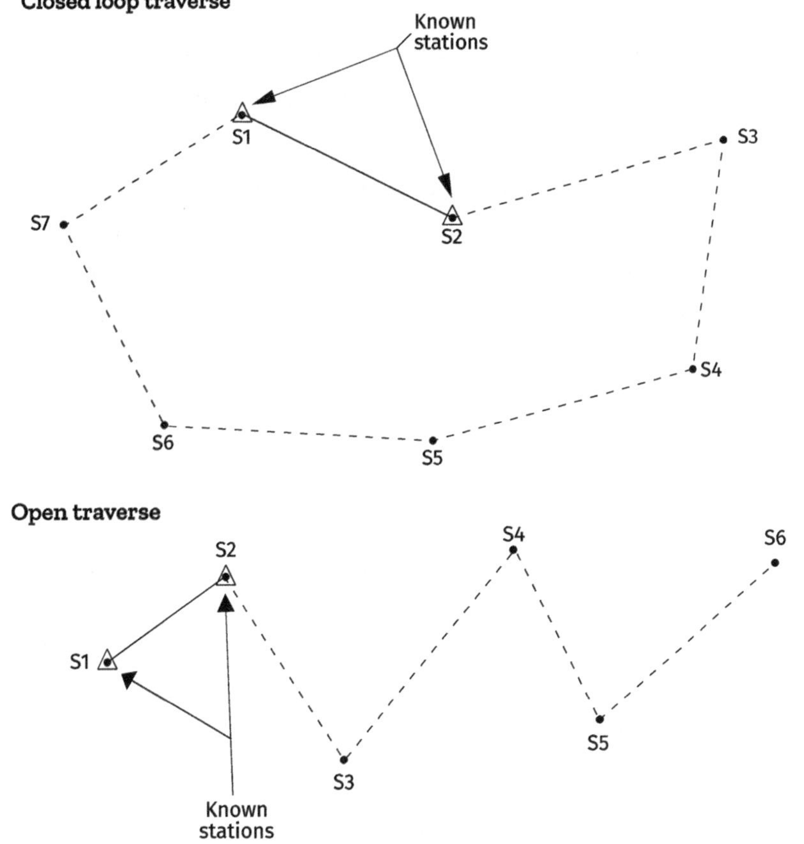

19.4 HOW DO I MEASURE AND RECORD THE TRAVERSE DATA?

The field measurements must be methodically recorded incorporating checks, and in such a way that the information is fully traceable.

Some total stations contain a programme to log and process the data, but for many instruments, a software licence is needed for this, so you may need to record the data by hand. In this case you can record the data either in your yellow field book as shown in Fig 19-3 or if you prefer to use a booking form.

The advantages of using your field book are:

> → the pages are waterproof and easier to work with in windy conditions than an A4 clipboard

> → it keeps all your survey work together in one place and forms part of the 'story' of your setting out activities

> → there is no need to design or source and print out special forms.

To minimise the risk of error, you should follow this set procedure and do it the same way every time.

A set of readings must be taken from every point in the traverse to measure the internal angle between the adjacent points and the distances to them from the set-up point (Fig. 19-4). Although it is not necessary to measure the points in the correct sequence, it is important that each one is clearly named and indicated on a plan before starting your traverse.

Instructions for measuring and recording traverse data:

1. Set up over a station

2. Sight to the point to your left, ensuring that you line up on the point using the correct method for measuring horizontal angles described in section **11.14**

3. Record the horizontal angle as a left face reading (LFR)

4. Sight to the point to your right

5. Record the horizontal angle as an LFR

6. Still pointing to the point to your right, change faces and sight to the point.

7. Record the horizontal angle as an RFR

8. Sight to the point to your left.

9. Record the horizontal angle as an RFR.

You have now measured one round of angles.

10. Return to the left face and reset the horizontal circle to zero at a random point

11. Re-centre your instrument over the point

12. If you are sighting to backsights, set up on tripods, re-centre them

13. Repeat steps 1 – 9

14. Sight to a prism on the point to your left. Record the distance once.

15. Sight to a prism on the point to your right. Record the distance five times.

Fig. 19-3: Traverse data booking example

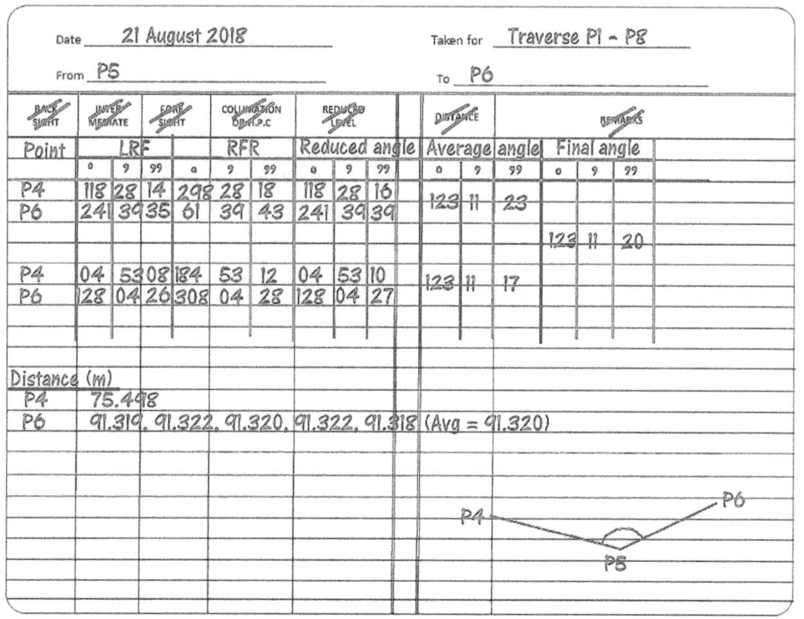

Date: 21 August 2018   Taken for: Traverse P1 – P8

From: P5   To: P6

Point	~~BACK SIGHT~~ LRF			~~FORE SIGHT~~ RFR			~~COLLIMATION OB H.P.C~~ Reduced angle			~~DISTANCE~~ Average angle			Final angle			~~REMARKS~~
	°	9	99	°	9	99	°	9	99	°	9	99	°	9	99	
P4	118	28	14	298	28	18	118	28	16	123	11	23				
P6	241	39	35	61	39	43	241	39	39							
													123	11	20	
P4	04	53	08	184	53	12	04	53	10	123	11	17				
P6	128	04	26	308	04	28	128	04	27							

Distance (m)
P4   75.498
P6   91.319, 91.322, 91.320, 91.322, 91.318 (Avg = 91.320)

Fig. 19-4: Traverse - measurements

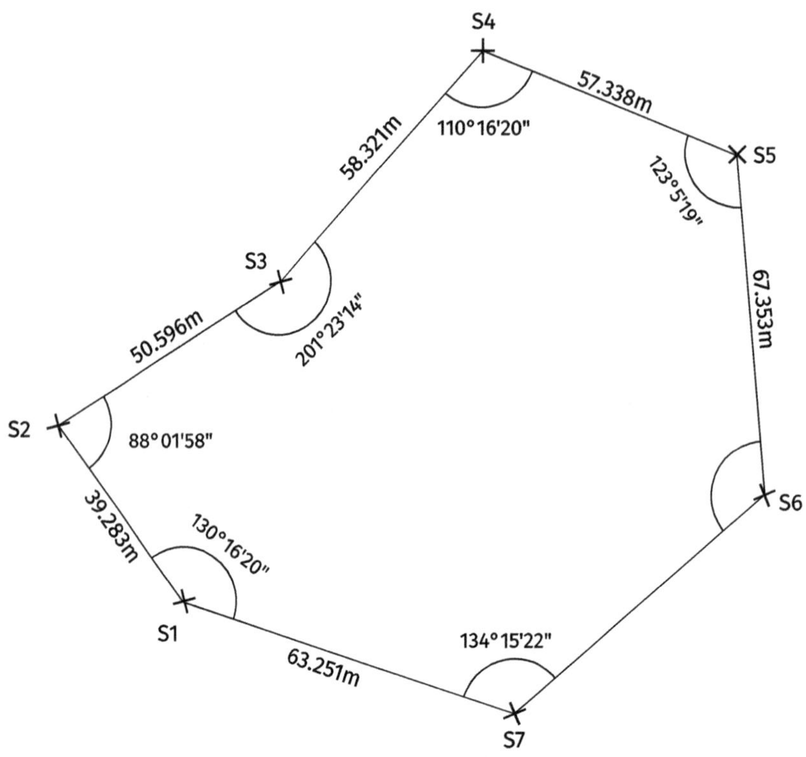

EXERCISE:

1. You have taken a set of readings at S3 and the results are as follows:

→ LFR S2 = 15°46'17", LFR S4 = 177°04'37", RFR = S4 357°04'33", RFR S2= 195°46'15"

→ LFR S2 = 168°11'30", LFR S4 = 329°29'50", RFR = S4 149°29'43", RFR S2= 348°11'22"

→ Dist to S2 = 75.498

→ Dist to S4 = 91.319m, 91.322m, 91.320m, 91.322m, 91.318m

   a. What is the final angle?

   b. What is the average distance from S3 – S4?

19.5 HOW DO I TAKE 5 CONSECUTIVE READINGS?

If your instrument is in tracking mode, it will automatically take a reading every second or so. Each time a new reading is taken, the value on the screen will flash up. Wait until the reading is steady within a few millimetres before recording 5 consecutive readings.

19.6 WHAT IF THERE IS A BIG VARIATION BETWEEN THE REPEATED DISTANCES?

The variation across the five distances should be no greater than 3mm. If you have a larger spread than this, discard the set of readings and repeat the measurement. If you are using a mini-prism, ensure that your assistant is holding it steady and take five more readings. If you are using a backsight set up on a tripod, high winds may be causing the problem. Ensure the tripod feet are pushed fully into the ground and repeat the measurement. If the problem persists, abandon the task until there is no wind or switch to a mini-prism if that is a viable option.

19.7 WHY DO I NEED TO RECORD THE DISTANCE 5 TIMES?

Recording the distance five times shows the prism was being held steady and over the point at the time of taking the reading. When you just take one reading, you just have a 'snapshot' which doesn't give any proof that the prism was being held steady.

19.8 WHAT IS THE ALLOWABLE ERROR BETWEEN LEFT FACE AND RIGHT FACE READINGS?

The theoretical difference between left face and right face readings taken to the same point is 180^00'00".

If the correct process for observing a round of angles is followed, any errors within the instrument will be cancelled out and so do not need to be taken account of.

If, however, there is a difference between the errors, then this indicates the presence of an external error, for example:

→ incorrect lining up of the cross-hair on the point

→ movement of the ground around the total station

→ a recording error

As a rule of thumb, if the difference between the LFR and RFR is more than 10" away from 180°, you should either discard the result or carry out further investigation into the reason for this. Follow these steps:

1. Carry out the total station checks in **chapter 12**

2. If you identify a collimation error of, say, 20" when you carry out your LFR/RFR check when taking your traverse readings, then increase the allowable error to 20" + 10"

3. It is ok to carry on as this error will be cancelled out in the calculations

4. If you have an LFR/RFR difference which is +/-10" different to this, then discard the results and repeat the measurement

5. If the error persists, investigate possible causes and minimise its effects.

19.9 HOW DOES ANGULAR ERROR TRANSLATE INTO A DISTANCE ERROR?

Fig. 19-5 illustrates how the distance error caused by an angular error increases with increasing distance from the set-up position.

An error of 10" propagated over a distance of 10m would translate into a positional error of 0.5mm.

The same 10" error over a distance of 10m would translate into a positional error of 5mm.

An error of 10" over a distance of 100m would translate into a positional error of approx. 50mm.

Fig. 19-5: How an angular error translates into a distance error

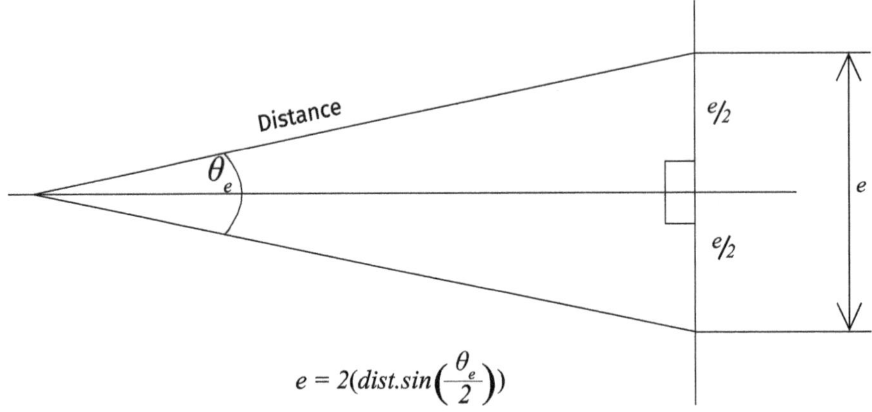

$$e = 2\left(dist.sin\left(\frac{\theta_e}{2}\right)\right)$$

19.10 HOW DOES A DISTANCE ERROR TRANSLATE INTO AN ANGULAR ERROR?

Fig. 19-6 illustrates how longer distances to backsights result in smaller angular errors.

An offset error of 10mm over a distance of 10 would subtend an angle of 00°03'26".

An offset error of 10mm over 100mm would subtend an angle of 00°00' 20".

An offset error of 10mm over 100mm would subtend an angle of 00°00' 02".

Fig. 19-6: How a distance error translates into an angular error

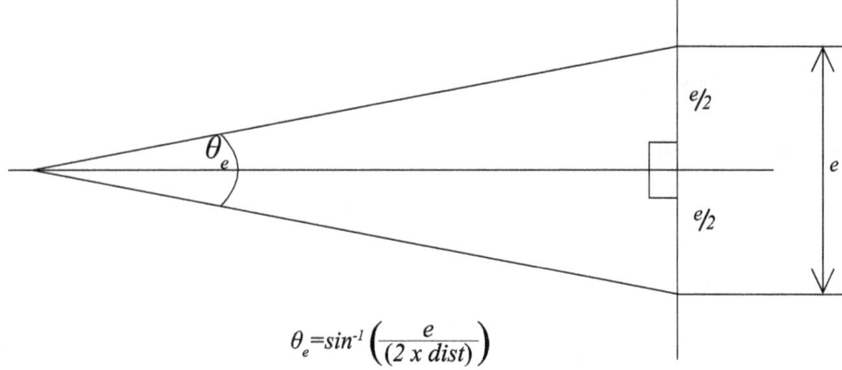

$$\theta_e = sin^{-1}\left(\frac{e}{(2 \times dist)}\right)$$

19.11 TRAVERSE TABLE

Once the traverse data has been gathered, the next step is to process the data to calculate the relative coordinates of the traverse stations. The data can be processed either by hand or using a spreadsheet or special software.

A traverse table is a convenient way of processing the field data and tabulating the calculations. It builds in checks and presents the information in a logical way. This makes it easier to spot any errors and for an independent person to check. Fig 19.7 shows a completed traverse table. The steps for completing the table are:

1. Complete the title boxes.
2. Draw an indicative diagram of the layout of the traverse.
3. Enter the coordinates of S1.
4. Enter the forward bearing of S1 to S2.
5. Fill in the forward bearings column from previous calculations (see worked example **'Calculating WCBs in a traverse'**).
6. Fill in the horizontal distances column from your measured data.
7. Calculate the total length of the survey.
8. Calculate change in eastings and change in northings (partial coordinates) for each leg (Fig. 19-8).
9. Calculate the misclosure in the eastings and northings by calculating the sum of the partial eastings and the sum of the partial northings. (Fig. 19-9)
10. Calculate the magnitude of the misclosure.

11. Calculate the coordinates of the points.

12. Calculate the adjustment to be applied to each point (using the Bowditch Adjustment as described in **Worked example for traverse calculations**).

13. Calculate the adjusted coordinates (Fig. 19-10).

Fig. 19-7: A completed traverse table

Survey Title	Control points S1 - S6
Date	21 May 2018
Observed	S Grant
Booked	S Grant
Calculated	S Grant
Checked	M Beetham

Station Name	Co-ordinates	
	Eastings (m)	Northings (m)
S1	65.000	135.000
S2	170.269	144.217
S3	207.758	133.278
S4	203.281	86.696
S5	133.053	89.900
S6	84.014	87.344

Station	Bearing
S1→S2	85°00'00"

Diagram	Station	Forward Bearing	Distance (m)	ΔE(m)*	ΔN(m)*	Eastings (m) / Adjustment(m) / Adjusted Eastings (m)	Northings (m) / Adjustment(m) / Adjusted Northings (m)	Station
	S1	85°00'00"	105.675	105.273	9.210	170.273 / -0.004 / 170.269	144.210 / +0.007 / 144.217	S2
	S2	106°15'24"	39.051	37.490	-10.932	207.763 / -0.005 / 207.758	133.278 / +0.009 / 133.287	S3
	S3	185°29'14"	46.808	-4.476	-46.594	203.287 / -0.006 / 203.281	86.684 / +0.012 / 86.696	S4
	S4	272°36'30"	70.298	-70.225	3.199	133.062 / -0.009 / 133.053	89.883 / +0.017 / 89.900	S5
	S5	267°00'47"	49.105	-49.038	-2.559	84.024 / -0.010 / 84.014	87.324 / +0.020 / 87.344	S6
	S6	338°14'59"	51.306	-19.012	47.653	65.012 / -0.012 / 65.000	134.977 / +0.023 / 135.000	S1

Total distance,	Misclosure,	Misclosure,
D = 362.243	ΔE = 0.012	ΔN = -0.023
e = √((ΔE)² + (ΔE)²) =	0.026	
Accuracy, 1:D/e =	1: 14,000	

*ΔE = Distance x sin(Forward Bearing)   *ΔN = Distance x cos(Forward Bearing)

Bowditch Adjustment

Adj. To E = (Running Distance x Misclosure (ΔE)) / Total Distance

Adj. To N = (Running Distance x Misclosure (ΔN)) / Total Distance

Fig. 19-8: Traverse - partial co-ordinates

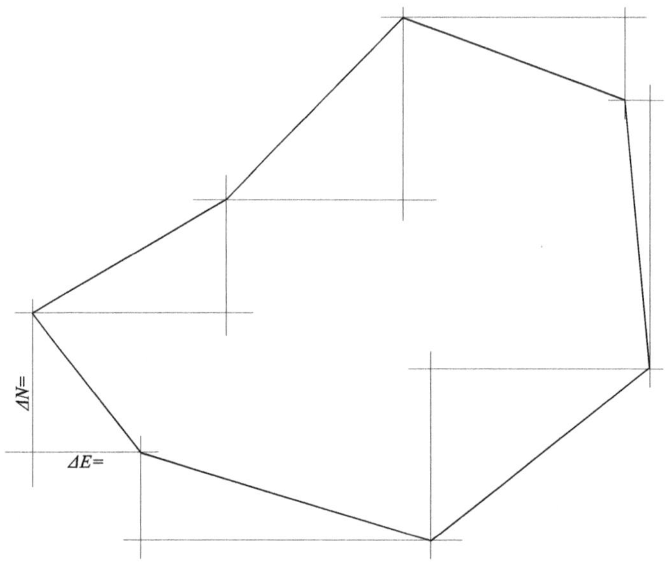

Fig. 19-9: Traverse misclosure

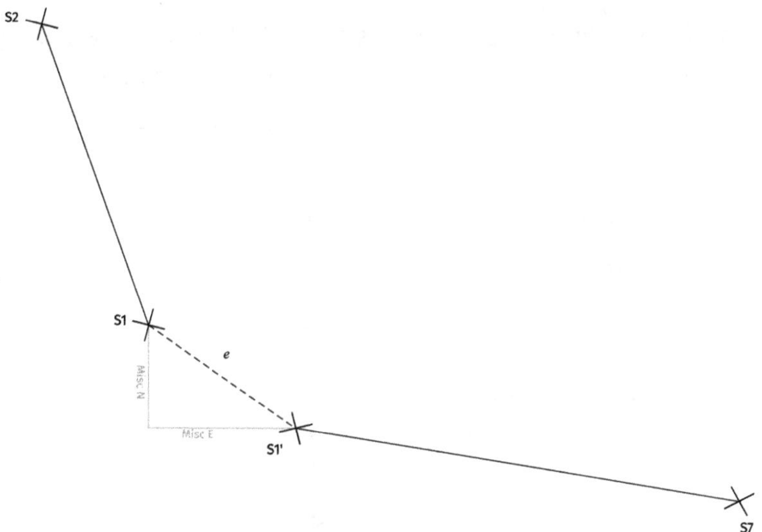

Fig. 19-10: Traverse - Adjusted co-ordinates

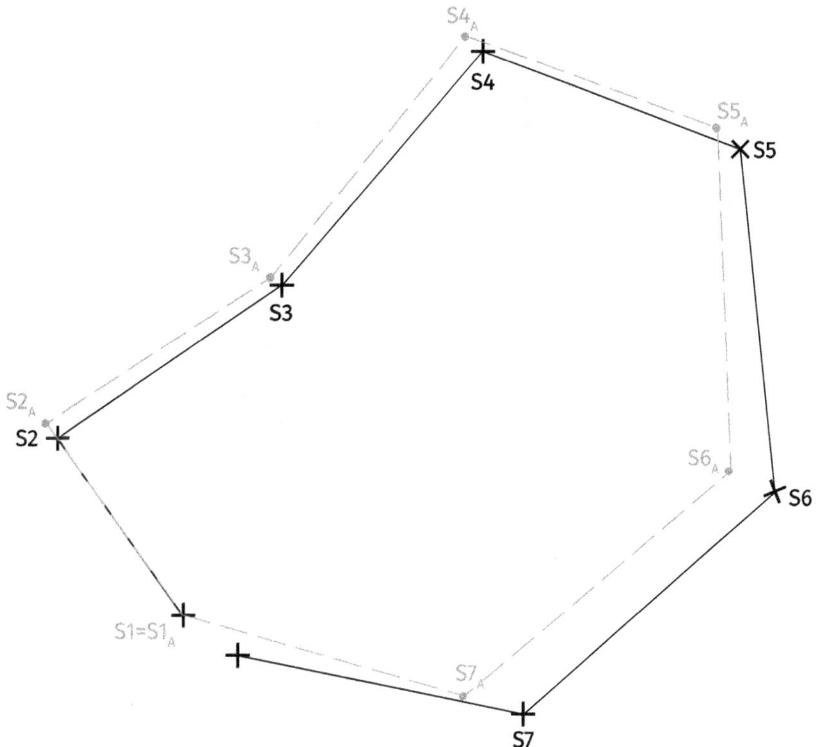

19.12 FORWARD BEARINGS

The 'forward bearing' of a line, also known as the 'whole circle bearing', is the clockwise angle from the start point to the end, measured from the north meridian line at the start point. The forward bearing from A to B is denoted as FB_{A-B}. Fig. 19-11 shows the forward bearing at each station of a traverse.

There are two stages to calculating the forward bearings for a traverse:

1. Establish the forward bearing from the first station to the second.

2. Based on this, calculate all other forward bearings.

Fig. 19-11: Traverse - Whole circle bearings

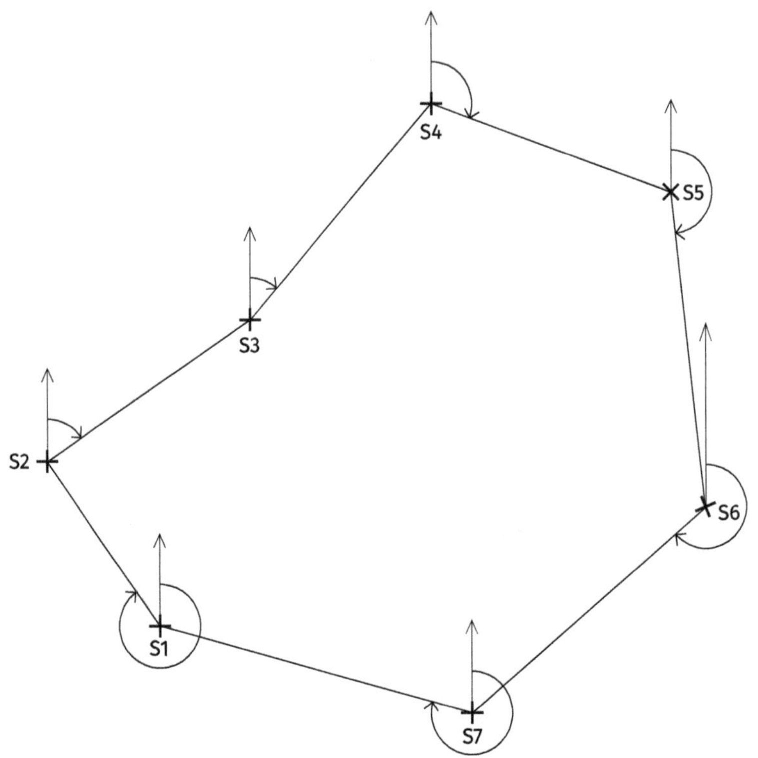

19.13 HOW DO I ESTABLISH THE FORWARD BEARING FROM THE FIRST STATION TO THE SECOND?

There are several ways to establish the forward bearing from the first station, ST1, to the second, ST2.

→ If you know the coordinates of ST1 and ST2, you can calculate the forward bearing (see section **28.36**)

→ If you are provided with the coordinate of ST1 and the forward bearing to ST2, you don't need to do anything. You already have the answer.

→ If you can identify ST1 and ST2 on a scale drawing you can either measure the changes in eastings and northings and calculate the forward bearing, or you can measure the forward bearing directly using a protractor.

EXERCISE:

2. The coordinates of the two starting points of a traverse are S1 = 800.466mE, 500.671mN and S2 = 722.389mE , 537.900mN. What is the whole circle bearing S1-S2?

19.14 HOW DO I MEASURE THE FORWARD BEARING FROM A DRAWING?

If you cannot ascertain the coordinates of S1 and S2, you can either measure the forward bearing with a protractor, or calculate it using the difference in eastings and the difference in northings. These methods are not suitable when the tolerance of the starting point is less than 100mm.

To find the forward bearing by measuring the difference in eastings and the difference in northings, the steps are as follows (Fig. 19-12):

1. Connect the two points, S1 and S2, with a straight line.

2. Create a construction line by extending the north arrow.

3. Create a construction line perpendicular to the north arrow extension line from S1 to the north arrow extension line.

4. Create a construction line perpendicular to the north arrow extension line from S2 to the north arrow extension line.

5. Create a construction line parallel to the north arrow extension line, through S1.

6. Create a construction line parallel to the north arrow extension line, through S2.

7. Using a ruler, measure the change in eastings and change in northings. If ST2 is to the right of ST1, the change is positive. If it is to the left, the change is negative. If ST2 is above ST1, the change is positive. If it is below it, the change is negative.

8. Calculate the forward bearing using the formula:

9. Forward bearing = $\tan^{-1}$ (northings/eastings).

The same process is used if the north arrow is lying vertically on the drawing. See Fig. 19-13.

To find the forward bearing using a protractor, the steps are as follows:

1. Connect the two points with a straight line.

2. Draw the north arrow at ST1 (either transferred from elsewhere on the drawing or using an arbitrary north if appropriate).

3. Use a protractor to measure the forward bearing.

Fig. 19-12: Change in E and N - Non vertical

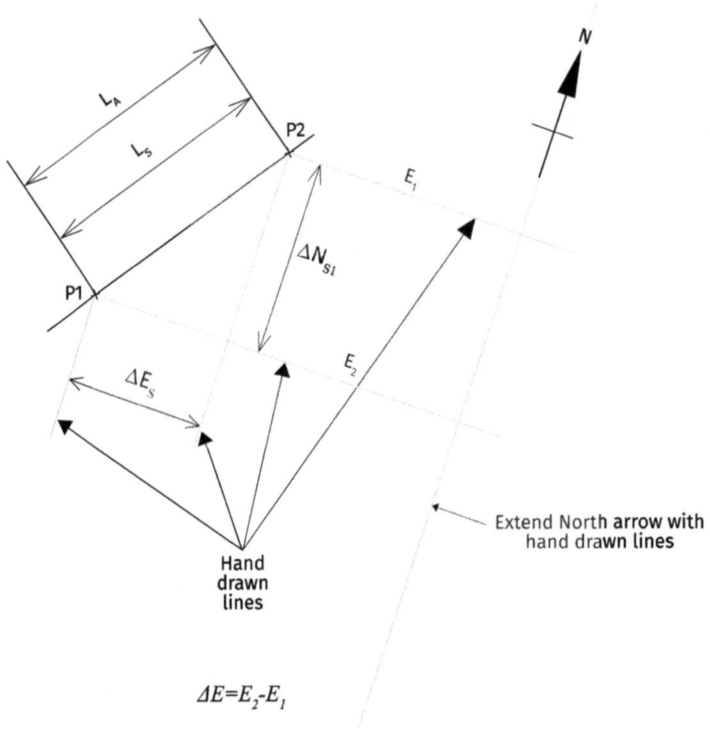

Fig. 19-13: Change in E and N from drawing

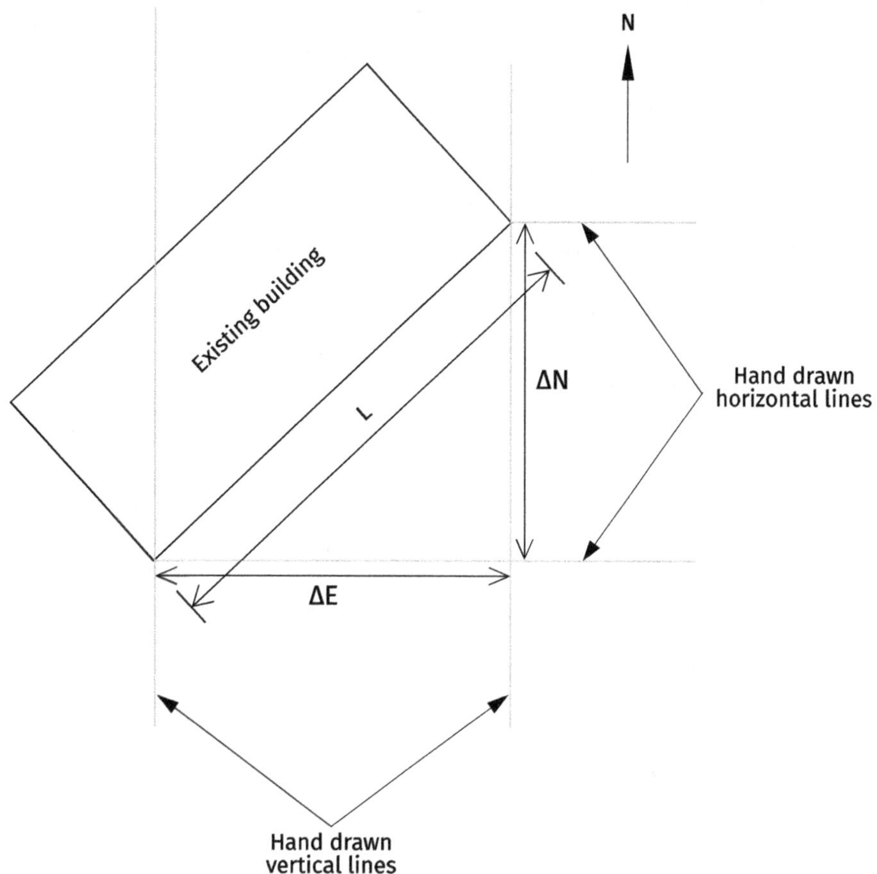

19.15 HOW DO I CALCULATE THE FORWARD BEARINGS FOR THE TRAVERSE?

Once you have established the forward bearing from ST1 to ST2, you can calculate the forward bearings for the remainder of the traverse either by inspection or by using a formula.

To calculate the forward bearing by inspection, the following geometry rules apply:

→ there are 360 degrees in a full circle

→ according to the law of opposite angles, A + B = 180 (Fig. 19-143)

To calculate the forward bearing at any given traverse station:

1. Draw a north arrow at the start of the line.

2. Draw on the line showing the forward bearing you need to calculate.

3. Mark the internal angle (as measured in the field).

4. Calculate the back bearing (the anticlockwise angle from the end point to the start point) and mark it on your diagram.

5. Calculate the forward bearing

For a full worked example see section **Worked example for calculating WCBs in a traverse**.

Alternatively, you can calculate the forward bearings using the formula:

→ Forward bearing = FB from the previous station +180 − internal angle

Fig. 19-14: Parallel lines geometry rule

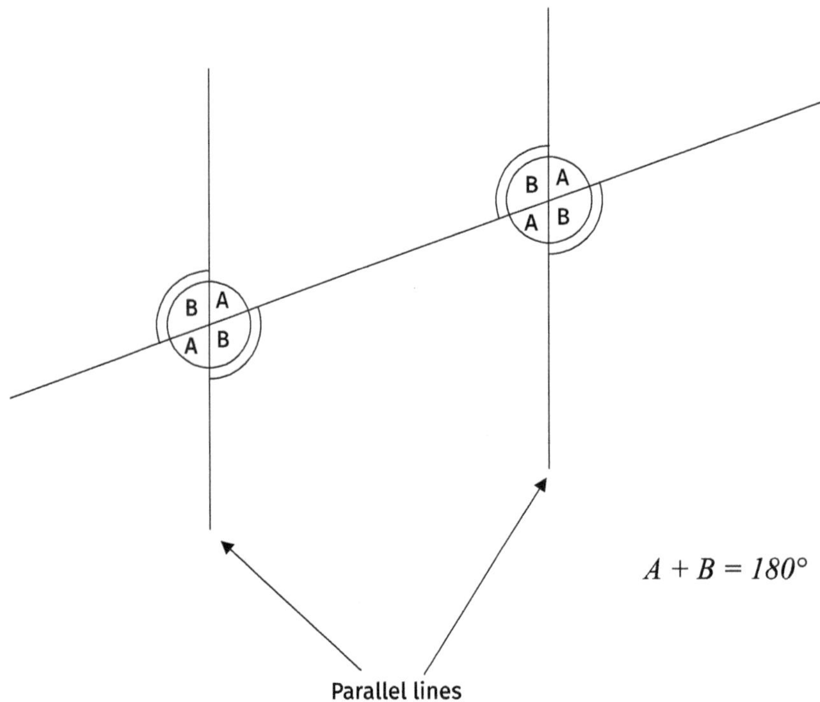

19.16 WHAT IS AN ANGULAR MISCLOSURE AND HOW DO I CALCULATE IT?

The difference between the sum of the internal angles measured using the total station and the theoretical sum of the internal angles is called the angular misclosure. The presence of this misclosure means that when you calculate the forward bearings round the traverse, when you get back to the starting point, there will be a difference between the original forward bearing and the answer calculated based on the measured angles. See Fig. 19-15.

→ From your field data, calculate the sum of your measured internal angles.

→ Calculate the theoretical sum of the internal angles using the formula:

→ Sum of internal angles = 180 x (n-2)

→ The error is the difference between the theoretical and the measured.

→ As an additional calculation check, once you have calculated the forward bearings of the traverse stations, continue back to the beginning, re-calculating the forward bearing of ST1 – ST2 based on the previous forward bearing. The difference between the re-calculated forward bearing and the original forward bearing should be the same as the error calculated in step 3. If it is not exactly the same, it means you have made a calculation error either in your adding up of the internal angles or your forward bearing calculations. Do not proceed until you have traced and rectified the error.

Fig. 19-15: Angular misclosure

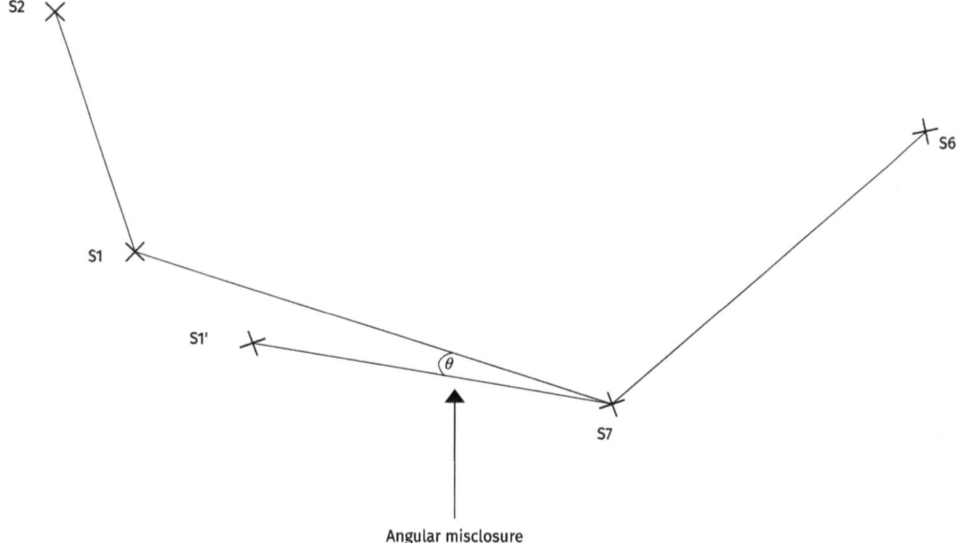

WORKED EXAMPLE FOR CALCULATING WCBS IN A TRAVERSE

In the diagram below, calculate the forward bearing for each leg of the traverse and calculate and check the angular closing error.

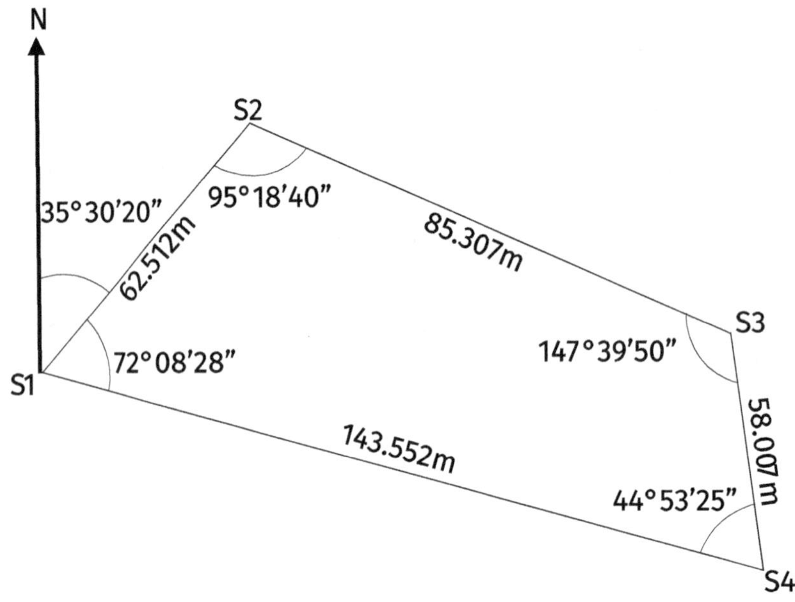

Forward bearing = 180 + (forward bearing of the previous point) – (Internal angle)

So:

FB_{S2-S3} = 180 + 35°30'20" - 95°18'40" = 120°11'40"

FB_{S3-S4} = 180 + 120°11'40"-147°39'50" = 152°31'50"

FB_{S4-S1} = 180 + 152°31'50" – 44°53'25" = 287°38'25"

FB'_{S1-S2} = 180 + 287°38'25" – 72°08'28" = 395°29'57" – 360° = 35°29'57"

To check the closing error, compare it to the error in the sum of internal angles:

The angular closing error = FB_{S1-S2} - FB'_{S1-S2} = 35°30'20" - 35°29'57" = 00°00'23"

Sum of internal angles = (n-2) x 180

Where n is the number of sides of the polygon.

In this case, n = 4, so:

Sum of internal angles = (4-2) x 180 = 360°

Sum of internal angles = 95°18'40" + 147°39'50" + 44°53'25" + 72°08'28" = 360°00'23"

Angular closing error = 360°00'23" - 360° = 00°00'23" ✓ which agrees with the previous calculation.

WORKED EXAMPLE FOR TRAVERSE CALCULATIONS

Where S1 = (120.550mE, 305.110mN), for the measurements given in the worked example above, complete the traverse table to determine the final adjusted coordinates of S2, S3 and S4.

Step 1: Complete the title blocks

Step 2: Enter the names of the stations at the start and end of each leg

Step 3: Complete the 'diagram' column with a diagram showing the leg of the traverse, the north arrow and indicatively mark on the forward bearing (whole circle bearing). See completed traverse table.

Step 4: Complete the 'distance' column using the distances measured

Step 5: Complete the 'change in Eastings' and 'change in Northings' columns by calculating the partial coordinates

$\Delta E_{S1-S2} = dist_{S1-S2} \times \sin(FB_{S1-S2}) = 62.512 \times \sin(35°30'20") = 36.306m$

$\Delta E_{S2-S3} = dist_{S2-S3} \times \sin(FB_{S2-S3}) = 85.307 \times \sin(120°11'40") = 73.733m$

$\Delta E_{S3-S4} = dist_{S3-S4} \times \sin(FB_{S3-S4}) = 58.007 \times \sin(152°31'50") = 26.757m$

$\Delta E_{S4-S1} = dist_{S4-S1} \times \sin(FB_{S4-S1}) = 143.552 \times \sin(287°38'25") = -136.802m$

$\Delta N_{S1-S2} = dist_{S1-S2} \times \cos(FB_{S1-S2}) = 62.512 \times \cos(35°30'20") = 50.888m$

$\Delta N_{S2-S3} = dist_{S2-S3} \times \cos(FB_{S2-S3}) = 85.307 \times \cos(120°11'40") = -42.904m$

$\Delta N_{S3-S4} = dist_{S3-S4} \times \cos(FB_{S3-S4}) = 58.007 \times \cos(152°31'50") = -51.467m$

$\Delta N_{S4-S1} = dist_{S4-S1} \times \cos(FB_{S4-S1}) = 143.552 \times \cos(287°38'25") = 43.502m$

Step 6: Calculate the positional misclosure by calculating the sum of the eastings and the sum of the northings

E = 36.306 + 73.733 + 26.757+ (-136.802) = -0.006m

ΣN = 50.888+ (-42.904) + (-51.467) + 43.502 = 0.019m

Step 7: Calculate the magnitude of the combined error, e

$e = \sqrt{(e_E^2 + e_N^2)}$

$= \sqrt{(0.006^2 + 0.019^2)}$

= 0.020m

Step 8: Calculate the coordinates of the stations based on the measured data and check that the misclosure corresponds with the closing error calculated in the ΣE and ΣN. Enter the answer into the top box in each row of the 'Eastings/Adjustment/Adjusted Eastings' column. Repeat for the 'Northings/Adjustment/Adjusted Northings' column.

$E_{S2} = E_{S1} + \Delta E_{S1-S2} = 120.550 + 36.306 = 156.856mE$

$E_{S3} = E_{S2} + \Delta E_{S2-S3} = 156.856 + 73.733 = 230.589mE$

$E_{S4} = E_{S3} + \Delta E_{S3-S4} = 230.589 + 26.757 = 257.346mE$

$E'_{S1} = E_{S4} + \Delta E_{S4-S1} = 257.346 + (-136.802) = 120.544mE$

Misclosure = $E'_{S1} - E_{S1} = 120.544 - 120.550 = 0.006m$ ✓ which agrees with ΣE

$N_{S2} = N_{S1} + \Delta N_{S1-S2} = 305.110 + 50.888 = 355.998mN$

$N_{S3} = N_{S2} + \Delta N_{S2-S3} = 355.998 + (-42.904) = 313.094mN$

$N_{S4} = N_{S3} + \Delta N_{S3-S4} = 313.094 + (-51.467) = 261.627mN$

$N'_{S1} = N_{S4} + \Delta N_{S4-S1} = 261.627 + 43.502 = 305.129mE$

Misclosure = $N'_{S1} - N_{S1} = 305.129 - 305.110 = 0.019m$ ✓ which agrees with ΣN

Step 9: Calculate the total distance of the traverse

Total distance = $dist_{S1-S2} + dist_{S2-S3} + dist_{S3-S4} + dist_{S4-S1}$

=62.512 + 85.307 + 58.007 + 143.552

= 349.378

Step 10: Calculate the accuracy of your traverse

Accuracy = $1: \frac{D}{e}$

Where D = total distance

And e = misclosure

Therefore, accuracy = $1: (\frac{349.378}{0.020}) = 1{:}17{,}468 < 1{:}10{,}000$ therefore OK

This equates approximately to an error of 1mm in every 17m

Step 11: Calculate the adjustment to be applied to eastings and northings using the Bowditch Adjustment. Enter the answer into the top box in each row of the 'Eastings/Adjustment/Adjusted Eastings' column. Repeat for the 'Northings/Adjustment/Adjusted Northings' column.

Bowditch Adjustment = $(\frac{Running\ distance}{Total\ distance})$ x misclosure

$$AdjE_{S2} = \left(\frac{65.512}{349.378}\right) \times -0.006 = -0.001m$$

$$AdjE_{S3} = \left(\frac{65.512+85.308}{349.378}\right) \times -0.006 = -0.003m$$

$$AdjE_{S4} = \left(\frac{65.512+85.307+58.007}{349.378}\right) \times -0.006 = -0.004m$$

$$AdjE_{S1} = \left(\frac{65.512+85.307=58.007+143.552}{349.378}\right) \times -0.006 = -0.006m$$

$$AdjN_{S2} = \left(\frac{65.512}{349.378}\right) \times -0.019 = 0.003m$$

$$AdjN_{S3} = \left(\frac{65.512+85.308}{349.378}\right) \times -0.019 = 0.008m$$

$$AdjN_{S4} = \left(\frac{65.512+85.308+58.007}{349.378}\right) \times -0.019 = 0.011m$$

$$AdjN_{S1} = \left(\frac{65.512+85.308+58.007+143.552}{349.378}\right) \times -0.019 = 0.019m$$

Step 12: Finally, apply the adjustment to calculate the final adjusted coordinates which is the value you will attribute to your stations.

$$E'_{S2} = E_{S2} + AdjE_{S2} = 156.856 - (-0.001) = 156.857mE$$

$$E'_{S3} = E_{S3} + AdjE_{S3} = 230.589 - (-0.003) = 230.592mE$$

$$E'_{S4} = E_{S4} + AdjE_{S4} = 257.346 - (-0.004) = 257.350mE$$

$$E'_{S1} = E_{S1} + AdjE_{S1} = 120.544 - (-0.006) = 120.550mE \checkmark$$

$$N'_{S2} = N_{S2} + AdjE_{S2} = 355.998 - 0.003 = 355.995mE$$

$$N'_{S3} = N_{S3} + AdjE_{S3} = 313.094 - 0.008 = 313.086mE$$

$$N'_{S4} = N_{S4} + AdjE_{S4} = 261.627 - 0.011 = 261.616mE$$

$$N'_{S1} = N_{S1} + AdjE_{S1} = 305.129 - 0.019 = 305.110mE \checkmark$$

The final adjusted coordinates are:

S1 = (120.550mE, 305.110mN)

S2 = (156.857mE, 355.995mN)

S3 = (230.592mE, 313.086mN)

S4 = (257.530mE, 261.616mN)

EXERCISE:

2. You have carried out a 5- station traverse.

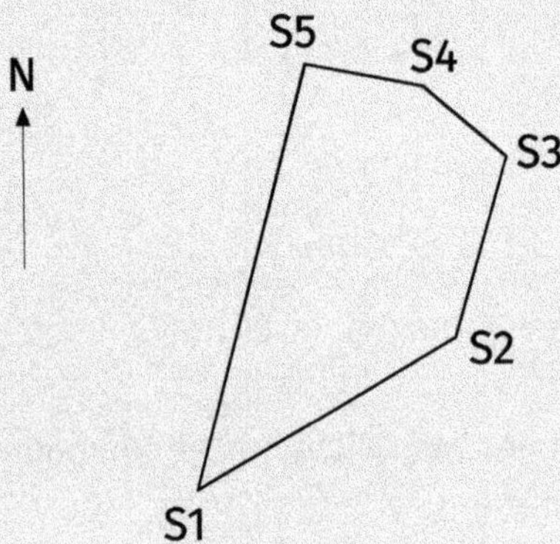

The results are as follows:

Internal angles: S1 = 49°22′56″, S2 = 126°26′20″, S3 = 100°56′53″, S4 = 155°59′28″, S5 = 107°14′12″

Distances: S1 – S2= 80.435m, S2 – S3 = 46.360m, S3 – S4 = 30.518m, S4 – S5 = 28.662m, S5 – S1 = 93.696m

    a. The coordinates of S1 are (106.743mE, 87.187mN) and S2 are (183.112mE, 112.438mN). What is the forward bearing S1 – S2?

    b. What are the forward bearings of the other legs of the traverse?

    c. What is the total distance of the traverse?

    d. What is the misclosure in the E and N?

    e. What are the final adjusted coordinates of S1 – S5?

    f. Rounded to the nearest 1000, what is the accuracy of the traverse?

19.17 HOW DO I KNOW THE COORDINATES OF MY STARTING POINTS?

To start your traverse, you need to have two physical points in the ground whose coordinates you know, or one physical point in the ground and the forward bearing to a reference point which you can sight to but may not necessarily be able to access directly.

Your two starting points may come in the form of:

→ surveying nails in the ground for which you are given coordinates on the setting out drawing

→ existing features whose coordinates you can reference in relation to design coordinates

In both cases, once you have identified your ST1 and ST2, you must go through the following process to verify them:

1. Based on the coordinates, calculate the forward bearing ST1 to ST2 (see section **28.29**)

2. Based on the coordinates, calculate the distance ST1 to ST2.

3. Set up the total station over ST1 and take a direct measurement to ST2.

4. Compare the distance calculated in step 2 and the distance measured in step 3. If there is any discrepancy, continue to step 5.

5. Recalculate the coordinates of ST2 based on the forward bearing calculated in step 1 and the distance measured in step 3. These recalculated coordinates supersede the original ones.

6. You now have the coordinates of your starting points ST1 and ST2.

19.18 HOW DO I DETERMINE THE COORDINATES OF AN EXISTING FEATURE?

To determine the coordinates of an existing feature in relation to design coordinates:

1. Identify a well-defined existing feature which is clearly marked on the drawing.

2. On the drawing, use a ruler to measure the change in eastings and the change in northings from a point at which design coordinates are given.

3. Apply the difference in eastings and northings to the known coordinate to calculate the coordinates of the existing feature.

19.19 HOW DO I DETERMINE THE COORDINATES OF A POINT WHICH IS OFFSET FROM AN EXISTING FEATURE?

If it is not possible to set the total station up over the existing feature you have chosen, for example if it is the end of a wall or the corner of a building, it may be necessary to put an offset nail in the ground at a convenient point and reference its position in relation to the existing feature as in Fig. 19-16. Follow these steps:

1. Choose the approximate location for your nail close to the existing feature. For example, the middle of the footway.

2. Choose the exact location of the nail in relation to the feature, for example, 3m along the wall and 2.5m offset.

3. Put the nail in the ground.

4. Plot the position of the nail onto your drawing.

5. Calculate the coordinates of the nail using the method in sections **28.25 - 28.29**.

Fig. 19-16: Coordinates of offset from an existing feature

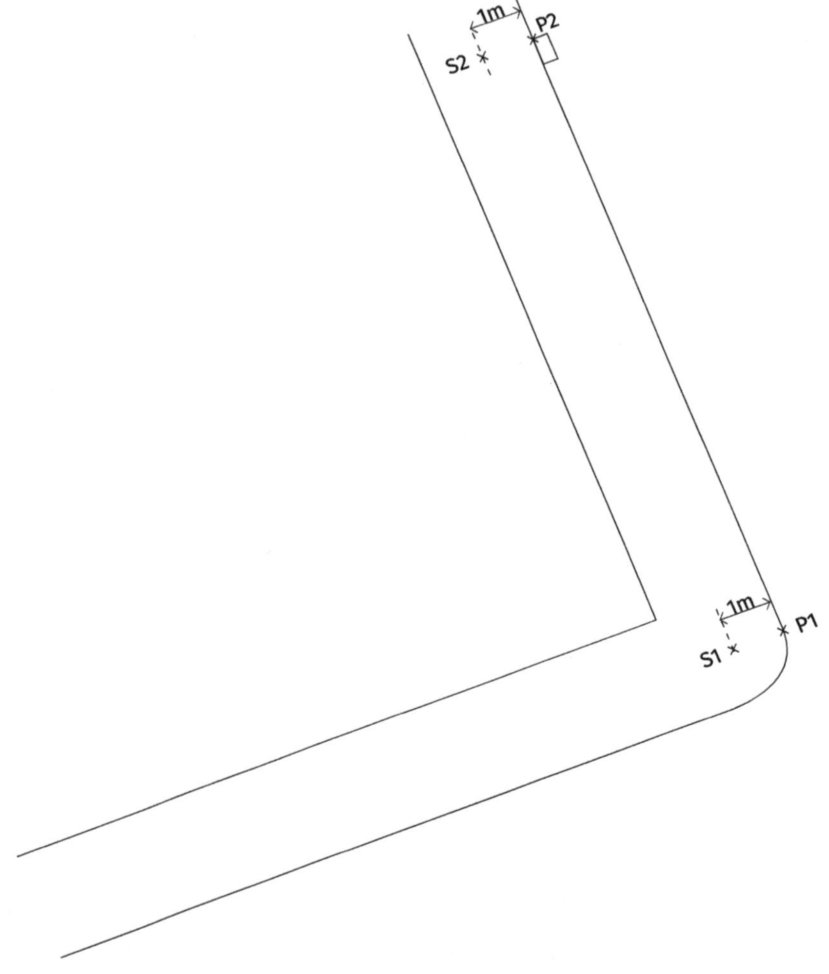

19.20 STATION COORDINATES ARE GIVEN ON THE DESIGN DRAWINGS. CAN I USE THESE?

Even if the design drawings indicate stations (usually a triangle with a dot at the centre) and provide coordinates for them, it is possible that these stations were installed as part of the initial topographical survey which was intended for design purposes only. It is not safe to assume that they are sufficiently accurate to use as control points for the construction phase of the works.

Sometimes the initial topographical survey is carried out months or even years before the construction commences. You may find that even though the stations are marked on the drawing, some or all of them may have been removed or obstructed by the time construction commences.

If the stations were established for design survey purposes, they may not necessarily be in suitable locations for the construction works. When deciding whether to incorporate these stations into your primary control network, consider section **18.9.**

If you do decide to incorporate some, or all, of these stations into your primary control network, you must include them in your traverse and recalculate them based on your own field data.

19.21 WHAT IS THE ALLOWABLE DISTANCE MISCLOSURE FOR A TRAVERSE?

The accuracy of a traverse is expressed as a proportion of the full survey length. You wouldn't expect to get the same error over a traverse length of 200m as you would over 2000m.

Guide required traverse accuracies are:

→ 1:10,000 for general civils projects (1mm in every 10m)

→ 1:20,000 for rail projects (1mm in every 20m)

→ 1:50,000 for specialist monitoring projects (1mm in every 50m)

19.22 WHAT IS THE DISTANCE BETWEEN THE TRAVERSE POINTS?

The distance between the points can be anything from 20m up to the maximum sight distance of the total station being used. The greater the distance, the more accurate the survey will be.

19.23 HOW CAN I IMPROVE THE ACCURACY OF MY TRAVERSE?

Traverse data can vary greatly in quality. In summary, the factors affecting accuracy are:

→ **quantity of readings** - the more measurements you take from each point, the closer the average measurement will be to the true value

→ **independence of readings** - the instrument should be re-levelled and re-centred in between each round of angles and distances. If this is not done, any centring errors will be repeated, achieving precision but not necessarily accuracy

→ **amount of experience** – the more practice you have, the better the quality of your readings

→ **amount of care** – even if you are under pressure of time, take as much time as you need to focus the object and the cross-hairs and to line up cross-hairs with care on the nail

→ **conditions** – wind, rain and wet ground conditions can affect the quality of your readings

→ **equipment** – use good quality, well-calibrated equipment. Carry out the full range of total station checks described in **Chapter 12** before commencing your traverse

→ **the process** – follow the process in section **19.4** to the letter

→ **checks** – carry out checks in the field, analysing and correcting any errors as you go along.

19.24 CAN I USE A SPREADSHEET FOR TRAVERSE CALCULATIONS?

The repetitive nature of the traverse table means that it lends itself well to a spreadsheet such as Excel. If you decide to use a spreadsheet, whether it is created by yourself or a third party, it is advisable to test the spreadsheet against a reliable set of calculations such as the worked examples in this chapter.

19.25 WHY AM I GETTING COMPLETELY DIFFERENT ANSWERS FOR MY REDUCED ANGLES TAKEN FROM THE SAME POINT?

If you are getting different answers between rounds of angles taken from the same point, possible reasons are that you have:

→ measured the external angle for one and the internal angle for the other. Fig. 19-17 shows internal and external angles in a traverse. To convert an external angle to an internal, subtract it from 360 degrees

→ sighted to a wrong point

→ taken the right readings but recorded them in the wrong order

→ made a calculation error

→ copied down the vertical angle from the display instead of the horizontal angle.

Fig. 19-17: Traverse - Internal vs external angle

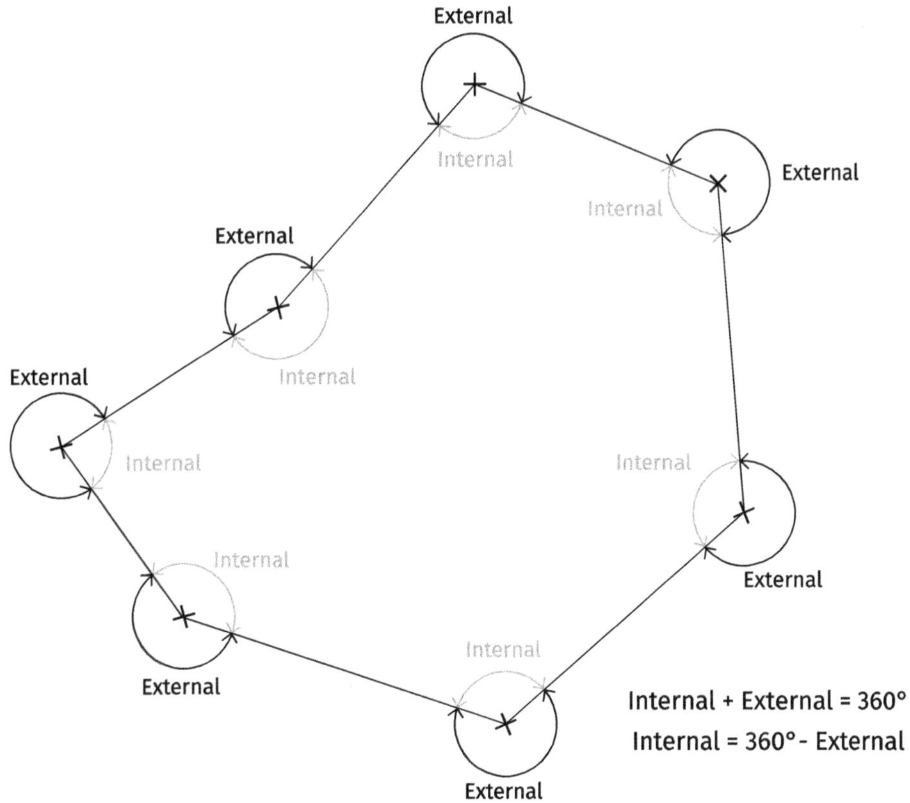

19.26　WHY AM I GETTING LARGE ERRORS BETWEEN LEFT FACE AND RIGHT FACE?

If you are getting large errors between your left and right face readings, it could be due to:

→ poor observing technique

→ poor visibility of the point you are sighting to

→ movement in the PGMs you are sighting to

→ sighting to the wrong peg on one of the faces

→ writing down some, or all, of the vertical angle instead of the horizontal angle

→ movement of the ground around your total station

→ the positioning of the pointer (e.g. nail or pencil) on the PGM.

19.27　MY LFR/RFR ANGLE CHECKS SEEM OK, BUT WHY AM I FINDING A LARGE MISCLOSURE IN MY TRAVERSE?

A large, unexplained misclosure could be due to:

→ the Electromagnetic Distance Measurement (EDM) settings being set to the wrong prism constant

→ mistaking an external angle for an internal angle (easy to do if it is close to 180°)

→ a combination of factors including poor observing technique, poor levelling and centring of the instrument and/or backsight targets, ground movement, peg movement and poor visibility.

19.28　HOW DO I ESTABLISH THE COORDINATES OF SECONDARY CONTROL POINTS?

Secondary control points must always be set out directly from primary control points. A secondary control point should never be set out by referencing another secondary control point as the internal errors will accumulate and become unacceptable.

As with primary control points, the locations of secondary control points should be strategically located to achieve optimum inter-visibility and flexibility.

Plan as far ahead as possible, taking into account the likely location of obstructions as the works progress. As many secondary control points as possible should be set out in each single exercise rather than one or two at a time on an as-needed basis. This will minimise the error between different set-up positions.

The procedure for establishing the coordinates of secondary control points is as follows (Fig 19-18):

1. Centre your total station over a primary control point using the 'Ori with coords' set-up function, or equivalent.

2. Backsight to a prism on a tripod or a mini-prism on primary control point.

3. Select the set-out function and change the display until it displays eastings and northings.

4. Sight to a prism or mini-prism on a primary control point other than your backsight point.

5. Press 'Dist' to display the measured coordinates of that point.

6. Compare the measured coordinates to the expected value. If they are less than 5mm error in the eastings and 5mm in the northings, continue to step 7.

7. Measure and record the coordinates of all of the secondary control points visible from your set-up point (retros and nails).

8. Move your total station to a different primary control point and repeat steps 1 – 7, if possible using three completely different points for your previous set-up, backsight and check.

9. Repeat steps 1-8 until you have at least three independent readings for every secondary control point, with a maximum spread of 3mm in eastings and 3mm in northings.

10. Calculate the averages of the eastings and northings of each target to give the final coordinates for each secondary control point.

Fig. 19-18: Establishing secondary control points from primary

Stage 1

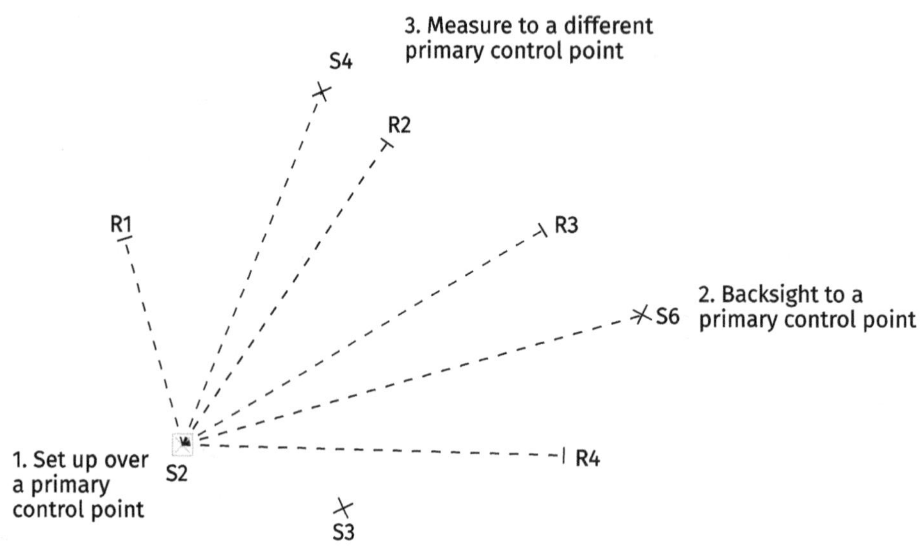

Stage 2

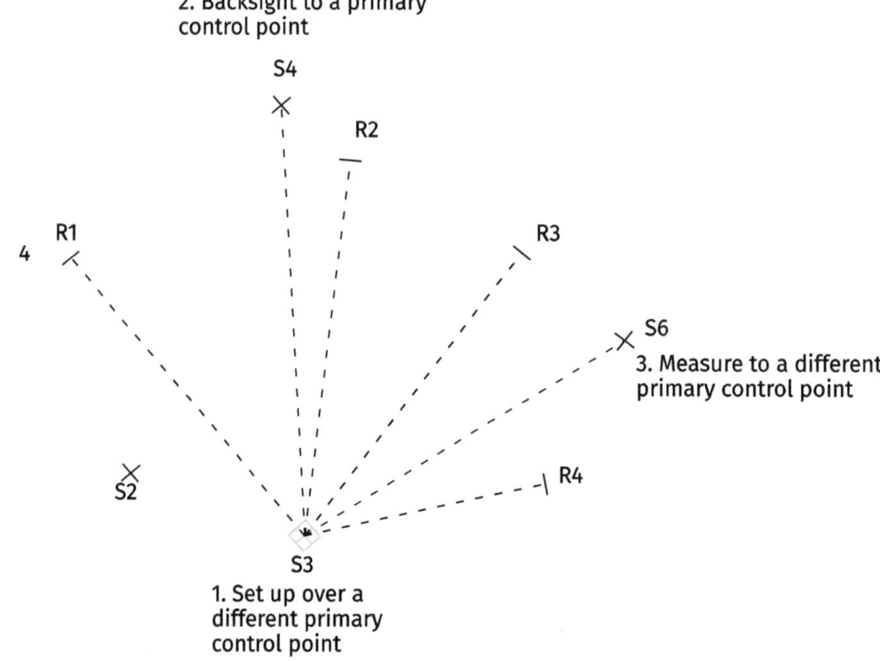

19.29 CAN I DETERMINE THE COORDINATES OF RETRO TARGETS USING A RESECTION SET-UP?

Although it is possible to do this, due to the errors inherent in the operation, a resection set-up should only be used for tertiary setting out, never for determining the positions of any control points.

The correct way to establish the coordinates of a retro target is by setting up over a primary control point, backsighting to another primary control point, checking to another primary control point and then measuring the coordinates of the retro as described in section **19.28**.

By determinig the coordinates of retro targets from a resection set-up, you will build in six layers of cumulative errors into your final setting out. The errors will build up as follows:

→ **cumulative error 1** - the errors involved in establishing the primary control points

→ **cumulative error 2** - the retro targets you use as backsights in your resection are secondary control points, therefore they will contain the error inherent in deriving them from your primary control points

→ **cumulative error 3** - it is not possible to set up directly over a retro target since it is on a vertical surface. Therefore, you would need to set up using a resection which would incorporate a further error into your set-up

→ **cumulative error 4** – there will be an angular and distance error built into the measurement of the coordinates of your new retros

→ **cumulative error 5** – you will set up a resection to carry out your final setting out. This will contain an error

→ **cumulative error 6** – there will be an error present in the final stage of your setting out.

19.30 IS THERE ANY SITUATION IN WHICH IT IS OK TO USE A RESECTION TO DETERMINE THE COORDINATES OF NEW RETRO TARGETS?

For accurate work, using a resection to determine the coordinates of new retro targets, either from primary control points or retro targets, is generally considered to be bad practice.

However, there are certain scenarios where it may be the best or only option available to you. For example:

→ on a site where there are no primary control points and you do not have time to install PGMs as outlined in section **19.4**

→ when all your primary control points have been obstructed or destroyed

→ for very low accuracy work such as earthworks

→ in an urgent situation where there is no alternative.

If you find yourself in any of these situations and you have no option, here are a few things to consider:

→ control point information should not be passed directly from one engineer to another. Whenever a retro target is set up, it should be entered into the Survey Management File. Control points established in this way should be flagged in the Survey Management File as 'Low Accuracy!' to reduce the risk of them inadvertently being used for tasks which require a higher level of accuracy

→ once you have set your instrument position and orientation using a resection, carry out extensive checking against other control points and as-built features before measuring your new retro targets

→ repeat the exercise several times, each time checking that the instrument is level, setting up the resection again, checking to known points and taking several readings to each new retro. The more sets of readings you take and the greater the number of independent set-ups, the better. Each time you set up a new resection, try and backsight to different control points rather than using the same ones each time

→ use as many backsights as possible for your resection, don't use the minimum of two

→ test your new control points extensively in relation to each other and to other control points

When you backsight to these low accuracy retro targets for setting out purposes:

→ once you have set up your resection, check to reliable control points and as-built features before you commence setting out

→ set out as many points as possible from a single set-up position. This will mean that the points are correct in relation to each other and minimise the errors between different set-ups

19.31 HOW DO I EXTEND MY PRIMARY CONTROL NETWORK TO TAKE IN A NEW AREA?

Primary control points should be positioned at the extents of the works, so setting out points are interpolated rather than extrapolated. In some cases, it is necessary to revise the area covered by the primary control point network. There are a couple of possible reasons for this:

→ the project is split into phases and there is no access to the subsequent phases at the time the primary control points are installed

→ the scope of work changes after the original primary control points have been installed and the area of the works needs to be extended.

To extend your primary control network, you must carry out a traverse, using two of your original control points as your baseline. Note the following:

→ verify your original control points - before you start, verify the accuracy of the two points you will start your traverse with

→ plan ahead – avoid extending your control network in small sections. Each time you install new control points, you are building in error and there is a risk of differential error between sections. To minimise error, ascertain the likely full extent of the work and aim to take in the whole area in one traverse

→ always start from original control points - if you extend your network, and then extend again later, use your original control points as starting points rather than using your extended network, otherwise your errors will accumulate.

In Fig. 19-19, S1 and S7 of the original traverse are used as the starting points for the adjacent traverse.

Fig. 19-19: Extending you control network

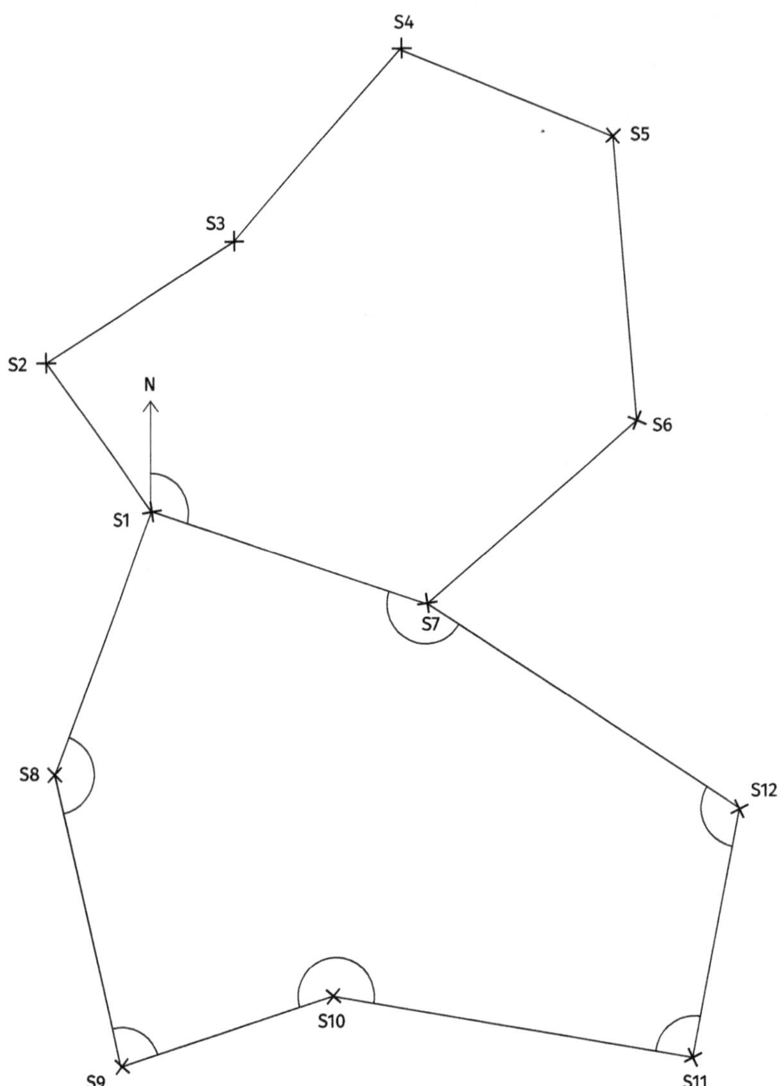

CHAPTER 20

VERTICAL CONTROL POINTS

20.1 PRIMARY VERTICAL CONTROL POINTS

Primary vertical control points can be established using either a level or a total station. If time allows it is prudent to verify the results of one method using another.

20.2 USING AN AUTOMATIC LEVEL TO ESTABLISH VERTICAL CONTROL POINTS

The process for establishing primary vertical control points using the level is known as transferring a Temporary Bench Mark (TBM). Primary vertical control points are usually situated around the perimeter of the works as shown in Fig. 20-1. The secondary vertical control points can be located at convenient locations throughout the works as shown in Fig. 20-2. They are determined in relation to the primary vertical control points using the same method.

Fig. 20-1: Primary control points - Vertical control

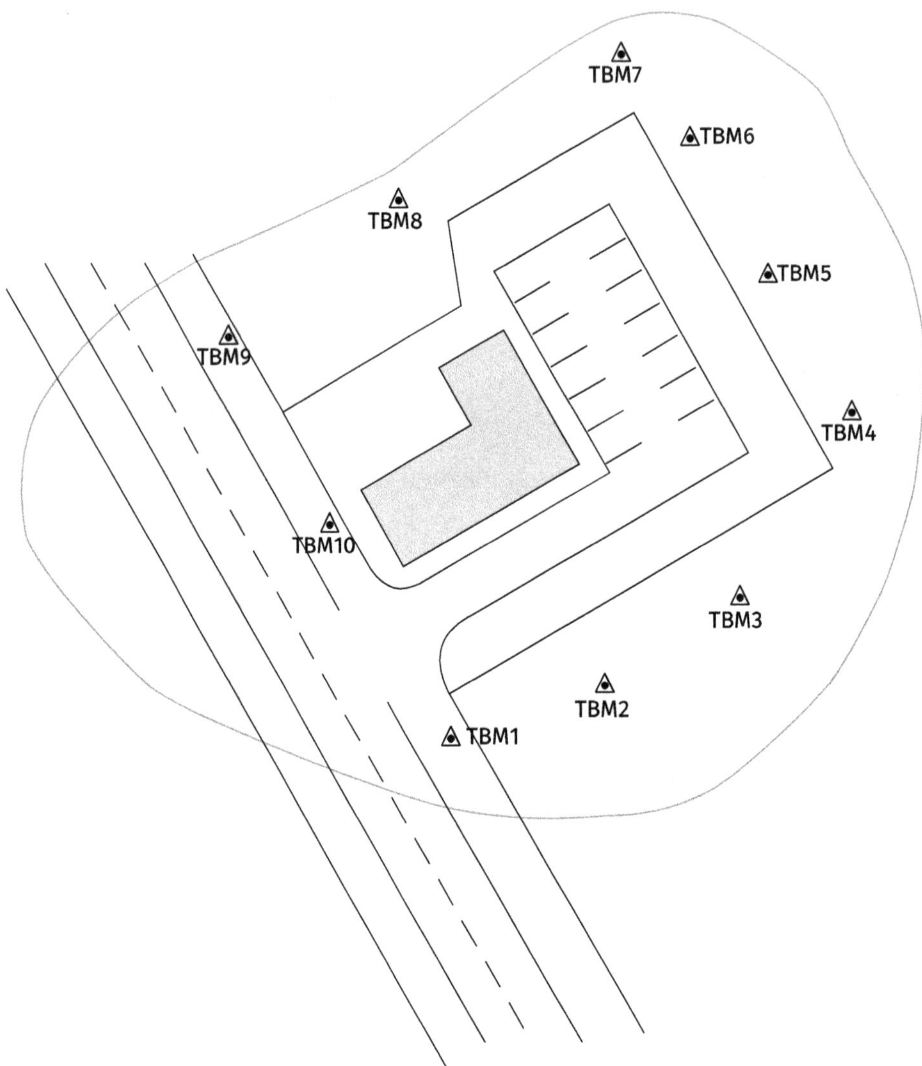

Fig. 20-2: Secondary control points - Vertical control

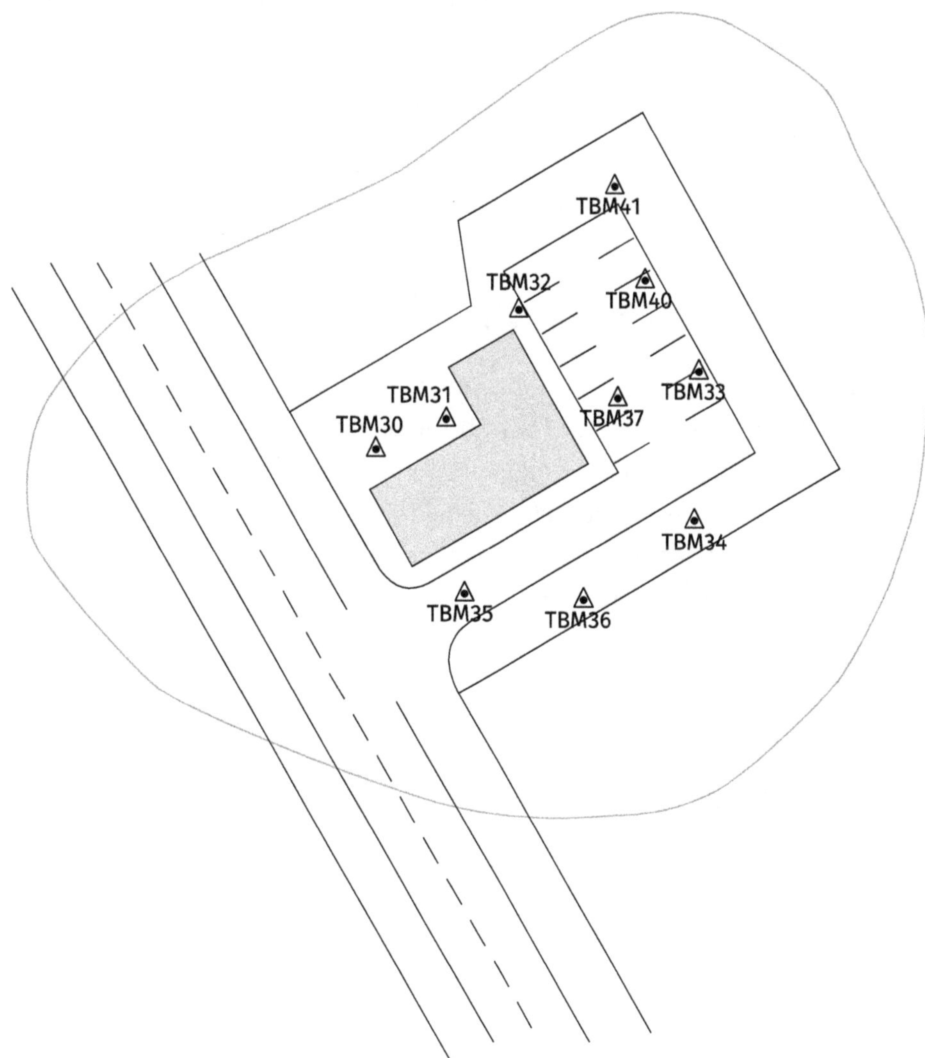

20.3 TRANSFERRING A TBM

Transferring a TBM means ascertaining the reduced level of a new TBM in relation to one which is already known. Transferring a TBM does not mean setting a physical mark at a reduced level to match that of an existing TBM.

There are several scenarios in which you will need to transfer a TBM. Examples include when:

→ you install your primary control points

→ you need to create a TBM close to the section of works you are working on

→ your existing TBMs are obstructed

→ your project is extended into a new phase of works and you need to extend your primary control point network

→ some of your setting out is not tying up correctly and you need to check the values of your existing TBMs.

20.4 HOW DO I TRANSFER A TBM?

The process for transferring a TBM involves measuring the reduced levels of intermediate change points between your original TBM and your new TBM as shown in Fig. 20-3. The reduced level of each change point is measured in relation to the previous one as shown in Fig. 20-4. The process is described below using the concept of change points, however in reality, each change point can also form a TBM. An example of how to record the information in your field book is shown in Fig. 20-5.

The process is as follows:

1. Choose two TBMs to use as your reference points. For the sake of this explanation, we will call them TBM1 and TBM2.

2. Choose the location of your new TBM or the TBM you would like to verify. We will call this TBM3.

3. Plan the locations of your change points. We will call these CP1, CP2 etc.

4. Set up your level between TBM1 and CP1, at a position where you can also sight TBM2.

5. Take a BS to TBM1.

6. Calculate the HPC.

7. Take an IS to TBM2.

8. Based on your staff readings, calculate the RL of TBM2. If correct, continue.

9. Take an FS to CP2. Do not calculate any RLs yet.

10. Subtract the FS from the BS and mark it on the diagram (Figure 20-3).

11. Pick up your level and move it so it is between the next two CPs.

12. Take a BS to your previous CP.

13. Take a FS to your next CP.

14. Subtract the FS from the BS and mark it on the diagram.

15. Repeat steps 11 – 14 until you have taken an FS to TBM3.

16. Now you should be between TBM3 and your final CP. You will now carry out the process, but in reverse, from TBM3 all the way back to TBM1, using the same CPs.

17. Raise or lower the height of collimation by adjusting the tripod legs or feet and re-levelling your instrument.

18. Take a BS to TBM3.

19. Take an FS to your final CP.

20. Subtract the FS from the BS and mark it on the diagram. This should match the level difference you measured between your previous BS and FS. If it does, continue.

21. Continue the process, until you get back to TBM1, each time checking that the level difference between your BS and FS match.

Now that you have taken all the measurements and checked their precision and accuracy, you can move onto the next stage of calculating the RLs of the CPs and TBMs. The process is as follows:

1. $RL_{CP1} = HPC - FS_{CP1}$

2. $HPC = RL_{CP1} + BS_{CP2}$

3. $RL_{CP2} = HPC - FS_{CP2}$

4. $HPC = RL_{P2} + BS_{CP3}$

5. Continue until all RLs have been calculated

The final step is to check your calculations. The check is:

$$\Sigma BS - \Sigma FS = \text{Calculated } RL_{TBM1} - \text{Original } RL_{TBM2}$$

The two sides of the equation should be equal. If they are not, then a calculation error is present.

Fig. 20-3: Transferring a TBM

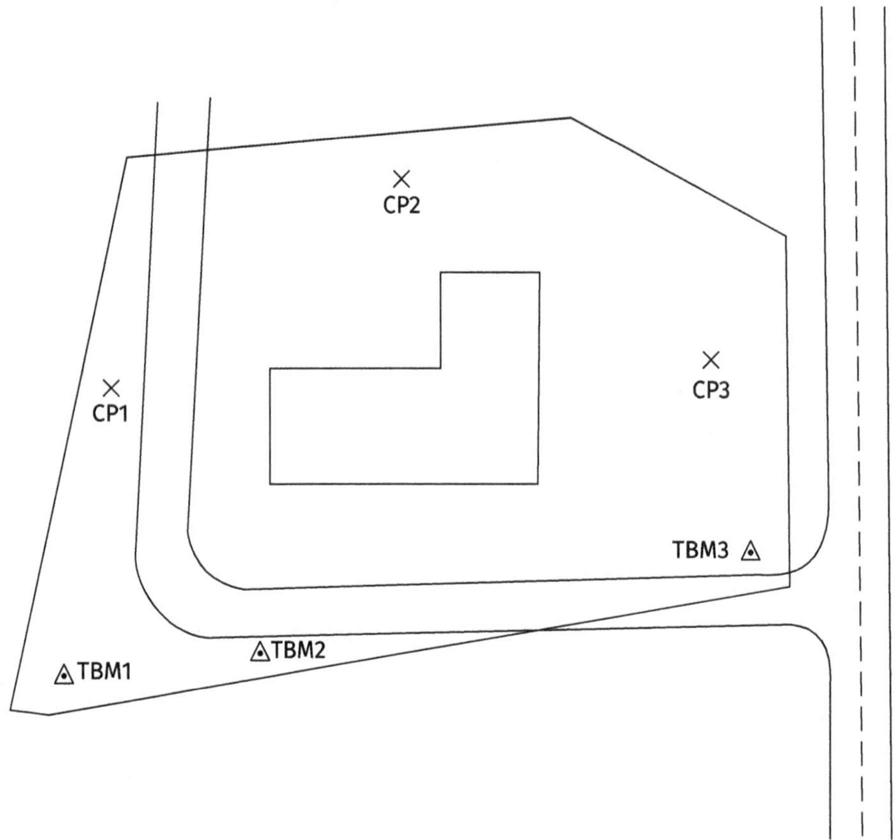

Fig. 20-4: Transferring a TBM - Each change point is measured in relation to the previous one

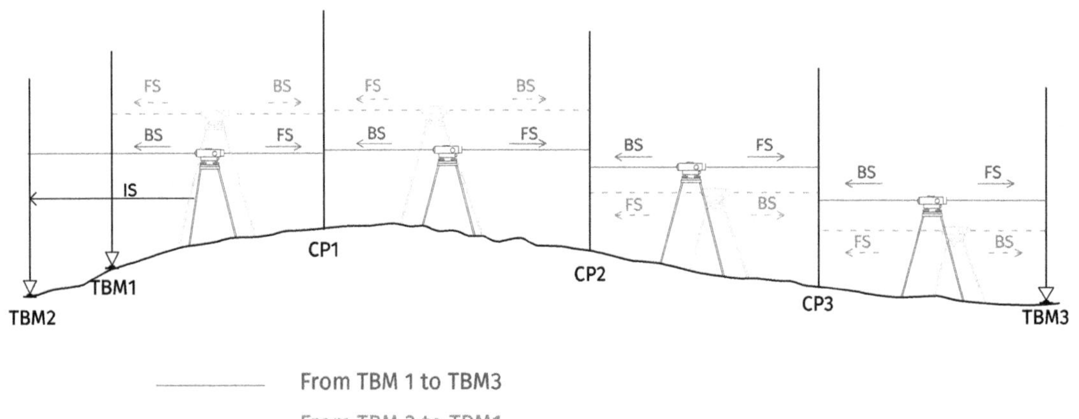

Fig. 20-5: Transfer a TBM - booking example

Date: 14 Feb 2019   Taken for: Establish TBM 3

From: TBM 1   To: TBM 3

BACK SIGHT	INTERMEDIATE	FORE SIGHT	COLLIMATION OR H.P.C	REDUCED LEVEL	DISTANCE	REMARKS
1.454			101.454	100.000		TBM 1
	0.608			100.846		TBM 2 (100.845 reqd, 1mm error ∴ OK)
1.572		2.221	100.805	99.233		Change Point 1 (CP1)
1.507		1.040	101.272	99.765		CP2
1.482		0.455	102.299	100.817		CP3
1.588		1.005	102.882	101.294		TBM 3
1.399		2.066	102.215	100.816		CP3
1.579		2.453	101.341	99.762		CP2
1.445		2.110	100.676	99.231		CP1
Σ=12.026		0.679		99.997		TBM 1 (3mm closing error)
		Σ=12.029				

Calculation Check

$$\Sigma BS - \Sigma FS = \text{Last RL} - \text{First RL}$$
$$12.026 - 12.029 = 99.997 - 100.000$$
$$-0.003 = -0.003 \checkmark \text{ Check OK}$$

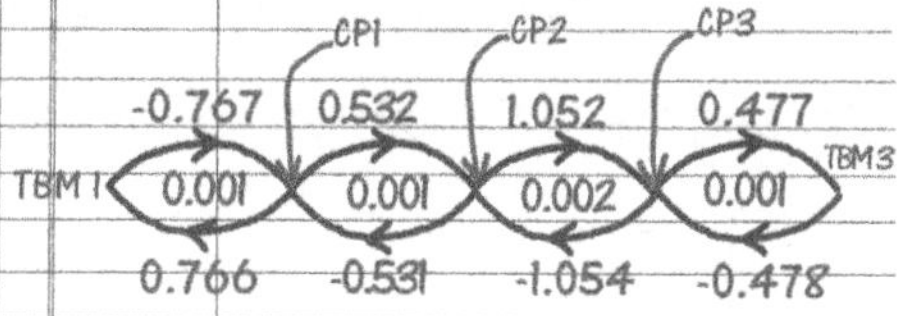

EXERCISE:

1. You have taken the following readings:

BS	IS	FS	HPC	RL	REMARKS
1.602				12.500	TBM1
	1.355				TBM2 (12.745 reqd)
1.533		1.567			TBM3
1.506		1.427			TBM4
1.550		1.828			TBM5
1.608		1.228			TBM4
1.612		1.713			TBM3
		1.649			TBM1

a. Draw the level difference diagram. What is the maximum error in level differences?

b. Calculate the RLs of the TBMs

c. What is the closing error?

d. Carry out the calculation check

20.5 HOW CAN I MINIMISE THE ERRORS WHEN TRANSFERRING A TBM?

It is essential to minimise any errors in the process of transferring your TBMs as errors in your TBMs will be transferred into the construction. Here are some measures you can take:

→ mark out the change points before you start the exercise

→ if your assistant will be deciding the change points, train them in how to choose a suitable point

→ locate change points on solid ground

→ choose change points which have a single point of contact between the point and the flat surface of the bottom of the staff

→ keep sight distances less than 40m.

→ for critical setting out, repeat the exercise several times

→ don't make sight distances too short, as the error increases with the number of set-ups.

20.6 HOW DO I USE THE TOTAL STATION TO ESTABLISH PRIMARY VERTICAL CONTROL POINTS?

To use the total station for measuring the reduced levels of your TBMs you can use either the tie- distance or survey functions. The process is as follows:

1. Measure to two known points to check that they tie in correctly.

2. Measure to the reduced levels of all the primary control points visible from your set- up position.

3. Repeat steps 1 and 2 until you have at least 3 measurements for each TBM with a maximum spread of 3mm.

4. Calculate the average RL for each point.

20.7 HOW DO I SET UP SECONDARY CONTROL POINTS FOR REDUCED LEVELS?

Secondary control points for reduced levels may take two different forms. You may decide to use either or both of these methods depending on the specific conditions of your site.

The methods are:

→ Attribute reduced levels to the same retro targets which you have set up for your secondary horizontal control points. This will allow you to set up a 3D resection (eastings, northings and reduced level) with your total station and use it for measuring and setting out levels.

→ Install a separate set of benchmarks on horizontal surfaces. This allows you to use your level and staff for measuring and setting out levels.

20.8 HOW DO I USED THE TOTAL STATION TO ESTABLISH THE VALUES OF SECONDARY VERTICAL CONTROL POINTS?

If your secondary control points are PGMs you can use the same method as described in section **20.4**. If your secondary control points are retro targets, use the method described in section **20.10**.

20.9 WHAT IS THE CORRECT PROCEDURE FOR TYING IN RETRO TARGETS TO PRIMARY TBMs?

There are two different methods suitable for establishing the reduced level of your retro targets in relation to your primary control points. You can:

→ use the tie distance/compute inverse function to measure the level distance between primary control points and retro targets or

→ use your primary control points to install a datum on a flat vertical surface directly below a retro target and then use a tape measure to determine the level difference. Once you have establish the level of one retro target, you can measure the other retro targets in relation to this one using the tie distance/compute inverse function or survey function.

20.10 HOW DO I USE THE TIE DISTANCE/COMPUTE INVERSE FUNCTION TO MEASURE THE REDUCED LEVEL OF RETRO TARGETS?

To use the tie distance/compute inverse function to determine the Reduced Level (RL) of retro targets, the procedure is as follows (this same method can be used for measuring the RLs of primary control points using a mini-prism):

PART 1 – CALIBRATE YOUR INSTRUMENT BY LINING-UP AGAINST TWO RELIABLE TBMS (FIG. 20-7):

1. Set up and level your instrument. You do not need to set its position and orientation. There may not be a position where you can view all your retros from a single set-up. If this is the case, find a position where you can see two TBMs and as many retro targets as possible. The remaining ones will be measured from different set-up positions.

2. Select the tie distance/compute inverse function.

3. Enter the reflector height (H_r) of your mini-prism.

4. Select the correct prism type and constant.

5. When prompted, sight to the mini prism held on TBM 1 and press 'Meas'.

6. You will then be prompted to sight to TBM2 and press 'Meas'. The level difference will be displayed on the screen. Make a note of it.

7. Exit the function.

8. Re-level your instrument and repeat steps 2 – 5. If you have a discrepancy of more than 2mm, repeat until you achieve several consistent readings.

9. If the level difference measured in step 7 corresponds with the expected difference, continue. You have now calibrated your instrument and your own accuracy against two reliable points.

PART 2 – MEASURE THE LEVEL DIFFERENCE BETWEEN YOUR TBM AND YOUR RETRO TARGETS (Fig. 20-8):

10. Repeat steps 1 – 6, checking. If the level difference is within 1mm of the expected value, continue.

11. Press 'New Point 2'

12. Set the reflector height (H_r) to 0mm.

13. Set the EDM type to 'Tape', checking the constant is correct.

14. Sight to a retro target giving priority to the vertical angle (see section **11.13**).

15. Press 'Meas'.

16. Make a note of the level difference between TBM 1 and the retro target, remembering to record whether it is negative or positive.

17. Repeat steps 14 - 16 for all the retro targets visible from that set-up position. Each time, you are measuring the level difference between TMB1 and the retro you are sighting to.

18. Exit the tie distance/compute inverse function.

19. Repeat steps 10 – 18. You now have two sets of results which should agree with each other within 2mm. If not, keep repeating steps 10 – 18 until you get several sets of consistent results.

PART 3 – MEASURE THE RETROS YOU COULD NOT SEE FROM YOUR INITIAL POSITION

20. Move your instrument to a position where you can sight a different sub-set of the retros.

21. Repeat steps 10 - 19, repeating any retros which you measured from your previous position.

22. Repeat steps 20 - 21 until you have a minimum of two correlating results for all your retro targets. Due to overlap, you may have many more results for some retros. The more data you have, the more accurate your final results will be.

PART 4 – CALCULATE THE REDUCED LEVELS OF THE RETROS

23. For each retro, take the average of all the level differences you have gathered, discarding any results which you believe are anomalies.

24. For each retro, add the averaged figure onto the reduced level of TBM1. You now have reliable reduced levels for your secondary control points.

Fig. 20-7: Tie Distance Level difference

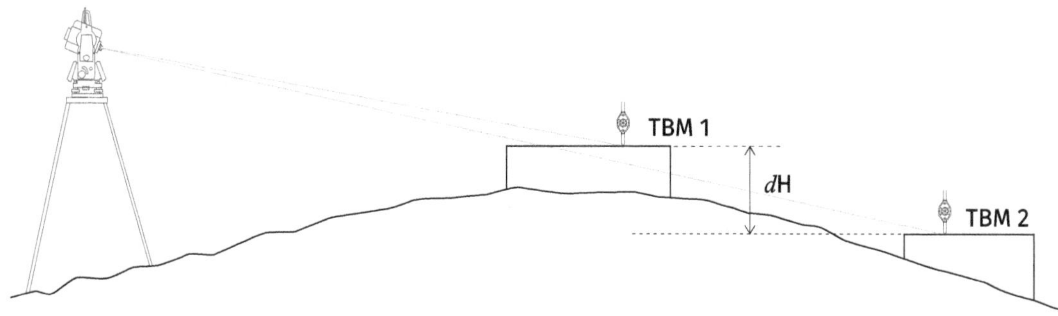

Fig. 20-8: Level difference between TBM and retro targets

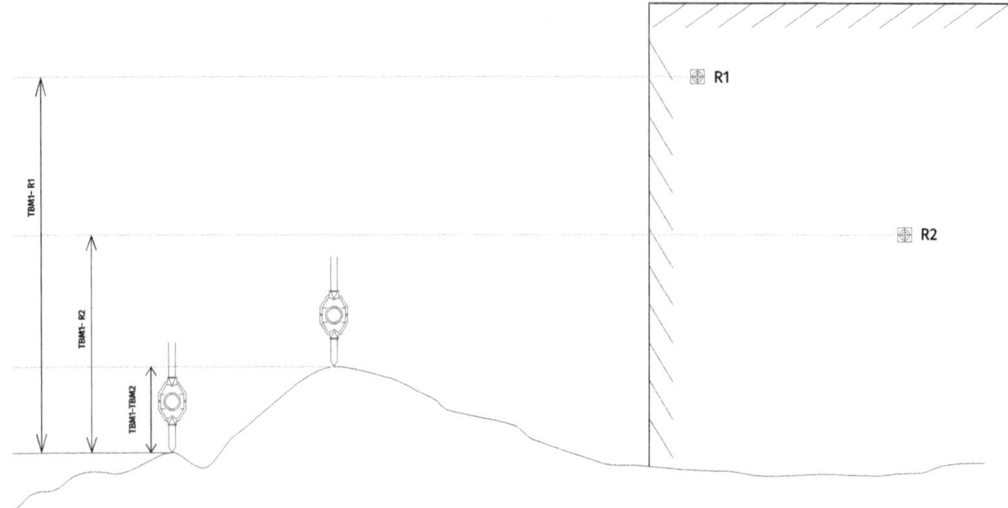

20.11 CAN I USE THE SURVEY FUNCTION TO DETERMINE THE REDUCED LEVELS OF THE RETROS?

It is possible to use the survey function to determine the reduced levels of the retros, however it is essential that once you have set up the instrument, you are able to check to other known TBMs to verify the accuracy of the set-up.

To use the survey function, you would have to set up the instrument in one of two ways:

→ **'Ori with coords'** – you would need to have a reliable reduced level for the point you are set up on. You would physically measure with a tape measure from the marker to the trunnion axis marker on the side of the total station. There is significant scope for error in this operation alone

→ **'3D resection'** – setting up the instrument using a resection incorporates errors which may be unacceptable when establishing control points. See section **11.56** on why a resection should not be used for establishing control points.

20.12 WHERE DOES THE INITIAL REDUCED LEVEL COME FROM?

If you are on a site where control points have already been established, your reduced level for your starting benchmarks will come from the control points in the Survey Management

File. Make sure that you are satisfied that they are reliable, verifying them yourself if necessary.

If you have arrived at a brand-new site, there may not be any TBMs in place for you to use. In this case, you will need to inspect the construction drawings and identify two existing features for which reduced level is provided. These may come in the form of:

→ the cover level of a manhole, inspection chamber or gully

→ a spot height whose location you can identify in relation to existing features such as a kerb, fence line or wall. You may need to scale off the drawing and translate this into a taped distance on the ground

→ the tie-in between existing and new carriageway.

Once you have selected two points which you think are reliable, use your level to measure one in relation to the other. If the points are more than 60m apart, use the process of transferring a benchmark described in section **20.4**. This exercise is merely to confirm that there is no gross error in the levels given on the drawings. The nature of the interface between the existing and the new will dictate the level of accuracy required for your first TBM. For example, at a road carriageway tie-in, there is a certain amount of play, whereas with a tie-in between cast-in pipes, a much higher level of accuracy will be required.

Once you are satisfied that you have a reliable TBM, you can create your second TBM following the process in section **20.13**.

20.13 HOW DO I SET UP MY SECOND TBM?

To establish your primary vertical control network, you will need two TBMs which are correct in relation to each other. These will be the reference points for installing the rest of your TBMs. Once you have set up your first TBM as in section **20.12**, you must create a second one approx. 30m away from your first one. To do this:

1. Set up your level so that you are equidistant from the two TBMs.

2. Take a BS and a FS.

3. Subtract the FS from the BS to find the level difference between the two points.

4. Change the height of collimation by adjusting the tripod legs or feet and level the instrument.

5. Repeat steps 2 - 4 until you have four level differences with a maximum spread of 2mm.

6. Discard any erroneous results.

Take the average of the four values and use the average to calculate the value of TBM 2.

CHAPTER 21

COORDINATES, BASELINES AND OFFSET LINES

21.1 HOW DO I SET UP A COORDINATE SYSTEM IF THE DESIGN DOES NOT PROVIDE ONE?

For steel-framed buildings or houses, setting out information is often provided as dimensions only, and no coordinates are given. If a design is provided without coordinates, it is most likely because they are not needed to set out the works. Consider how you can set out the building using a baseline and measurements from existing features using the Reference Line function.

If, after careful consideration, you feel you do need a coordinate system to set out the works, a local coordinate system can be set up by choosing an arbitrary origin and arbitrary north (Fig. 21-1). You can then use this system to attribute coordinates to any point on the drawing (Fig. 21-2). This works best for buildings where most walls are parallel and perpendicular to each other. Follow these steps:

1. Decide which axis of your building will lie along the horizontal (eastings) axis.

2. Choose an arbitrary coordinate for the bottom left hand corner of your building. Make the coordinates large enough to ensure that no part of your building will be negative. Make the easting different to the northing so that you can distinguish between them, reducing the risk of getting them confused with each other, for example 100.000mE, 300.000mN.

3. Select a point which you want to calculate the coordinates of.

4. Determine the distance from your starting point along the eastings axis and add it onto the eastings of your first point. This will give you the eastings of your new point.

5. Determine the distance from your starting point along the northings axis and add it onto the northings of your first point. This will give you the northings of your new point.

6. Repeat steps 3 – 5 for all the points which you need to know the coordinates of.

The more coordinates you calculate, the higher the risk of error. Where possible, identify the coordinates of only two points per building, either on the building or offset from it, and use the Reference Line function to set out the building in relation to your baseline.

If you have been provided with a single coordinate system for a group of structures and they are not convenient to work with, you can use the same principle to set up local coordinate systems for each individual structure or building. If you do this, you will also have to attribute corresponding coordinates to each control point for each local grid. There is lots of scope for error here, so plan carefully before making any decisions about how you will organise your setting out data.

Fig. 21-1: Coordinates for a building on a gridline (Part1)

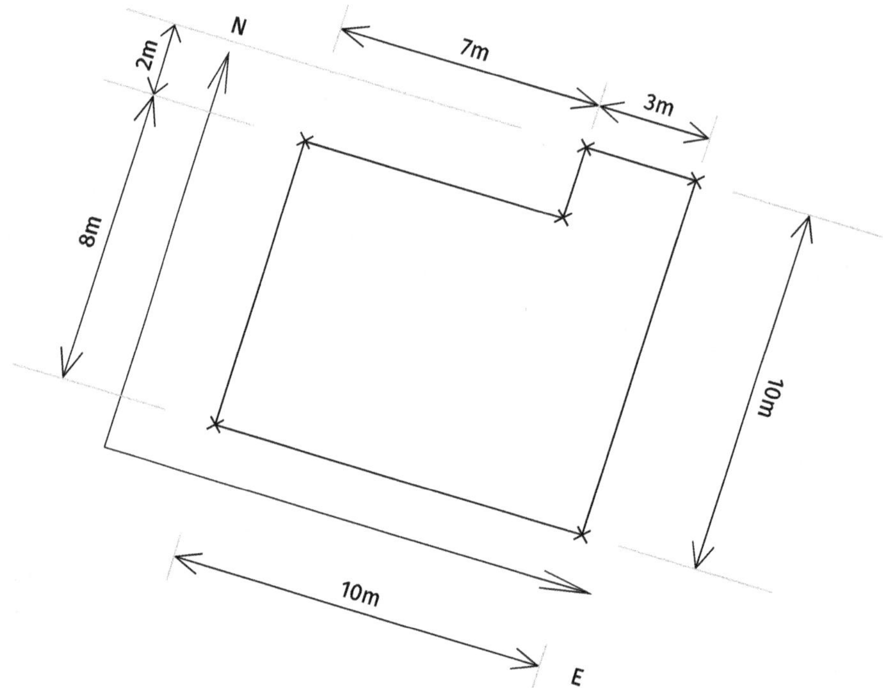

Fig. 21-2: Coordinates for a building on a gridline (Part2)

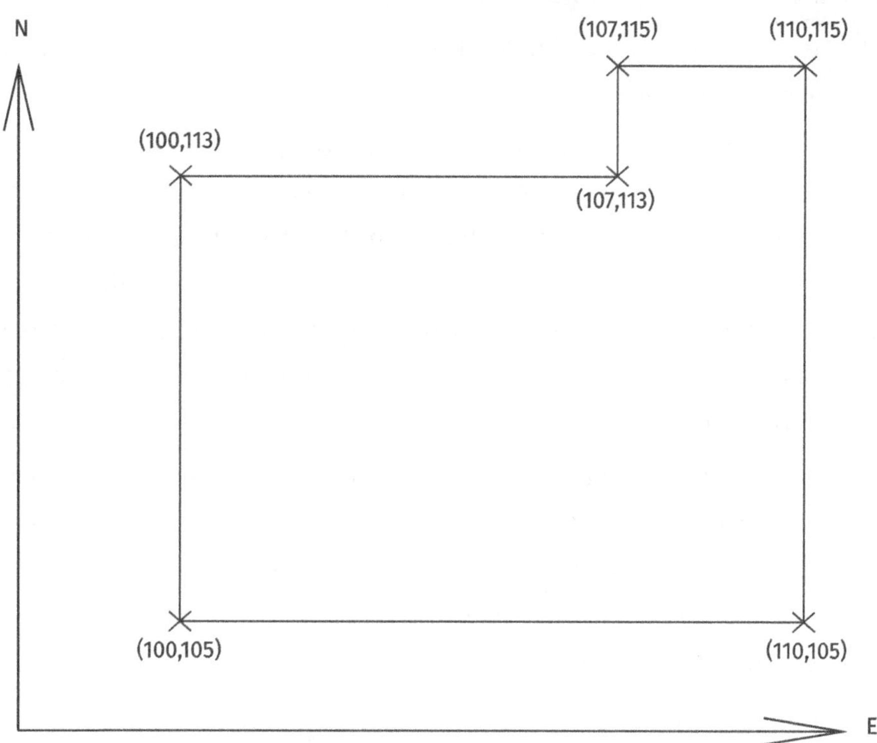

21.2 WHAT ARE THE BENEFITS OF COORDINATES OVER GRIDLINES?

The main benefit in converting your setting out information into coordinates is that you can use the stake-out function for setting out your points.

21.3 WHAT ARE THE DRAWBACKS OF USING COORDINATES?

Be careful not to over-rely on coordinates as they do have some drawbacks:

→ contractually, the dimensions given on the design drawings are correct. If you make a mistake when calculating the coordinates, you are responsible

→ contractually, the Designer is obliged to provide the minimum information required, with no repeated information. If you request coordinates from the designer, unless there is provision in the contract, they probably have no obligation to provide them.

Therefore, if they differ from the dimensions on the drawing then the Contractor may bear the risk of any errors

→ if you are provided with coordinates, it is your responsibility to cross-check them all and query them if any do not tie in with the dimensions you are given

→ if the Designer provides you with a .dfx file to enable you to extract the coordinates you need, exercise extreme caution. Unless the contract states otherwise, it is the dimensions and coordinates provided on the paper drawings which relate to the contract. If you use coordinates from an electronic file and they turn out to be wrong, it is fully the Contractor's responsibility. The exception to this rule is complex projects where there is provision in the contract for exchange of data files, digital information

→ when working with coordinates rather than dimensions, it is not as obvious when you have made an error. Once you have set out adjacent points using the stake-out mode, you should carry out a direct distance measurement between the two points and cross-reference it with the dimensions given on the drawing

→ there is a higher risk of confusion or error

→ it takes time and effort to calculate coordinates if you don't have access to the appropriate survey software.

21.4 WHAT IS A BASELINE?

All surveying and setting out is based on the principle of a baseline.

A baseline is a notional line joining two defined points. It can be shown on a drawing, but does not physically exist on the ground, however the two end points of the baseline are clearly marked physically on site.

The two end points of the baseline form a reference line from which other points are measured or set out. See Fig. 21-3. The positions of the start and end points are known in relation to the works and may be defined by:

→ their coordinates

→ their relationship to gridlines or

→ their relationship to existing features

The orientation of the line is known in relation to either north or design gridlines and existing features.

You can use any two of your primary control points to form a baseline. A traverse is effectively a series of baselines joined together, whose positions are known in relation to a master baseline, formed by the two starting points of your traverse.

You can also create a local baseline for each building or structure you are setting out as shown in Fig. 21-4. To provide the greatest accuracy, the baseline should be as long as possible. When setting out a house, building or structure, choose the longest side as your baseline.

Fig. 21-3: Baseline - Control points

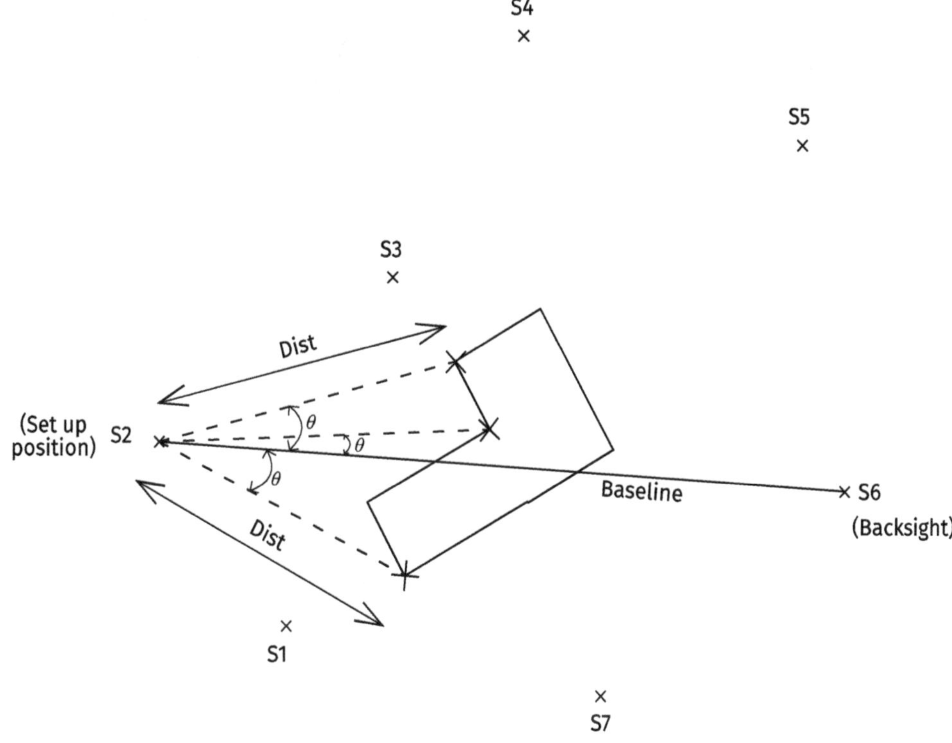

Fig. 21-4: Baseline - Offsets from a gridline

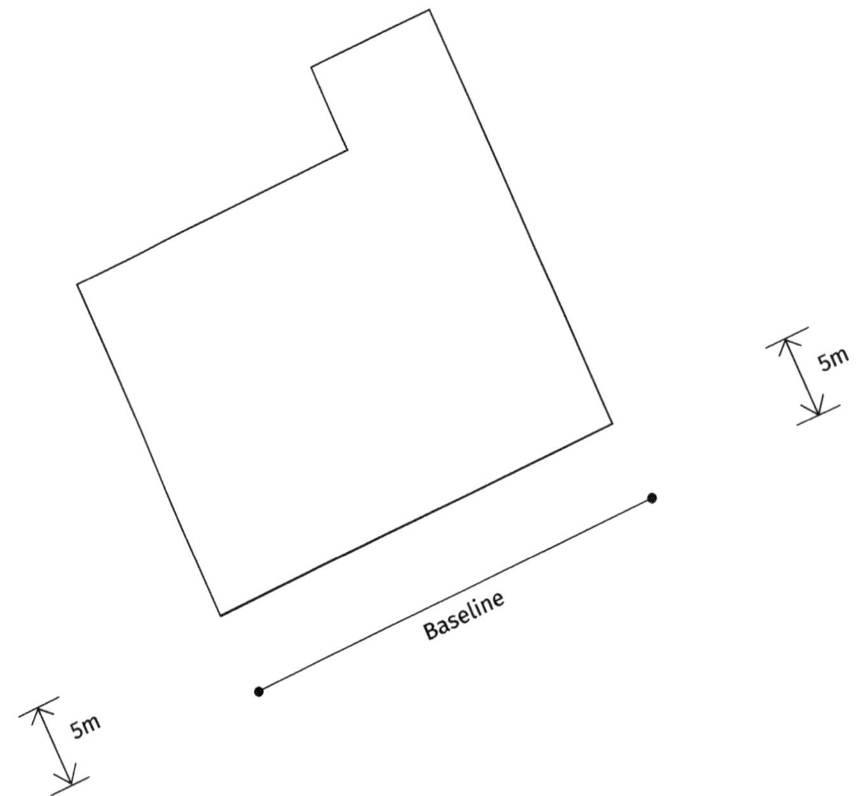

21.5 HOW DO I SET OUT OR MEASURE FROM A BASELINE USING A TAPE MEASURE?

Using a baseline and a tape measure is the traditional method for setting out relatively small construction elements. Fig. 21-5 shows the principle. The total station has largely precluded the use of this method of setting out, however it is a perfectly adequate method for small scale projects such as housing plots, plinths, slabs, strip foundations, landscaping or manholes and is still used.

To set out points from a baseline using a tape measure:

1. Plot the start (A) and end (B) points of the baseline on the design drawing.

2. Scale or calculate the distance from A and B to the point being set out, P1.

3. Extend the tape from A to the approx. location of P1 and spray an arc.

4. Extend the tape from B to the approx. location of P2 and spray an arc.

5. Point P1 is the intersection of the two arcs.

6. Repeat steps 2 – 5 for a second point, P2, on the element you are setting out.

7. Adjust the position of P2 until the line P1 – P2 is exactly the right length.

8. P1 – P2 can now be used as an accurate local baseline for the element you are setting out.

Fig. 21-5: Baseline - Using a tape measure

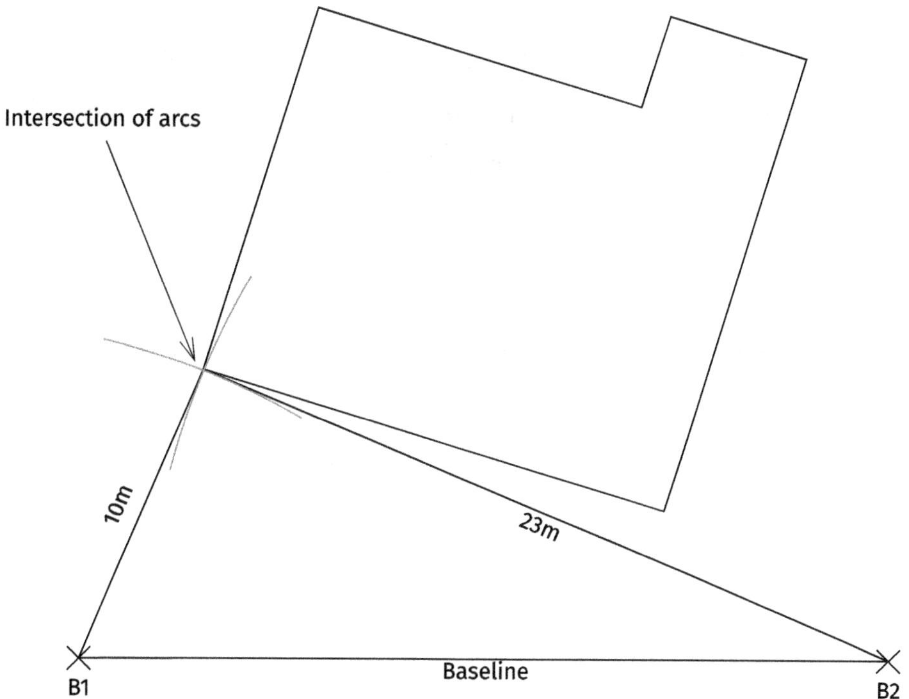

21.6 HOW DO I SET OUT POINTS FROM A BASELINE USING TWO TAPE MEASURES?

1. Plot the start (A) and end (B) points of the baseline on the design drawing.

2. Scale or calculate the distance from A and B to the point being set out, P1.

3. Extend two tapes from A and B simultaneously.

4. Pull the two tapes taut until they cross at the point where both tapes are at the correct distance from A and B respectively.

5. Point P1 is the intersection of the two tape measures.

6. Repeat steps 2 – 5 for a second point, P2, on the element you are setting out.

7. Adjust the position of P2 until the line P1 – P2 is exactly the right length.

8. P1 – P2 can now be used as an accurate local baseline for the element you are setting out.

21.7 HOW DO I DETERMINE THE POSITION OF AN EXISTING POINT IN RELATION TO A BASELINE USING A TAPE MEASURE? (Fig. 21-6):

1. Measure the distance from B1 to the point you are trying to determine, P.

2. Measure the distance from B2 to the point you are trying to determine, P.

3. Use the sine rule to calculate the angle PB1B2.

4. If the bearing B1B2 is known, calculate the bearing B1D.

5. The coordinates of P can now be calculated.

Fig. 21-6: Measuring co-ordinates in relation to a baseline

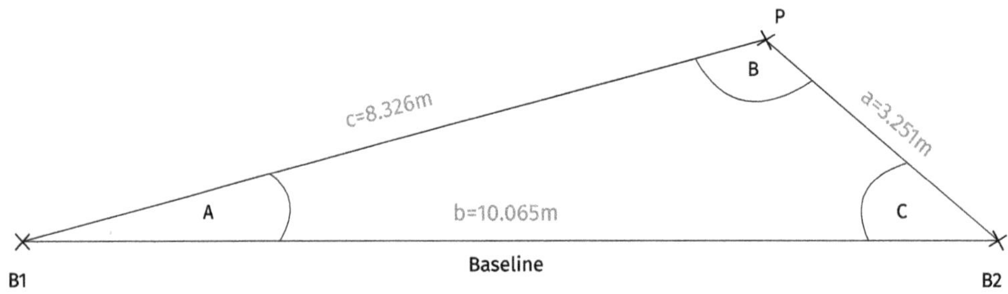

21.8 HOW DO I USE A BASELINE WITH A TOTAL STATION?

Each time you set up your total station, you are either using or creating a baseline. The following are examples of activities which rely on the baseline principle:

→ setting up over a point of known coordinates and sighting to another point of known coordinates

→ setting up a resection by sighting to two known points

→ using a chainage and offset system

→ setting up over the intersection of two gridlines, sighting along one gridline and turning off 90 degrees

→ carrying out a traverse

→ setting out by bearing and distance.

21.9 WHAT ARE OFFSET LINES?

It is often necessary to mark out offsets from the actual position you are setting out, to prevent your setting out marks being destroyed by the construction process. For example, if you were to mark out the centreline of a strip foundation, the first job would be to excavate to formation level, destroying the marks you made. However, if you were to mark out a line which was 2m offset from the centreline, once the excavation had been completed, your original markers would still be in position, allowing subsequent elements such as the shuttering and reinforcement to be set out from your original marks.

21.10 WHAT SIZE OF OFFSET SHOULD I USE?

A common offset distance is 2m, however some scenarios may require a smaller or larger offset. The offset must be small enough to be convenient for the works, but large enough to allow for the appropriate clearance and working room. If in doubt, ask the foreman or tradespeople who will be working to your marks what offset distance will work well for them.

21.11 HOW DO I SET OUT OFFSETS?

There are two approaches you can use to set out offset lines. The first is to set out the correct lines, and then measure the required offset, perpendicular to the line. The second approach is to calculate the position of the offset line, and set out the offset lines directly.

To mark your offsets, hammer two timber pegs into the ground, at a chosen distance back from the start of the line, and at equal offsets from the centreline. The two pegs should be oriented square to the offset line. Attach a profile board to the pegs and tap a nail vertically into the top edge of the board, exactly on the centreline. You can do this using a total station or a tape measure. Repeat at the other end of the offset line being set out. A string line can be attached between the two nails, so that construction elements can be set out from the string line.

21.12 WHY MIGHT I NEED TO EXTEND A LINE?

It is not always possible to set out the full length of a line in one section. It may be necessary to set out a short section of a line as a temporary measure, then extend the line to its full length at a later stage. This situation arises when access to a part of the site is not available during the initial stages of setting out or when a slab must be poured in sections because of obstructions or other activities taking place. As the slab is extended, it may be necessary to project the line of shutters and walls to the next section.

21.13 HOW SHOULD I EXTEND A LINE BEYOND ITS ORIGINAL END POINTS?

Before deciding which method to use to extend your line, consider whether it's more appropriate to establish your new end-point in relation to your primary control points, or to extend an existing line.

If you decide that it's necessary to extend your existing line, never transit the telescope (i.e. do not rotate it through the vertical circle) to project a line, as trunnion axis errors are more likely and usually more severe than horizontal collimation errors. Instead, use one of the following options:

→ **Reference Line function** - use the Reference Line function to input the two original points (either by coordinates or by sighting to them) and create a third point on the line at the required distance from the start point.

→ **project the line through two points** - set up your total station over one point, sight to the other point on the line, and project the line in the same direction.

→ **rotate through 180 degrees** - set up your total station over one point, sight to the other point and rotate through 180 degrees to extend the line in the opposite direction. Repeat using the opposite face. The mid-point between the two marks is the correct position. Avoid using a short backsight distance and long foresight distance. This will exacerbate the effects of any centring, observing and collimation errors. The longer the distance to the backsight, and the shorter the foresight, the better. Only use this method if no other options are available as this method is most likely to result in a kink in the line.

CHAPTER 22

PROFILE BOARDS

22.1 PROFILE BOARDS

Profile boards are an arrangement of timber posts and boards used as reference points for setting out positions and levels of construction elements. They comprise vertical timber posts fixed into the ground with a horizontal board attached (Fig. 22-1). The boards are set to fixed reduced levels and positions in relation to the desired level and position of the element being set out.

Profile boards, sometimes known as sight rails or boning rods, are used for controlling levels and may also be positioned to define offsets. These profile boards are used for setting levels of elements such as:

→ excavations

→ fill material

→ road construction

→ drainage, including excavation of trench and crown of pipe.

Positional profile boards are used for marking offsets from horizontal lines (Fig. 22-2). A nail or 'V' is used to mark the position of the offset so a string line can be pulled between the two points and a tape measure used to reference the offset line. This type of profile board is typically used for setting out foundation elements including:

→ edges of excavations

→ centrelines of strip foundations and walls

For the purpose of this book, the term 'profile boards' refers to those used for level control.

Fig. 22-1: Profile board

Fig 22-2: Taping - Building offsets

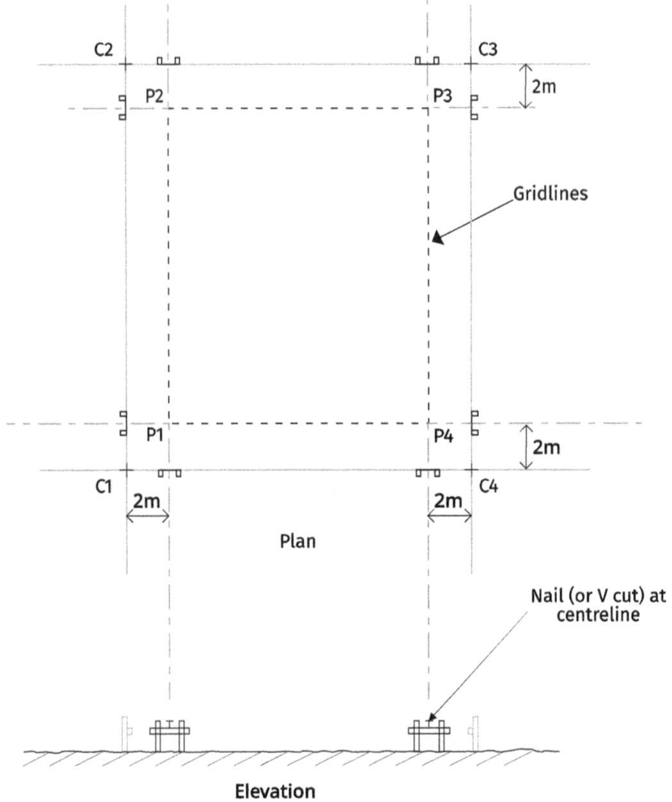

22.2 HOW DO PROFILE BOARDS WORK?

The principle of profile boards is that a Reference Line is set up at a fixed parallel distance above the plane being set out. A traveller, a free moving timber post of fixed length with a cross-head, is 'sighted in' between the fixed profile boards (Fig. 22-3). This is known as boning in.

To check if a point on the surface being set out is at the correct level, the observer adjusts their eyeline until the tops of the fixed profile boards are lined up. The traveller is then held with its base in contact with the surface being measured. If the top of the traveller is visibly above or below the line of sight, then the excavation or fill needs to be adjusted accordingly. When the traveller is exactly in line with the line of the two profile boards, the surface is at the correct level.

Fig. 22-3: Profile boards for a level excavation

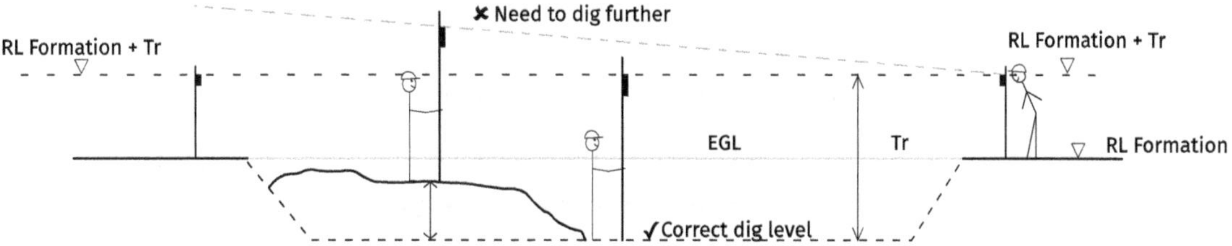

22.3 WHAT ARE THE ADVANTAGES OF SETTING UP PROFILE BOARDS AND BATTER RAILS?

Although there are a range of alternatives available for setting out levels and slopes, such as the optical level, laser level, total station and GNSS, there are many situations where profile boards and batter rails are the most suitable method of control. Here are some of the advantages of profile boards and batter rails over the other methods:

→ **they are cheap** – the basic equipment needed is an optical level, a tape measure and the timber.

→ **they are time-efficient** – once the engineer has set up the profile boards or batter rails, they are left in place for use by the operatives. Although the engineer is still responsible for the setting out, they do not need to be constantly present. This frees them up to work on other tasks.

> → **they are quick** – once they are set up, they take seconds to use.

> → **they are simple** - once they have been set up by the engineer, they are very simple to use. There is no equipment needed and there are no numbers involved so there is relatively little chance of confusion.

22.4 WHAT ARE THE DRAWBACKS OF USING PROFILE BOARDS AND BATTER RAILS?

There are some drawbacks to profile boards and batter rails:

> → **they require some learning on the part of the engineer** – although the principles and calculations involved are surprisingly simple, some engineers avoid using profile boards and batter rails because they do not have the necessary skills and understanding.

> → **they may be destroyed** – because they need to be positioned close to the works area, they are vulnerable to being knocked or even completely destroyed by excavators and other plant.

> → **they may look right even if they are not** – if the profile boards or batter rails are not set up correctly, it may not be obvious. Independent checking is strongly advised.

> → **they are only suitable for certain depths** – profile boards are not suitable for depths greater than, say, 2m as the traveller then would be too long and unmanageable.

> → **they are limited to low-accuracy tasks** – due to the accumulation of errors in setting up the boards, staff verticality, boning in and local inconsistencies in the surface being measured, profile boards and batter rails are only suitable for low-accuracy tasks.

22.5 WHAT IS A TRAVELLER?

A traveller is a timber post with a cross head attached at a specified distance from the base of the post. The post is typically 40mm x 40mm, up to 2m long and the board is usually 90mm and anything from 500 – 900mm long.

22.6 HOW DO I KNOW WHAT THE TRAVELLER LENGTH SHOULD BE?

The vertical offset from the feature being set out to the top of the profile board is chosen so that the top of the board is at a convenient height for the user to sight through. Since the profile boards are offset from the dig area and are set up before excavation commences, this is usually 1 to 1.5m above existing ground level. If they are higher than this, they will be too high to sight through. If they are too low the user will need to lie on the ground to sight through. Trial and error is used to find a traveller length that will suit the height of the profile boards on the ground, but will not be too long and unmanageable.

22.7 WHAT DISTANCE SHOULD PROFILE BOARDS BE OFFSET?

Where the excavation is long and narrow, for example in the case of a drainage trench, to avoid being destroyed, the profile boards must be offset from the centreline of the dig. Even when the profile boards are offset in this way, it is possible to line in the top of the traveller.

The most common offset used is 2m, however you should use your engineering judgment for each scenario and be guided by the requirements of the ganger or foreman. Profiles should be positioned where they are least likely to be damaged by plant. Consider the following:

→ the depth of the dig

→ the slope of the sides of the dig

→ where the castings need to go

→ approach routes for dumpers and other plant.

22.8 HOW DO I USE PROFILE BOARDS OVER A LARGE AREA?

For large area excavations, a grid of profile boards should be set up. In this case, it's not possible to offset the profile boards so the boards must be positioned within the excavation itself. A mound is left around the boards and removed at the end, once the whole area has been taken down to the correct level. See Fig. 22-4.

Fig. 22-4: Profile boards for large area

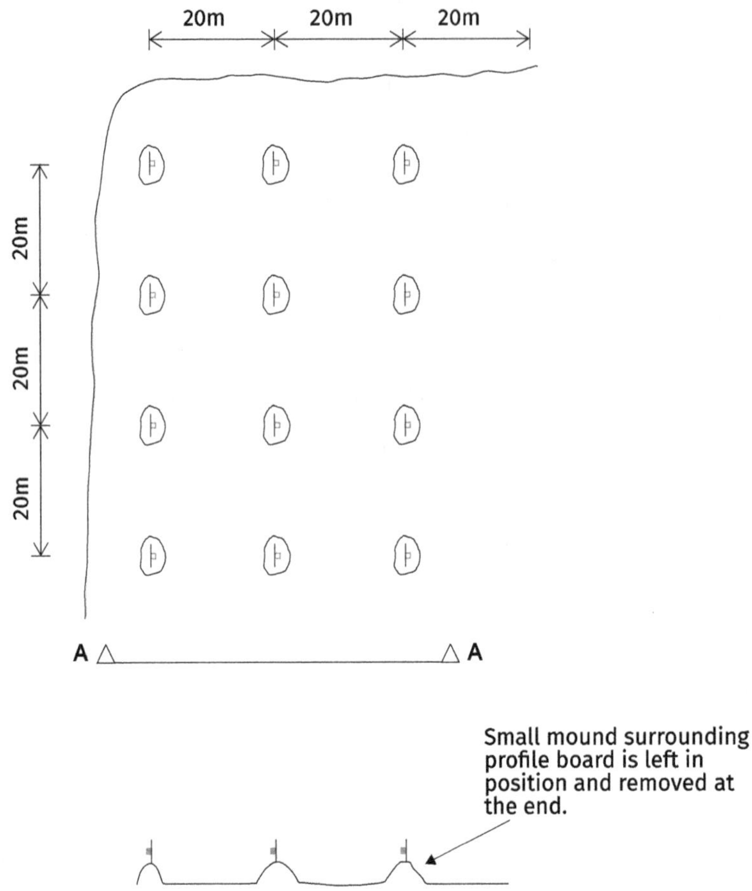

22.9 HOW FAR APART SHOULD PROFILE BOARDS BE?

As a rule of thumb, the maximum distance between any two profile boards should be 20m or the distance from the profile to the next change in grade, whichever is shorter. However, you should use your engineering judgment, taking account of:

→ convenience

→ accuracy

→ obstructions

22.10 WHAT IS THE MAXIMUM TRAVELLER LENGTH?

As a rule of thumb, the maximum traveller length should be 3m, however there may be cases where it is appropriate to relax this limit or reduce it (e.g. if there is a risk of contact with overhead services)

22.11 WHAT ARE DOUBLE-HEADED PROFILES USED FOR?

Double-headed profile boards are most commonly used for the upper layers of road construction where there are opposing cross-falls on each side of the carriageway. The method is suitable for symmetrical or asymmetrical arrangements and can be used on straight and curved stretches.

For accurate setting out, the profile boards should be set up approx. 200mm above the required level and a string line and tape measure used instead of a traveller.

22.13 WHAT IS A DOUBLE-HEADED TRAVELLER USED FOR?

A double-headed traveller is used to set out several levels or layers from one set of profile boards. Two or more boards can be attached to the traveller for different stages of the construction; for example, formation level, top of pipe bedding and crown of pipe. Each board should be marked clearly with the element to which it corresponds.

22.14 WHAT ARE STEPPED PROFILES?

In some cases, the shape of the existing ground level does not allow for a single length of traveller to be used for the full length of the run. An example of this is a drainage run when the existing ground level falls at a faster rate than the pipe. If the traveller length was consistent, for part of the stretch, the profile boards would be too high to sight through. To solve this problem, a step in the level of the profile boards is introduced and the traveller length is amended accordingly.

22.15 PROFILE BOARDS FOR LEVEL EXCAVATION

An example of how to record your setting out is shown in Fig. 22-5. The process for setting up profile boards for a level excavation is as follows:

1. Determine formation level from the construction drawing.

2. Determine existing ground level at the extents of the excavation, either from drawings or by taking a level survey.

3. Calculate reduced level of the top of the profile boards at 1m above existing ground level.

4. Calculate the traveller length. This will be the level of the top of profile board from step 3 minus the formation level in step 1.

5. Round the traveller length up or down to the nearest 0.5m so that the profile boards will be 1-1.5m above existing ground level.

6. Calculate the required reduced levels of the profile boards based on the traveller length calculated in step 5.

7. Calculate the height of each profile board above existing ground level. If any of them fall outside of the 1 – 1.5m range, choose a different traveller length and recalculate the profile boards. If there isn't a solution for a single traveller it may be necessary to use a stepped profile arrangement as in section **22.14**. If the existing ground level will be taken down to a lower level, allow for this when deciding the traveller length.

8. Knock 1.2m or 1.5m long pegs into the ground at suitable locations.

9. Survey the tops of the pegs using the automatic level and calculate their reduced levels.

10. Calculate the positions of the top of the profile boards in relation to the top of the pegs by subtracting the required reduced level of the board from the reduced level of the top of the corresponding peg (Fig. 22-6).

11. Attach the profile boards at the correct heights and annotate clearly with a marker pen.

12. Construct the traveller to the correct length.

Fig. 22-5: Profile boards for level excavation - booking example

BACK SIGHT	INTER MEDIATE	FORE SIGHT	COLLIMATION OR H.P.C	REDUCED LEVEL		DISTANCE	REMARKS
							Date: 30 March 2018 Taken for: Dig level
							From: TBM1 To: Strip foundation, Gridline A
1.703			51.803	49.100			TBM 1
	2.686			49.117			TBM 2 (49.117 reqd, 1mm error ∴ OK)
	1.332			50.471		↓0.221	Peg 1 (50.250 Reqd)
	1.151			50.652		↓0.402	Peg 2 (50.250 Reqd)
		1.703		49.100			TBM 1 (0mm closing error ∴ OK)

Fig. 22-6: Close-up of profile board

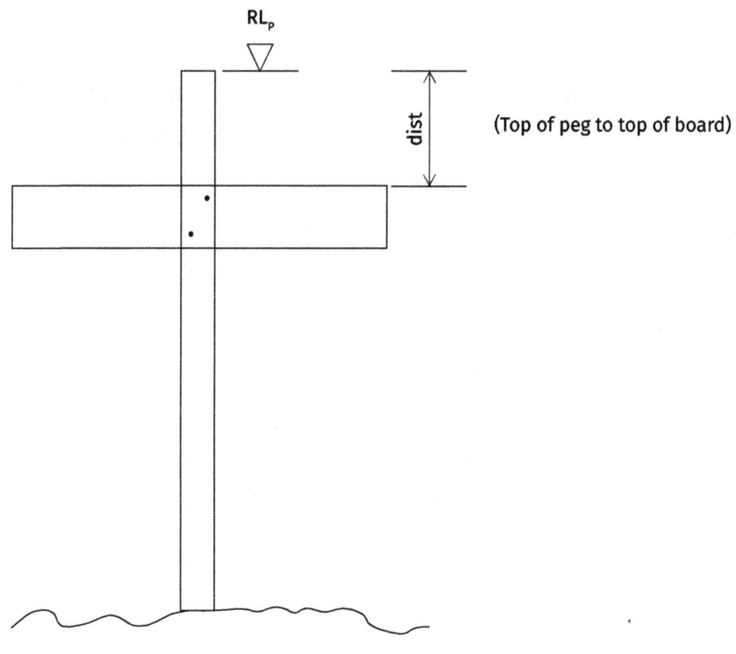

22.16 PROFILE BOARDS FOR A SLOPING EXCAVATION

The principle of using profile boards to set out a sloping or graded excavation is the same as for a level excavation, except the tops of the boards create a line of sight at a fixed distance above a sloping line rather than a level one.

For a sloping excavation, the relative position of the profile board along the run must be correct, otherwise there will be a sudden gradient change in the dig.

22.17 HOW DO I SET OUT PROFILE BOARDS FOR A DRAINAGE RUN?

Profile boards positions can be set out using a total station, GNSS or a tape measure. For an example of how to record your setting out, see Fig. 22-7. The process for setting out profile boards for a drainage run using a level and tape measure is as follows:

1. Determine the full length of the run.

2. Calculate the level change between the beginning and the end.

3. Calculate the fall of the run (length/level change). This is expressed as 1: X , the ratio of the horizontal to vertical distance.

4. Determine the reduced level of formation at each end of the run.

5. Divide the run into 20m lengths starting at the start point (chainage 0). The final section will be the remainder. If this is very small, say, less than 10m, it can be incorporated into the final section.

6. Determine the level of the existing ground along the alignment of the run.

7. Calculate reduced level of the top of the profile board at chainage 0, at 1m above existing ground level.

8. Calculate the traveller length. This will be the level of the top of profile board from step 7 minus the formation level at chainage 0 in step 4.

9. Round the traveller length up or down to the nearest 0.5m so that the profile board will be 1-1.5m above existing ground level.

10. Based on the chosen traveller length, determine the profile board levels at the beginning and the end points.

11. Calculate the height of each profile board above existing ground level. If any of them fall outside of the 1 – 1.5m range, choose a different traveller length and recalculate the profile boards. If there isn't a solution for a single traveller it may be necessary to use a stepped profile arrangement as in section **22.14**.

12. Using the fall calculated in step 3, calculate the profile board levels at the intermediate points as shown in the worked example at the end of this section.

13. Set out the positions of the beginning and end points of the run at the centreline of the pipe.

14. Ascertain which side the digger will be sitting and which side the castings will be placed. Decide which side of the alignment to construct the profile boards. Consult with the ganger or foreman to do this.

15. Set out 2m offsets at the beginning and end points and knock in 1.2 or 1.5m long pegs at each offset. Ensure that the pegs are vertical and orientated square to the alignment.

16. Knock in the intermediate long pegs at 20m intervals lining them in by eye (see section **22.18**)

17. Carry out a basic level survey to calculate the reduced level of the top of each peg.

18. For each peg, calculate the position of the top of the profile board in relation to the top of the peg by subtracting the required reduced level of the board (calculated in step 12) from the reduced level of the top of the corresponding peg.

19. Using a spirit level to ensure the peg is vertical and the board is horizontal, attach the profile boards at the correct heights and annotate clearly with a marker pen.

20. Construct the traveller(s) to the required length.

Fig. 22-7 Profile boards for drainage run - booking example.

Date: 15 December 2018
Taken for: Profile boards (Formation +2m)
From: MH S1
To: MH S2

BACK SIGHT	INTER MEDIATE	FORE SIGHT	COLLIMATION OR H.P.C	REDUCED LEVEL	DISTANCE	REMARKS
1.492			15.392	13.900		TBM 1
	1.842			13.550		TBM 2 (13.550 reqd, 0mm error ∴ OK)
	0.585			14.807	↓0.164	Ch.0 (14.643 Reqd)
	0.729			14.663	↓0.220	Ch.20 (14.443 Reqd)
	1.204			14.188	↑0.055	Ch.40 (14.243 Reqd)
	0.947			14.445	↓0.382	Ch.58 (14.063 Reqd)
		1.494		13.898		TBM 1 (2mm closing error ∴ OK)

WORKED EXAMPLE FOR PROFILE BOARDS OF SLOPING EXCAVATION

Calculate the RL of the profile boards at each end of the drainage run and at 20m intervals along the pipe for the following:

Distance MH1 – MH2 = 58m

Dig level at MH1 = 11.643m, Dig level at MH2 = 11.063m

EGL at MH1 = 13.423m, EGL at MH2 = 12.995m

Step 1: Draw a sketch

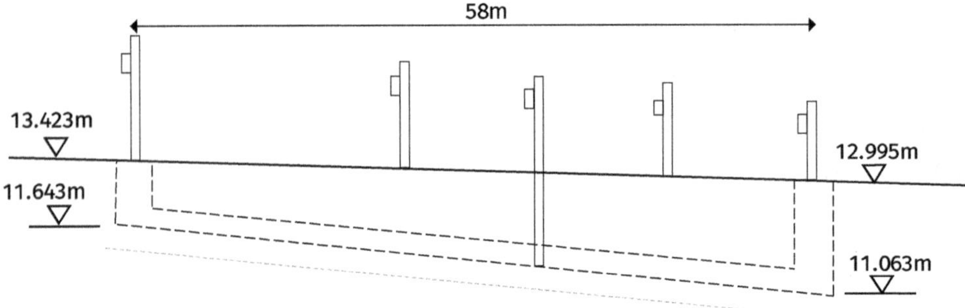

Step 2: Decide the traveller length

The traveller length is chosen based on the specific conditions. The profile boards should be at a convenient height for sighting through, say between 1 and 1.5m above the existing ground level, and the traveller should be a manageable length, a maximum of say 3m.

So that the profile boards are a minimum of 1m above ground level:

At MH1, RL_{Min} = EGL + 1m = 13.423 + 1 = 14.423m

For initial traveller length:

Traveller length = RL_{board} – RL_{dig}

   = 14.423 – 11.643

   = 2.78m

Traveller length should be rounded up to the nearest 0.5m which would make the traveller length 3m.

Step 3: Calculate the required RL of the profile board at MH2:

At MH2, required RL of profile boards = formation level + traveller length

   = 11.063 + 3.000

   = 14.063m

Step 4: Check that the profile boards are at a convenient height above EGL at MH2.

Height above EGL = RL of the profile board − EGL_{Min}

= 14.063 − 12.995

= 1.068 < 1.5m therefore OK.

Step 5: Calculate the gradient of the pipe.

Length of drainage run = 58m

Level change = 11.643 − 11.063 = 0.58m

Gradient of pipe = 1: $\left(\dfrac{Length}{Level\ change}\right)$ = 1: $\left(\dfrac{58}{0.58}\right)$ = 1:100

Therefore, fall (mm/m) = 1 $\left(\dfrac{1}{100}\right)$ x 1000 = 10mm/m

Step 6: Calculate RL of profile boards at MH1 and MH2.

At MH1, RL_{board} = 11.643 + 3 = 14.643m

At MH2, RL_{board} = 11.063 + 3 = 14.063m

Chainage	Level change from Ch.0		RLboard (m)
0	(10 x 0)/1000 = 0	14.643 - 0	14.643
20	(10 x 20)/1000 = 0.2m	14.643 − 0.2	14.443
40	(10 x 40)/1000 = 0.4m	14.643 − 0.4	14.243
58	(10 x 58)/1000 = 0.58m	14.643 − 0.58	14.063 ✓

EXERCISE:

1. For a drainage run between two manholes, MH1 and MH2, the data is as follows:

   → Invert Level (IL) at MH1 = 18.180m, MH2 = 17.853m

   → Pipe thickness = 30mm

   → Depth of pipe bedding material = 200mm

   → Length of the run is 49m

   → A 2m traveller is used.

   a. What is the level change?

   b. What is the fall in mm/m run, percentage and as a ratio?

   c. What are the required RLs of the profile boards at Ch.0m, Ch.20m, Ch.40m and Ch.49m?

22.18 HOW DO I SET OUT PEGS IN A STRAIGHT LINE BY EYE?

To set out pegs in a straight line:

1. Put in the first and last peg.

2. Have your assistant measure out the required interval using a tape measure and hold the peg vertically at the correct distance (the back face of one peg to the back face of the next one). Either you can hold the zero end of the tape in position or they can tap a nail into the peg to hook the tape onto.

3. Step back several metres from the first peg and crouch down, close one eye and move side to side until the last peg is in line with the first one

4. Direct your assistant to move their peg left or right until it is lined in between the first and last peg.

5. Have them knock the peg into the ground.

6. Move to the next intermediate peg to hold the tape in position so the subsequent interval can be measured off.

7. Return to your original position several metres back from the peg at chainage zero to line the peg in laterally as in step 4.

8. Repeat steps 6 and 7 until all intermediate pegs are in position. The pegs should all be in a straight line.

22.19 ARE PROFILE BOARDS RE-USEABLE?

Timber pegs and boards are re-used where possible and disposed of once they become worn out. Adjustable plastic profile boards are available but are more expensive to purchase.

22.20 CAN I USE GNSS FOR SETTING UP PROFILE BOARDS?

GNSS is a very quick and convenient way of setting up profile boards. The onboard software will provide you with positions of the pegs and guide you to the level at which the board should be attached.

22.21 IS THERE A STANDARD WAY OF COLOUR-CODING AND MARKING UP PROFILE BOARDS AND BATTER RAILS?

There is a standard convention for colour-coding and marking up profile boards and batter rails, however it is rarely adhered to nowadays. The full details of the system can be found in the CIRIA publication 'Setting out procedures for the modern built environment'.

For general site purposes, the main reason for the use of different colours is to avoid confusion when using double-headed travellers, stepped profiles or double-headed profiles on road cross-sections. The profile board cross-head should be sprayed the same colour as the corresponding traveller cross-head.

When colour is not needed for differentiation purposes, boards can be sprayed white to improve visibility and so that relevant information can clearly be marked on with marker pen.

When marking up profile boards, include information such as:

→ **chainage** - the chainage of the profile board

→ **offset** - the offset of the post and the line it is offset from, e.g. centreline of pipe or kerb

→ **height** - the height above a reference plane, e.g. 2m above dig.

When marking up batter rails, include:

→ **chainage** - the chainage of the profile board

→ **offset** - the offset of the post and the line it is offset from, e.g. centreline of pipe or kerb

→ **height** - height above the face of the batter

→ **distance for level control** - distance to toe or top of batter (from a point marked by a nail)

→ **slope** - the slope of the batter, e.g. 1 in 3.

WORKED EXAMPLE FOR PROFILE BOARDS FOR LEVEL EXCAVATION

Determine the required RL of the profile boards for the following:

Required formation level (dig level) = 137.230m

Maximum (EGL_{Max}) = 138.050m

Minimum existing ground level (EGL_{Min}) = 137.865m

Step 1: Draw a sketch.

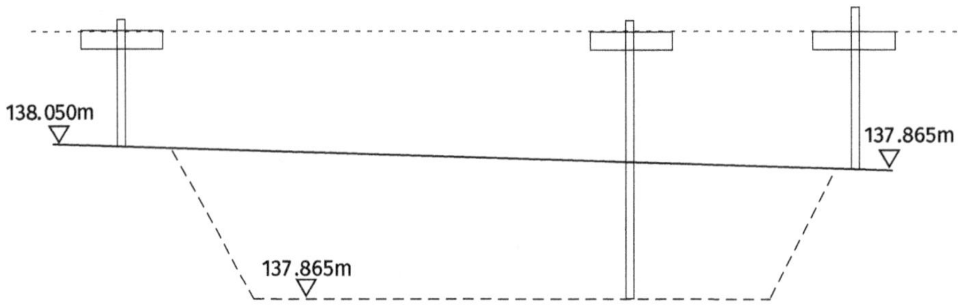

Step 2: Decide the traveller length

The traveller length is chosen based on the specific conditions. The profile boards should be at a convenient height for sighting through, say between 1 and 1.5m above the existing ground level, and the traveller should be a manageable length, a maximum of say 3m.

So that the profile boards are a minimum of 1m above ground level, the profile board needs to be at a minimum RL of EGL_{Max} + 1m = 139.050m.

If the RL of the board is min 139.050, then the traveller length would be 139.050 − 137.230 = 1.82m. Traveller length should be rounded up to the nearest 0.5m which would make the traveller length 2m.

Step 3: Calculate the required RL of the profile boards.

Required RL of profile boards = formation level + traveller length

= 137.230 + 2.000

= 139.230m.

Step 4: Check that the profile boards are at a convenient height above EGL

Check that for the chosen traveller length, the profile board at the minimum level will not be more than 1.5m above ground level:

Height above EGL = RL of the profile board − EGL_{Min}

= 139.230 − 137.865

= 1.365 < 1.5m therefore OK.

EXERCISE:

2. You have been asked to set out batter rails for the following arrangement:

    Existing ground level (EGL) = 20.500m and can be assumed to be level.

    Formation level = 16.500m

    The batters are at a slope of 1:2

    The distance from the CL of the dig to the toe of the batter is 4.3m

    P1 is 0.5m back from the top of the batter and P2 is 0.5m back from that.

    a. What is the plan distance from the CL of the dig to top of the batter?

    b. What are the RLs of the board at P1 and P2

CHAPTER 23

BATTER RAILS

23.1 WHAT ARE BATTER RAILS?

Batter rails, also known as sight rails, are an arrangement of timber posts and boards used for setting out the slopes of cutting and fill-in earthworks. They comprise vertical timber posts fixed into the ground with a sloping board attached which creates a sight line, either coincident with the projected line or a specified distance above it.

23.2 HOW DO I SET UP BATTER RAILS FOR CUTTING ON LEVEL GROUND?

In cutting, batter rails can be used with (Fig. 23-1) or without (Fig 23-2) a traveller. The position of the batter rails determines the position of the final excavation and must be in the correct relative position. There are two methods for setting up batter rails on level ground (Fig. 23-3); the levelling method and the taping method.

THE STEPS FOR SETTING UP BATTER RAILS FOR CUTTING ON LEVEL GROUND USING THE LEVELLING METHOD ARE:

1. Determine the slope of the batter

2. Determine the Reduced Level (RL) of the existing ground level (by taking a backsight to a TBM and a foresight to a point on the ground)

3. Determine the RL of the bottom of the excavation from drawings

4. Based on the slope identified in step 1, calculate the horizontal distance from toe of the batter to the top of the batter.

5. Determine the horizontal distance from the centreline of the excavation to the toe of the batter

6. Decide if a traveller will be used. If so, decide the length (0.3m – 0.5m recommended)

7. Choose the offset from the top of the batter to the inside peg, P1 (0.5m recommended)

8. Choose the distance from P1 to the outside peg, P2 (0.5m recommended)

9. Calculate the distances from the centre peg to the edge peg (step 4 plus step 5)

10. Calculate the distances from the centre peg to P1 and P2

11. Based on the slope and the traveller length, calculate the RLs of the top of the board at P1 and P2

12. Calculate the distance from the top of the board at P1 to the toe of the batter

13. Bang the edge pegs, P1 and P2 into the ground at their correct locations using a tape measure/total station/GNSS as appropriate

14. Take staff readings to determine the RL of the tops of P1 and P2

15. Determine the distance down from the top of each peg to the top of the board

16. Mark the distance down and attach the board at the required positions

17. Take a staff reading to check the that RLs of the board at P1 and P2 are correct

18. Tap a nail into the top of the board at peg 1, leaving it protruding so a tape measure can be hooked onto it

19. Mark up the board with its chainage, the distance from the nail to the toe of the batter and traveller length

20. Construct the traveller if you are using one

THE STEPS FOR THE TAPE MEASURE METHOD ARE:

1 – 10. The same as the levelling method above

11. Based on the slope and the traveller length, calculate the distance from EGL to the top of board at P1

12. Based on the slope, calculate the level difference between the RL of the board at P1 and P2

13. Calculate the distance from the top of the board at P1 to the toe of the batter

14. Bang the edge peg, P1 and P2 into the ground at their correct locations using a tape measure/ total station/ GNSS as appropriate

15. On peg 1, measure up from the existing ground level (EGL) using a tape measure and make a mark at the correct height.

16. Transfer a horizontal line from P1 to P2 and on P2, measure and mark the level difference calculated in step 12

17. Tap a nail into the top of the board at peg 1, leaving it protruding so a tape measure can be hooked onto it

18. Mark up the board with its chainage, the distance from the nail to the toe of the batter and traveller length

19. Construct the traveller if you are using one

Fig. 23-1 Batter rails in cutting (traveller)

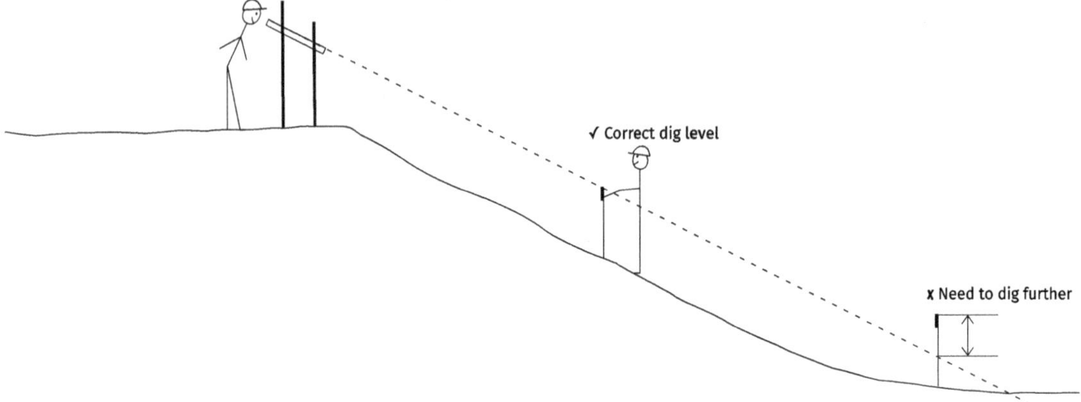

Fig. 23-2: Batter rails in cutting (no traveller)

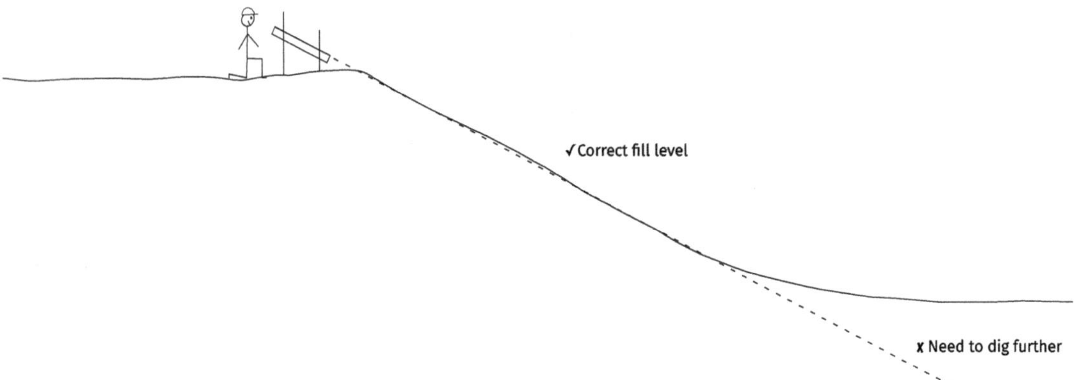

Fig. 23-3: Batter rails on level ground

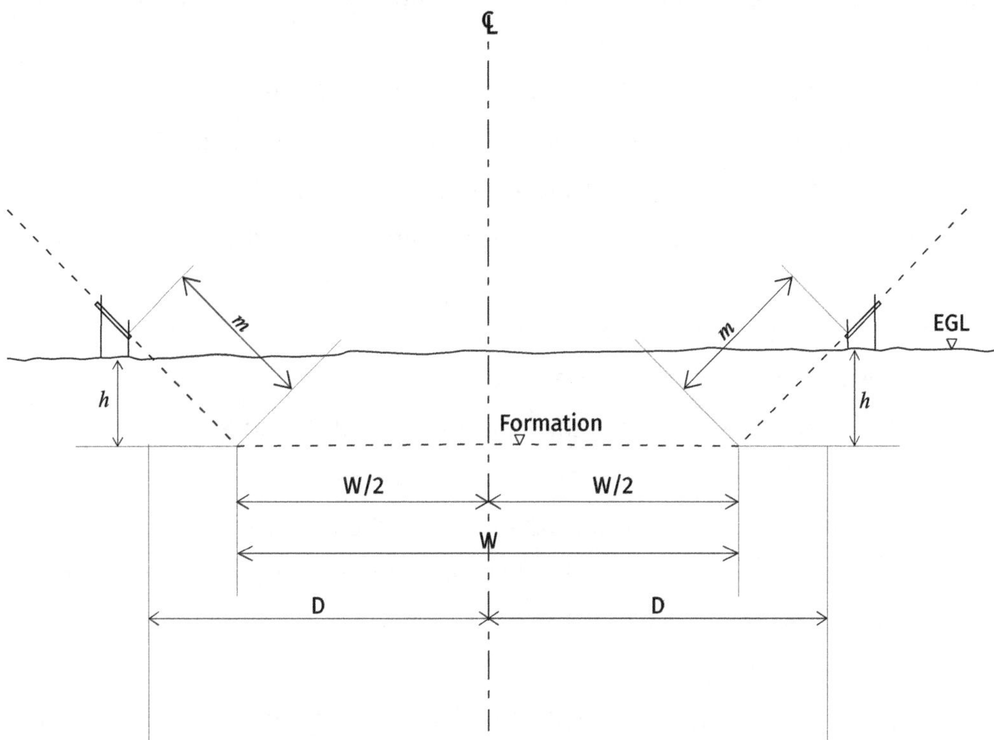

23.3 HOW DO I SET UP BATTER RAILS FOR FILL ON LEVEL GROUND?

A traveller is needed for all fill exercises. See Fig. 23-4. Additional rails can be placed on the constructed batters once some fill has been placed. The steps for setting up batter rails for fill, on level ground using the levelling method are:

1. Determine the slope of the batter

2. Determine the RL of the existing ground level (by taking a staff reading)

3. Determine the RL of the top of the batter

4. Determine the horizontal distance from the centreline of the fill to the top of the batter

5. Decide the traveller length. 1.5m is usually suitable

6. Choose the offset from the top of the batter to the inside peg, P1 (0.5m recommended)

7. Choose the distance from P1 to the outside peg, P2 (0.5m recommended)

8. Based on the slope identified in step 1, calculate the horizontal distance from toe of the batter to the top of the batter

9. Calculate the distances from the centre peg to the edge peg, P1 and P2

10. Based on the slope of the fill, calculate the projected RL at P1 and P2

11. Based on the traveller length, calculate the RLs of the top of the board at P1 and P2

12. Calculate the distance from the top of the board at P1 to the top of the traveller when it is at the top of the batter

13. Bang the edge pegs, P1 and P2 into the ground at their correct locations using a tape measure/total station/GNSS as appropriate

14. Take staff readings to determine the RL of the tops of P1 and P2

15. Determine the distance down from the top of each peg to the top of the board

16. Mark the distance down and attach the board at the required positions

17. Take a staff reading to check the that RLs of the board at P1 and P2 are correct

18. Tap a nail into the top of the board at peg 1, leaving it protruding so a tape measure can be hooked onto it

19. Mark up the board with its chainage, the distance from the nail to the toe of the batter and traveller length

20. Construct your traveller

The tape measure method described in section **23.2** can also be used for setting up batter rails for fill. Instead of setting the boards in relation to the RL of the top of the peg, a tape measure can be used to set the correct RL in relation to the EGL.

Fig. 23-4: Batter rails for fill

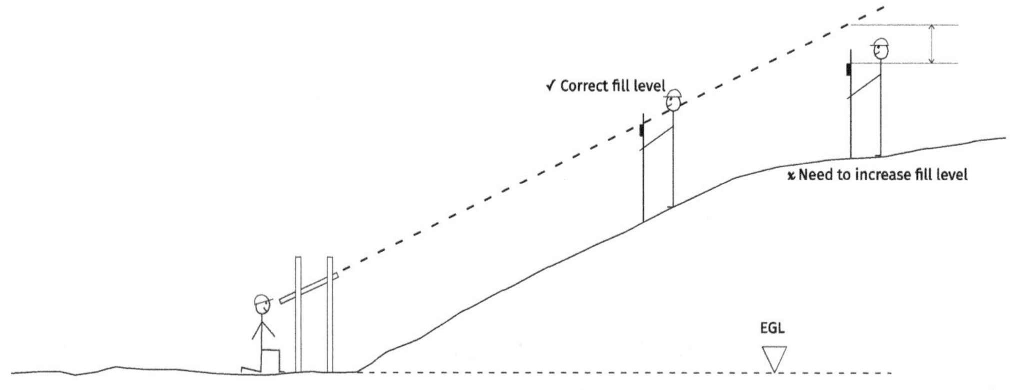

23.4 HOW DO I SET UP BATTER RAILS ON SLOPING GROUND?

When the existing ground is sloping, the arrangement is asymmetrical (Fig. 23-5) and the maths is a lot more involved. The associated maths is outwith the scope of this book.

Fig. 23-5: Batter rails on sloping ground

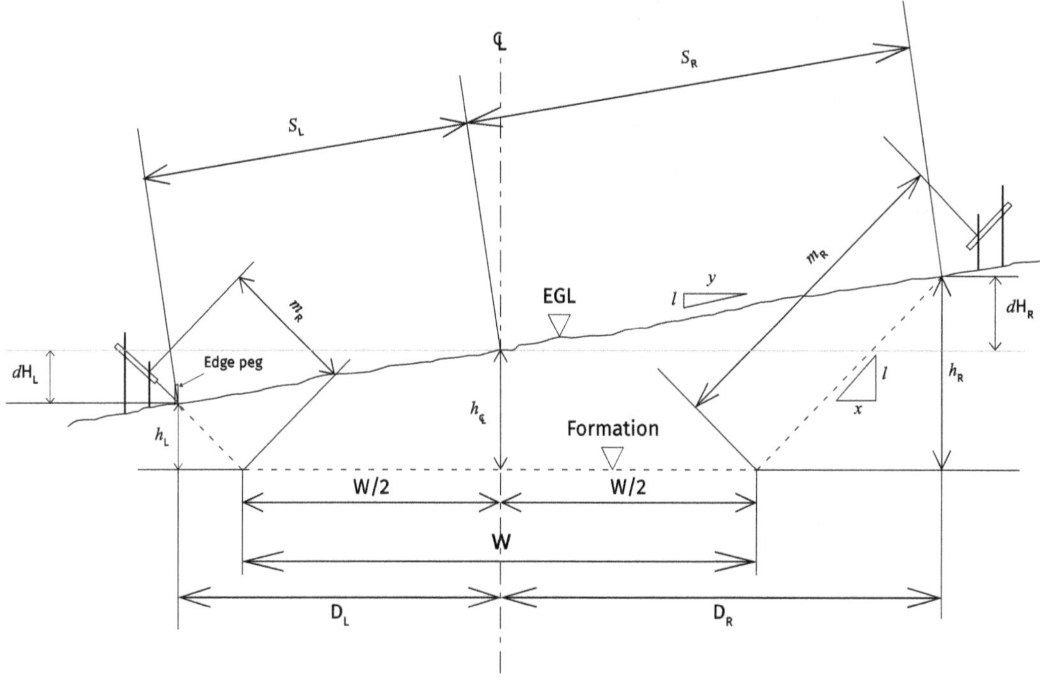

23.5 HOW CAN I USE BATTER RAILS FOR LEVEL CONTROL?

Batter rails can be used for approximate level control by providing tape measure distance from a nail at a fixed position. For cutting, the level set should be set a fixed distance, say, 200mm, above formation level. Appropriate level control, for example profile boards or laser level, should be installed for the final stage of the dig.

23.6 HOW DO I SET OUT THE POSITION OF PROFILE BOARDS AND BATTER RAILS?

The positions of profile boards and batter rails can be set out using a total station, GNSS or a tape measure.

23.7 ALTERNATIVES TO BATTER RAILS

It is not always necessary to construct batter rails, particularly for relatively short slopes or where the earthworks are for temporary stockpiles or excavations. Here are some quick and simple alternatives for situations when batter rails are not appropriate:

→ **use a template** – create a plywood template comprising a right-angled triangle cut to the correct slope. The template is placed on the face of the batter with a spirit level placed on the horizontal surface.

→ **by eye** - experienced ground workers and machine drivers can judge a slope by eye

→ **level change over slope distance** - convert slope distance into plan distance based on level difference between two points. Use the level to check vertical distance per horizontal distance.

WORKED EXAMPLE FOR BATTER RAILS ON LEVEL GROUND

In the arrangement on the opposite page, calculate:

a) the distance, D1, from the centreline (CL) to the edge peg (EP)

b) the distance, D2, from the CL to Peg 1

c) the RL of the board at Peg 1

d) the RL of the board at Peg 2

e) the distance, L1, from the top of the board at Peg 1 to the toe of the batter

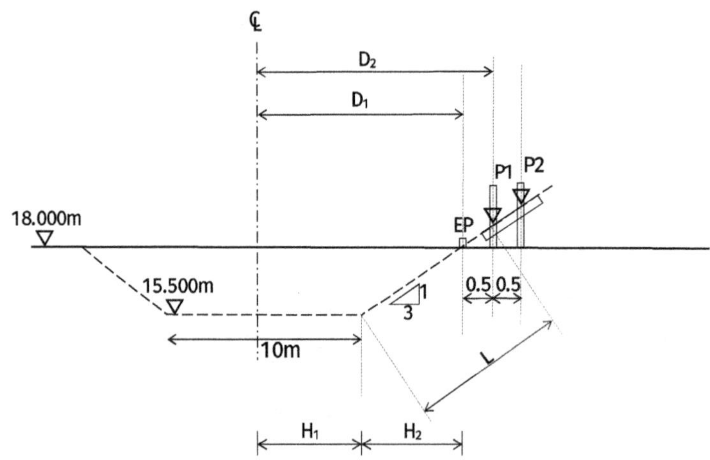

Answers:

a. The distance from the CL to the EP

Calculate the horizontal distance from the CL to the toe of batter, H_1.

$H_1 = 10/2 = 5m$

Calculate the horizontal distance from the toe of batter to the top of batter, H_2.

If $RL_{EGL} = 18.000m$

and $RL_{dig} = 15.500m$

then the vertical distance, $V_1 = 18.000 - 15.500 = 2.500m$

If the slope of the batter is 1:3, then the horizonal distance, H_2 is 3 x the vertical distance.

Therefore $H_2 = 3 \times 2.500m = 7.500m$

So the distance from the CL to the EP, $D_1 = H_1 + H_2 = 12.500m$

The distance from the centreline to Peg 1

$D_2 = D_1 + 0.5m$

$= 12.500 + 0.5 = 13.000m$

The RL of the board at Peg 1

For a slope of 1:3, for a horizontal distance of 0.5m, the level change, V2, will be 0.5 x = 0.167m.

$RL_{P1} = RL_{EGL} + V2$

$= 18.000 + 0.167$

$= 18.167m$

The RL of the board at Peg 2

For a slope of 1:3, for a horizontal distance of 0.5m, the level change, V3, will be 0.5 x = 0.167m.

$RL_{P1} = RL_{P1} + V3$

$= 18.000 + 0.167$

$= 18.337m$

The distance, L1, from the top of the board at Peg 1 to the toe of the batter

$H3 = 7.500 + 0.5 = 8.000m$

$V4 = 2.500 + 0.167 = 2.617m$

L1 is the hypotenuse of the triangle H_3 and V_4

Therefore $L1 = \sqrt{(H_3{}^2 + V_4{}^2)}$

$= \sqrt{(8.002 + 2.1672)}$

$= 8.288m$

EXERCISE:

1. You have been asked to set out batter rails for the following arrangement:

    Existing ground level (EGL) = 20.500m and can be assumed to be level.

    Formation level = 16.500m

    The batters are at a slope of 1:2

    The distance from the CL of the dig to the toe of the batter is 4.3m

    P1 is 0.5m back from the top of the batter and P2 is 0.5m back from that.

    a. What is the plan distance from the CL of the dig to top of the batter?

    b. What are the RLs of the board at P1 and P2?

CHAPTER 24

CONSTRUCTION ACTIVITIES

24.1 TOPSOIL STRIP

Setting out the topsoil strip (Fig. 24-1) will be one of the first setting out tasks on a new site. The topsoil strip often covers the footprint of the area of the works.

Related tasks include:

→ calculating the volume of topsoil to be stripped, stored, reinstated or sold. The topsoil will either need to be exported off-site or a suitable storage found for the duration of the project so that it can be reused at the end of the job. The area of topsoil can be calculated from drawings, or the area marked out and measured using the area function in the total station. Once the area is known, it is multiplied by the depth, to find the volume to be stripped.

→ surveying the final area stripped for payment/cost control purposes

→ marking out the perimeter of the topsoil strip with spray paint or timber pegs, clearly showing which is to be stripped as well as identifying areas of topsoil which must be protected from construction traffic during the works (it costs money if topsoil is unnecessarily damaged!)

→ marking out the areas where topsoil is to be reinstated

24.2 WHAT IS TOPSOIL?

Fig. 24-1: Topsoil strip

Topsoil is the layer of fertile ground at the surface of undisturbed terrain. It contains the roots of grass and other vegetation and is typically 50 – 300mm deep. The layer of soil underneath the topsoil is called the subsoil.

The depth of topsoil may have been determined as part of the site investigation carried out at the design stage, or for smaller projects, it may be determined at the outset of the works by digging trial holes.

Topsoil is a valuable commodity and is treated separately from other excavated material. Its value depends on its quality, location and accessibility.

One of the first activities at the start of any construction site is to 'strip' the topsoil from the entire footprint of the works. The topsoil is then either taken off site or stockpiled on the site for re-use at the end of the job.

The specification will contain details about required quality; the design may specify a depth.

If topsoil is required to be reinstated to certain areas on completion of the works, the required amount will be retained and stockpiled on site. Stockpiles should be strategically located where they won't get eroded or washed away and where fine silts won't be transported into watercourses.

Surplus topsoil is removed from site and may be sold depending on its value. This may not be economically viable for small quantities, in which case, it may be used as fill material if required.

24.3 HOW DO I MARK OUT THE AREA TO BE STRIPPED OF TOPSOIL?

1. Plan the topsoil strip. The perimeter of the area to be stripped may be specified on the construction drawings. If not, mark out on the Site Layout drawing, the footprint of the substructures, any temporary cuttings and embankments, stockpile storage areas, crane hardstanding areas, haul roads and site compound. Draw a single outer perimeter extending 1m beyond the total footprint. This can be done by hand or in AutoCAD if relevant.

2. Mark out the perimeter of the topsoil strip on the ground. You can do this by using the set-out or the Reference Line function in the total station, by using a tape measure or pacing to measure from existing features such as buildings, fence lines and kerb lines or a combination of all or any of these methods. Accuracy is not critical, so the required accuracy may vary from 300mm – 2m. You will need to use your engineering judgment here or ask for guidance from the works manager. The perimeter can be marked with spray paint lines, or simply pegs at the vertices. The machine drivers will be able to eye-in the line between the pegs.

24.4 AREA CALCULATIONS FOR TOPSOIL STRIP

There are two ways you can calculate the area of the topsoil strip:

→ CAD – draw on the perimeter and determine the area of the shape

→ Calculate the area by hand (see section **28.16**)

→ Use the area measurement function in the total station (see appendices for instructions)

Note: Do not rely on the Bill of Quantities for this information. It is ok to use it as a rough cross-check once you have completed your own calculations independently, but not as a reliable figure for construction phase purposes.

24.5 VOLUME CALCULATIONS FOR TOPSOIL STRIP

The steps are as follows:

1. Calculate the total area of the topsoil strip using one of the three methods in section **28.16**

2. Calculate the total volume of topsoil to be stripped by multiplying the area by the depth (either given on the drawings or from the depth of topsoil identified in trial holes).

3. From the drawings, calculate the area of topsoil to be reinstated at the end of the works. The depth of reinstated topsoil (typically 150mm) will be given on the drawings or in the specification. Use this information to calculate the volume to be stockpiled for the duration of the works. Take into account any expansion and compaction co-efficients.

24.6 EXCAVATIONS

There are several different components to setting out an excavation:

They are:

→ topsoil strip

→ the positions of the edges of the excavations

→ the batters (slopes) of the sides of the excavation

→ the level of the bottom of the dig (either level or graded)

→ 'benching'

→ access and egress to excavations

→ slopes in cutting or fill

24.7 HOW DO I KNOW WHAT GRADIENT TO SET THE SLOPING SIDES (BATTERS) OF THE EXCAVATION TO?

There are two scenarios when a batter would need to be set to a specific slope:

→ **scenario 1** - excavations and mounds forming part of the construction process, which will be backfilled or removed before completion of the works, e.g. excavations and stockpiles

→ **scenario 2** - permanent cutting and fill which form part of the completed works, e.g. landscaping and road-cutting and embankments.

For the temporary batters in scenario 1, the Contractor will most likely be responsible for determining a safe slope. These will be estimated in advance based on anticipated soil properties, safe slopes and footprints, but may be subject to change based on actual ground conditions once the excavations have started. If you are unsure, consult the Site Manager.

For the permanent batters in scenario 2, these slopes will be shown on the design drawings. If the slopes are not given explicitly, you can calculate them by identifying the level change over a given horizontal distance.

24.8 WHAT SETTING OUT IS INVOLVED IN EXCAVATIONS?

The steps involved in setting out excavations are:

1. Carrying out the calculations to determine the positions of shoulders and toes of batters based on the required slopes (allowing for working room and clearances as necessary).
2. Marking the edges of the dig with spray paint.
3. Setting up profile boards or a laser level to control the level of the bottom of the excavation.
4. Setting up batter rails to control the slopes of the sides of the excavation.
5. Checking the excavation with an automatic or laser level.

24.9 WHAT IF THE DIG LEVEL IS TOO HIGH OR TOO LOW?

Once earth has been dug out, it loses strength, so in the case of accidental over-digging resulting in the dig level being too low, the excavated material cannot simply be put back in again. This means additional blinding concrete or capping must be used. This increases construction costs since per unit volume, the cost of blinding concrete or crushed stone is significantly higher than the cost of digging out earth.

If the dig level is too high, and only discovered when it is time to pour the blinding this will cause delays as extra digging will need to take place before the blinding can be poured.

24.10 BLINDING CONCRETE

Blinding is a thin layer of non-structural concrete which provides a flat, stable surface on which to begin the construction of reinforced concrete foundations. The blinding creates a clean, dry, sound surface for setting out and constructing formwork and steel reinforcement.

The purpose of the blinding is to take up any undulations in the formation (the exposed upper surface of the excavation). It acts as a seal to the underlying material and prevents dirt and mud interfering with the structure. The upper surface will have a smooth(ish) finish and the underside will depend on the nature and quality of the formation dig. The thickness of the blinding is specified on the construction drawings and is typically a minimum of 50mm.

Blinding concrete (typically C30) is cheaper and much more workable (free flowing) than structural grade concrete (C50).

There is no minimum or maximum area for blinding and each section is usually poured between several days and a few weeks before it is needed. In wet conditions where there is a risk of the sides of the excavation being washed in, the blinding is placed as late as possible.

Because the blinding is only 50mm in depth, no formwork is needed. The edges of the blinding may be the extents of the excavation or may extend approx. 1m beyond the footprint perimeter of the substructure, whichever is most convenient and cost-effective.

It is ok to create cold joints in blinding. It is not necessary that each section is placed in one continuous pour. To avoid wastage and having to dispose of surplus concrete from other pours, it can instead be redirected to parts of the site where blinding will soon be required.

Blinding concrete does not require curing as it is too shallow for bleeding to occur and its structural strength is not critical.

24.11 HOW SHOULD I SET OUT BLINDING?

BLINDING LEVEL METHOD 1: SET THE TOP OF BLINDING PINS TO THE TOP OF CONCRETE LEVEL

This method involves tapping a grid of steel pins into the formation until the top of the pin is level with the required level of the blinding. A spacing of 2m is typical, however this is a guide. Accurate setting out of the position of the pins is not necessary. Pacing out the positions is sufficiently accurate.

The advantages of this method are:

→ the pins are not sticking up afterwards so do not cause a trip hazard or clash with the steel reinforcement in the bottom of the slab

→ there is no need to saw them off as they are not sticking up

→ it is easy to check the level of all the pins once they are set to the right level, as the required staff reading will be the same on top of every pin

The disadvantage of this method is that once the blinding covers the top of the pin, the pin is hidden and there is no visual check as to how much higher the top of concrete is above the correct level.

Steps:

1. Set up your optical level and determine the RL at random spot heights across the area of the blinding in relation to benchmarks. Identify the lowest formation RL.

2. Calculate the maximum blinding depth which will occur at the lowest formation RL.

3. Tap steel pins into the ground in a 2m grid (or whatever interval the foreman has requested), leaving at least 100mm of steel pin above the blinding level. E.g. if the maximum blinding depth is 70mm, tap in the pins so they are protruding from the ground at least 170mm.

4. Set up your level and determine the HPC.

5. Calculate the required staff reading for the correct blinding level.

6. Mark the staff at the required reading with a pencil line or electrician's tape.

7. Have your assistant hammer the pin into the ground, holding the staff on top of the pin after every few taps. Have the pin tapped into the ground until the mark on the staff is level with horizontal cross-hairs.

8. Once all of the pins have been tapped into the ground, set up the level at a different HPC, recalculate the required staff reading, mark the staff and check all of the pins.

BLINDING LEVEL METHOD 2: TOP OF BLINDING PINS PROTRUDING

This method involves tapping a grid of steel pins into the formation until the top of the pin is between 50mm and 200mm above the required blinding level and placing a strip of electrical tape round the pin such that the bottom of the tape is set to the top of the blinding level.

The advantage of this method is that the pins and the tape remain above the level of the concrete giving a visual check as to the accuracy of the concrete pour.

The drawbacks of this method are that the pins are left sticking up out of the blinding. The protruding pins need to be sawn off as they cause a temporary trip hazard and may obstruct the bottom layer of steel reinforcement.

1. Set up your optical level and determine the RL at random spot heights across the area of the blinding in relation to benchmarks. Identify the lowest formation RL.

2. Calculate the maximum blinding depth which will occur at the lowest formation RL.

3. Tap steel pins into the formation in a grid. A spacing of 2m is typical, however this is a guide. Accurate setting out of the position of the pins is not necessary. Pacing out the positions is sufficiently accurate. Leave at least 100mm of steel pin above the blinding level. E.g. if the maximum blinding depth is 70mm, tap in the pins so they are protruding from the ground at least 170mm.

4. Set up your optical level and determine the HPC by backsighting to a TBM.

5. Take a staff reading to the top of a pin.

6. Calculate the RL of the top of the pin.

7. Calculate the distance down the pin to blinding level.

8. (Option 1): If you have a reliable assistant, have them measure the correct distance from the top of the pin and place electrician's tape around the peg with the underside of the tape at the correct level.

9. (Option 2): If you don't have a reliable assistant, decide on a referencing system and allocate each pin a unique reference code. For each pin, record the required distance down each pin in the distance column of your yellow field-book. Once you have taken staff readings to all pins, apply electrician's tape to each of them with the underside at the correct level.

10. Using a tape measure, check the distance to the underside of the electrician's tape on each pin.

BLINDING LEVEL METHOD 3: SET SCREED RAILS

This method involves fixing screed rails (parallel sets of scaffolding tubes) in place, with their crowns level with the top of the blinding. A 'screed' (a timber beam with vertical handles at each end) is then 'tamped' (dragged/tapped up and down) across the tops of the top of the tubes. The screed rails act as a physical fixed guide for the screed.

The advantages of this method are:

→ the screed rails are fixed in position and can be checked in advance of the blinding pour, reducing the risk of error

→ the screed spans between the screed rails, minimising the chance of localised dips and bumps

→ the screed rails remain in position for the duration of the concrete pour, which means the as-built levels of the blinding can be checked whilst the concrete is green.

The disadvantage of this method is that it takes time to prepare.

The steps are as follows:

1. Have the screed rails fixed approx. 3m apart (or slightly less than the length of the screed) at the approx. RL of the blinding.

2. Set up your level and determine its HPC.

3. Calculate the required staff reading for the correct blinding level.

4. Check the levels of the screed rails and have them adjusted to exactly the right level.

5. Ensure they are fixed firmly in position, so that they do not move with the force of the screed.

6. Once all the screed rails have been fixed to the correct level, set up the level at a different HPC, recalculate the required staff reading and check the level at the beginning and end of each section of screed rail.

24.12 HOW DO I CHECK BLINDING?

Once the blinding has gone off, the as-built levels should be checked and marked on the blinding prior to steel-fixing stage. This allows the steel-fixers to set the packers to the underside of the steel correctly. Since the reinforcement bars are pre-fabricated to a fixed slab depth, if the bottom layer of steel is set to the correct level, then the top layer will also be the correct level (although it goes without saying that this should be checked once the steel has been fixed in position).

If the blinding levels are not marked prior to the steel-fixing stage, and the blinding is assumed to be correct when it is not, then any localised high or low points will be transferred to the top layer of steel. The depth of concrete cover to the top layer of steel is critical, so if the steel is too high or too low, then further work will be required to adjust it to the right level after it has been fixed. This may require re-fixing of the steel or a crane to lift the steel cage so that extra packers can be inserted. It is much easier to set the levels correctly in the first place than to adjust once the steel is in place.

To check blinding levels:

1. Set up your level and determine its HPC.

2. Calculate the required staff reading for the correct blinding level.

3. Have your assistant hold the staff at intervals in a grid of approx. 2m (pacing is sufficiently accurate).

4. At each point, shout out to your assistant the actual level in relation to the required level. Have them mark the spot with spray paint. If the level is exactly right, they should mark it with a tick or '0'. If it is too high or too low, they should mark the amount with a 'H' or 'L' accordingly. So 10mm too high would be marked 10H and 15mm too low would be marked 15L.

24.13 WHAT CAN GO WRONG WITH BLINDING?

In terms of setting out, there are several things that can go wrong with blinding:

→ the as-built level of the blinding is too high

→ the as-built level of blinding is too low

→ the blinding is too low in some places and too high in others

→ the blinding is set out in the wrong position.

24.14 IS IT BETTER TO AIM TO SET BLINDING SLIGHTLY LOW OR SLIGHTLY HIGH?

It's better to set it to the right level! However, high blinding is more disastrous than low. When the blinding is set too low, the slack can be taken up either by adding additional blinding concrete or by increasing the cover to the underside of the steel reinforcement of the base slab. When it is set too high however, the effects are more serious. At worst, remedial work could involve having to break out the incorrect blinding or requiring the levels for the rest of the works to be redesigned.

The best-case scenario is that there is scope to reduce the cover to steel at the underside of the slab, or to raise the reduced level of the entire project by a constant amount. In this latter case, if there are no issues with tying into existing features, the benchmarks can be adjusted accordingly to avoid recalculating design levels.

24.15 STRIP FOOTINGS

A strip footing, also known as a strip foundation, continuous foundation or ground beam, is a long, narrow, shallow concrete foundation which supports a wall or series of columns.

Strip footings are usually specified when the bearing capacity of the soil is good, and the load is relatively low or when it is cheaper to construct a strip footing than a series of pad foundations. Some strip footings include reinforcement (Fig. 24-2), and some consist of concrete only. The edges of strip footing may be formed with formwork, shuttering or stiff plastic sheets against the vertical edges of the excavated soil (Fig. 24-3).

Fig. 24-2: Strip footings including reinforcement

Fig. 24-3: Plastic sheets against the vertical edges of the soil

24.16 WHAT IS INVOLVED IN SETTING OUT A STRIP FOOTING?

1. Once the topsoil strip has been completed, set out the gridlines and/or offsets on pegs or H-profiles as described in section **21.11**.

2. Determine if there are any voids, box-outs or drainage service ducts affecting the foundation.

3. Using a combination of a tape measure, string line and/or the Reference Line function in the total station, mark the outline of the trench, or an offset from the outline of the trench, by referencing the gridline and offset pegs set up in step 1.

4. Mark out the edges of the trench with spray paint. If the ground is not suitable for spraying, mark the corner points with pegs or steel pins.

5. Mark the locations of any stepped level changes or increased depth of the bottom of the foundation.

6. Set the dig levels using a laser level or profile boards.

7. When the excavation is complete (Fig. 24-4), check the as-built dimensions and levels.

8. Based on both the drawings and as-built trench dimensions, calculate the required volumes of blinding and structural concrete needed for each pour.

9. If blinding is being used (Fig. 24-5), mark out the positions if any recesses to allow for service ducts entering from below. Set the level using a laser level.

10. Once the blinding has gone off, check that the level is within tolerance.

11. If the footing is to be constructed in sections, mark the positions of any stop-ends.

12. If the footing is to contain reinforcement, at approx. 1m spacing, mark up how high or low the blinding is in relation to required level (as describe in section **24.12**).

13. If the footing is to contain reinforcement or requires shuttering to be constructed, mark either the centreline and ends of the footing or mark the corners of shutters with nails in the blinding. The joiners and steel-fixers can work to these reference marks to construct the reinforcement and shuttering.

14. Check the as-built positions of the reinforcement and shuttering.

15. Once the shuttering is constructed, set out the positions of anchor bolts if relevant (see section **24.22** or **24.35**).

16. Set the top of the concrete level with marks on the shuttering and/or the laser level.

17. Record the as-built positions of the extents of the footing and the positions of bolts.

18. Once the footings for the entire building are complete, record the top-of-concrete levels at centres of bolt groups.

Fig. 24-4: Excavation complete

Fig. 24-5: Blinding poured for pad foundation

24.17 ANCHOR BOLTS

Anchor bolts (also called holding-down bolts) are long vertical bolts, which attach universal steel columns to concrete foundations (Fig. 24-6), resisting uplift forces in the structure.

The bolts are integral with the concrete base, meaning that they must be set in position as part of the base construction.

The bolts can be installed by drilling holes in the correct positions once the base has been poured, or more commonly, by casting them into position during the placement of the base concrete. When casting them into position, the lower part of the bolt is cast into the concrete during pouring of the base slab (Fig. 24-7). Sacrificial waxed cardboard cones, known as bolt cones, surrounding the upper part of the bolt, are stuffed with rags before the base slab concrete is poured. Once the concrete has gone off, the rags and bolt cones are removed, leaving a gap between the bolt and the concrete. This means that there is a small amount of 'play' in the tops of the bolts. The bolts are secured in their final position using epoxy grout once the structural steel is in position and tightened up.

The steel columns and beams of steel-framed buildings fit together like a giant jigsaw, with very little play between the component parts. As such, a great deal of skill, care and engineering judgment is required for setting out the bolt positions in the base slab.

As well as ensuring the required tolerances between column spacings are met, consideration must be given to how the columns being set out will tie in with any interfacing structures.

Fig. 24-6: Anchor bolts

Fig. 24-7 Bolts in concrete

24.18 WHAT IS A BOLT GROUP?

'Bolt group' is the collective term for all the bolts of the base plate of a single structural steel column. Column base plates are pre-drilled to accommodate the bolts.

24.19 WHO SETS OUT THE ANCHOR BOLT POSITIONS?

The main contractor is responsible for ensuring the bolts are in the correct positions and the top of concrete is set and poured to the correct reduced level.

Typically, once the base slab has been constructed, the main contractor will mark out the column centrelines (on both axes) and provide the steel-erector with information about the lowest and highest points of the top of the concrete. The main contractor is responsible for the final checking once the steelwork has been erected. Any lack of fit due to poor setting out of the bolt or column positions is the main contractor's responsibility.

24.20 WHAT ACCURACY IS REQUIRED FOR BOLT POSITIONS?

Since universal beams are manufactured to within a tolerance of +/- 3mm and act as fixed length spacers between the columns, accuracy is critical when setting out bolt positions.

According to the National Structural Steelwork Specification for Building Construction 6th Edition, bolt positions are required to have a relative position accuracy of +/- 10mm when bolt cones are used, or +/-3mm when set in a fixed position. This can be used as a rule of thumb, however any tolerances given in the specification for your contract would take precedence over these figures.

24.21 HOW DO I SET OUT ANCHOR BOLTS?

There are two common ways to position cast-in anchor bolts in the concrete base. Both methods involve creating a 'bolt-box'. This consists of a plywood template of the bolt group arrangement, which holds the tops of the bolts in the correct relative positions (Fig. 24-8). The underside of the bolts, which will be submerged in the concrete, are protected by cardboard cones which are narrow at the bottom and wider at the top. The cardboard cones are held in place by a washer plate and a nut, and are stuffed with rags to prevent concrete entering the void. The short section of bolt protruding from the bottom of the cone, the washer plate and the nut are covered with protective tape so that the bolt remains free to pivot from that point. Immediately prior to erecting the steel columns, the template and cardboard cones are removed. Due to the void created by the cones, there is sufficient movement at the top of the bolts for them to fit exactly into the holes drilled in the base plate of the column. Fig. 24-9 and Fig 24-10. show a bolt box in position before and after the concrete is poured. Once the columns are in position, the voids surrounding the bolts (Fig. 24-11) are filled with grout.

The first method is known as 'hanging the bolts' and involves securing the bolt-box in the correct position prior to the placement of the surrounding concrete (Fig. 24-12 and Fig. 24-13). The second method is known as 'floating (or plunging) the bolts'. In this method, the concrete is poured, then the bolt box is placed into the wet concrete at the correct position.

Do not rely on measuring from shutters to set bolt positions unless you have verified their position.

Avoid cumulative errors when setting out bolt-boxes by measuring all distances from a single point (Fig. 15-7)

Use the tie distance/compute inverse function or a tape measure liberally for checking between as-set-out positions. Don't just check to the adjacent columns, but to all other column positions. Check between the extents of rows, columns and diagonals as well as intermediate columns.

Fig. 24-8: Bolt Box Section

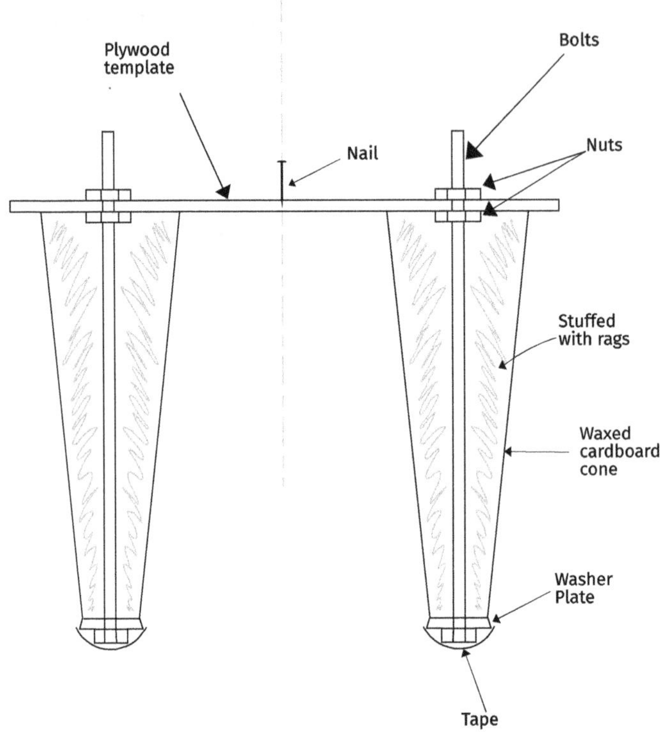

Fig 24-9: Before concrete is poured

Fig 24-10: After concrete is poured

Fig 24-11: Voids surrounding the bolts

Fig 24-12: Bolt Box Level

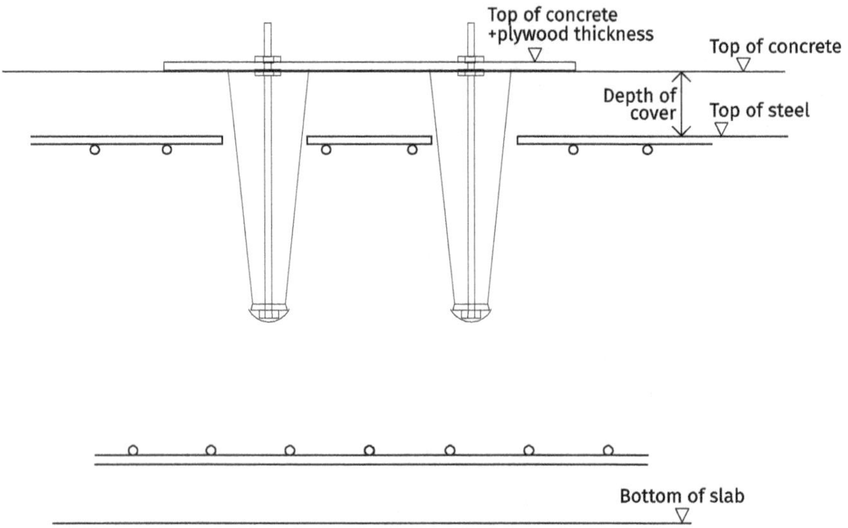

Fig. 24-13: Bolt box secured in the correct position

24.22 WHAT IS THE PROCEDURE FOR HANGING BOLTS?

When using the hanging method to set out bolts, the bolt-box is secured in the correct position before concrete is poured. The bolt box may be fixed to the surrounding reinforcement (Fig 24-13), or to transverse timber struts attached to either the shuttering or pegs driven into the adjacent ground. (Fig.(s) 24-13,24-14, & 24-15)

Rigorous checking is carried out as the pour reaches its final level, to ensure that the placement of the concrete has not caused any movement of the bolt-box.

Sections of steel reinforcement may need to be cut out to facilitate the positioning of the bolt cones. If this is the case, and the construction drawings do not cater for this scenario, permission should be sought in writing from the designer.

The procedure for hanging bolts is as follows:

1. Create a diagram of the bolt group arrangement for the joiners, including details of the bolt diameter and length and the quantity of bolt-boxes required. The joiners can then construct the bolt boxes (Fig. 24-16). The plywood template will have the centrelines of the bolt group marked on.

2. If it has not already been done, set out the relevant gridlines and offsets as described in section **21.11**.

3. Either side of where the template will be located, mark the centreline of the bolt group, in both the longitudinal and transverse directions (Fig. 24-17 Steps 1 and 2). There are several ways to set out these centreline marks. See section **24.23**.

4. Have the joiners fix the template in the correct position (Fig. 24-17 Step 3), matching up the marks on the centreline of the template with the centreline marks on the reinforcement.

5. Check the as-set-out position of each bolt in relation to the centreline marked on the template.

6. As an additional check, once the template is secured in place, use a tape measure to measure the perpendicular distance from the adjacent shutter to the centreline of the bolt group at each edge of the template. Mark this distance on the shutter. This means that the position of the template can easily be checked without the total station being set up during the pour.

Fig. 24-14: Hanging Bolts Plan

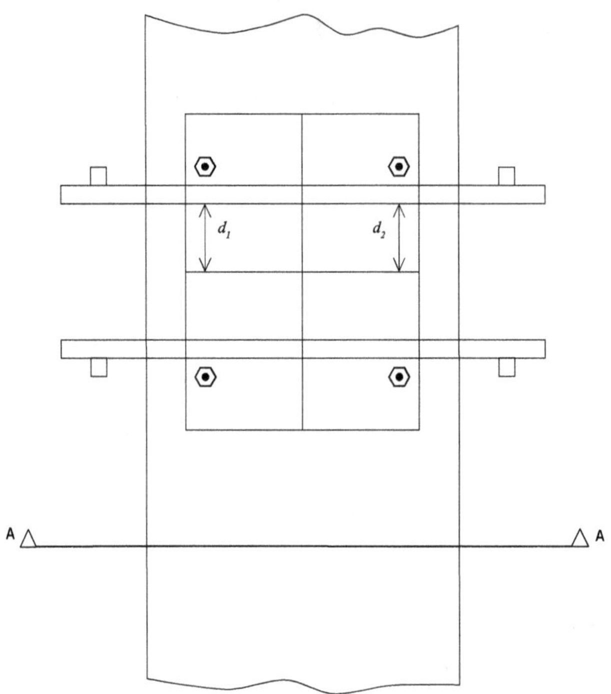

Fig. 24-15: Hanging Bolts Section

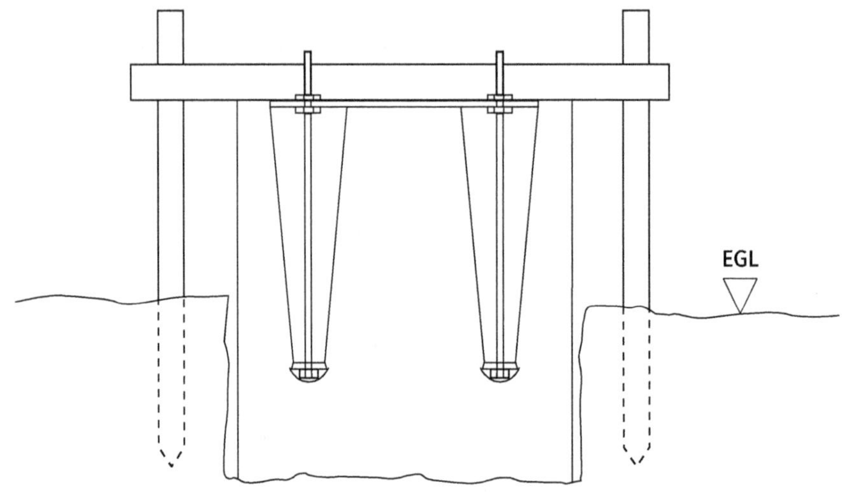

Fig. 24-16: Bolt boxes

Fig. 24-17: Fixing Bolts

Step 1

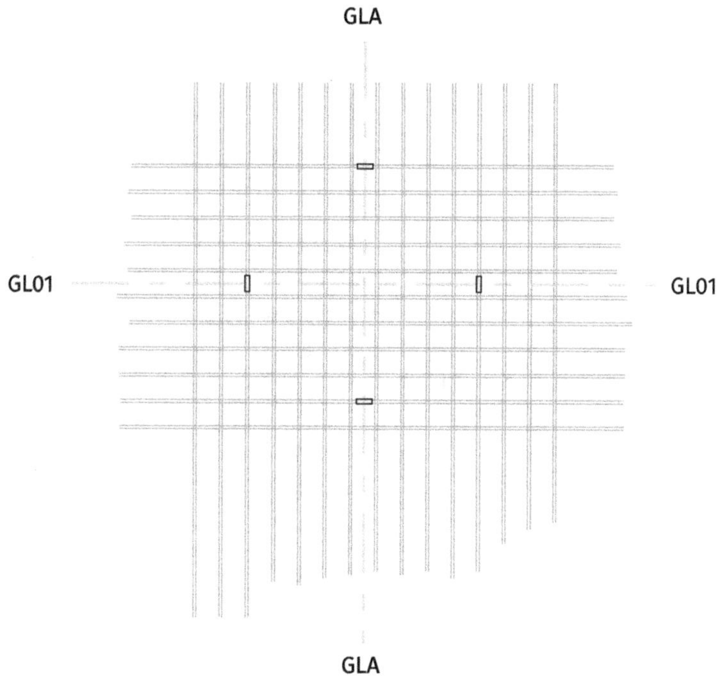

Step 2

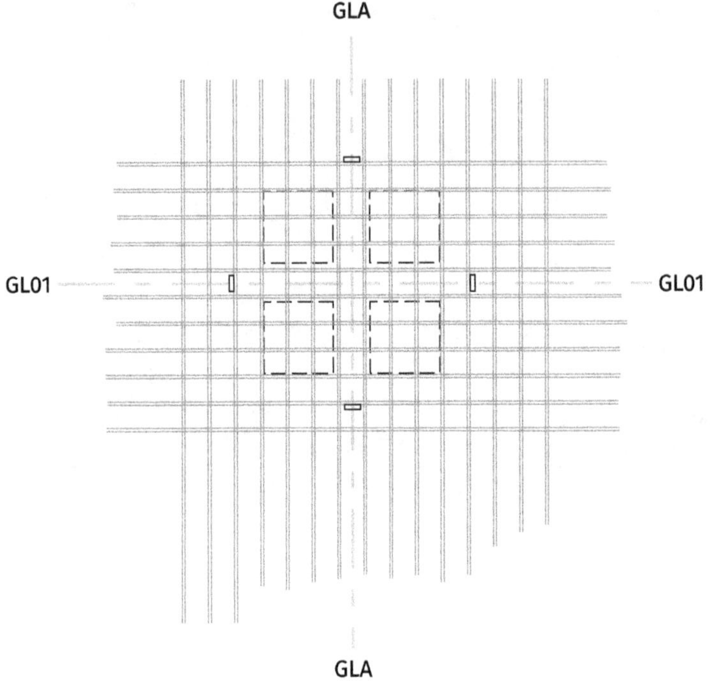

Step 3

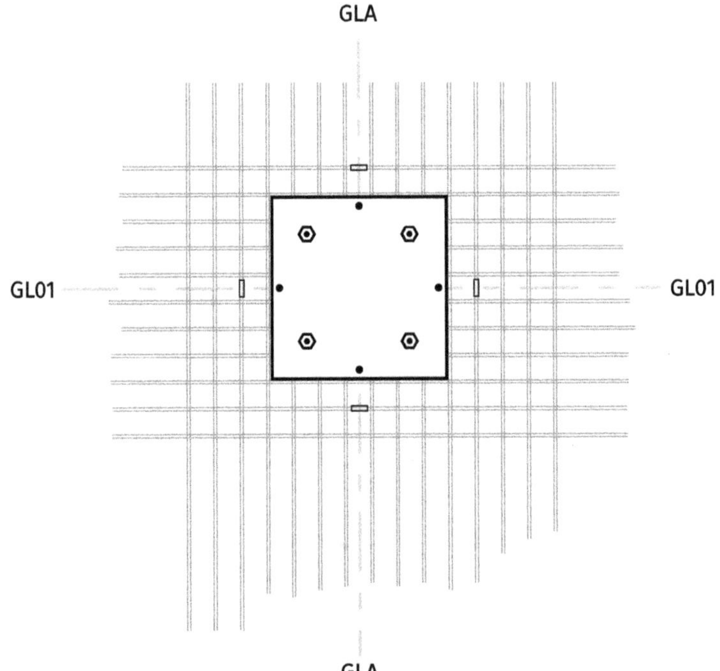

24.23 HOW TO SET OUT X-AXIS CENTRELINE MARKS ON REINFORCEMENT

METHOD 1- SET UP THE TOTAL STATION ALONG THE CENTRELINE:

1. Install protruding nails on the X-axis centreline of the template, at the near and far edges.

2. Install Nail 1 on the centreline of the X-axis at least 10m back from the closest bolt group (using one of the methods described in section **21.11**).

3. Install Nail 2 on the centreline of the X-axis at the far end of the line of bolt groups. This may be in a peg, H-profile, blinding or the top edge of the shuttering.

4. Set up the total station over Nail 1.

5. Line up the single vertical cross-hair on Nail 2.

6. At each location where a centreline mark is required, have your assistant make a patch of yellow wax crayon on the reinforcement bar or wrap some electrician's tape round the bar.

7. Have your assistant move the tip of their pencil left or right along the reinforcement until it is line with the vertical cross-hair.

8. Have them make a mark at this point and then hold the tip of their pencil on the mark so that you can check it is in the correct position.

9. Set out the Y-axis marks.

10. Re-check both axes once the template has been secured in place. Adjust if necessary.

11. Re-check the template position in both axes within 5 minutes of the concrete being poured to the correct level.

METHOD 2: REFERENCE LINE FUNCTION:

1. Set up a baseline as described in section **21.4**.

2. Set up and level your total station at a convenient point where you have visibility of all the bolt-groups to be set out.

3. Input the baseline into the Reference Line function by sighting to the mini-prism at Point 1 and Point 2 of the baseline.

4. If you don't have a reliable baseline set up for step 3, set up the position and orientation of the instrument. If you use the resection function for this, check extensively in relation to bolts set out from previous set-ups.

5. Define the line you wish to set out, in relation to the baseline (Fig. 24-17).

6. Have your assistant move the prism left or right until the prism is on the correct line.

7. Have them make a mark at this point.

8. Y-axis marks can also be set out in the same way, from this same set-up position.

9. Re-check both axes once the template has been secured in place. Adjust if necessary.

10. Re-check the template position in both axes within 5 minutes of the concrete being poured to the correct level.

Fig. 24-18: Setting out bolts from a baseline

24.24 HOW TO SET OUT Y-AXIS CENTRELINE MARKS

The Y-axis of each bolt group should be marked on the reinforcement and on each edge of the shuttering so that the template position can be checked after the top layer of steel has been covered by concrete.

This allows the joiners to adjust the template in the Y direction, before, during and after concrete pouring, without the need for an engineer to set up the total station on that line. If there are no suitable shutters, H-profiles or pegs should be put in place with nails on the Y-axis.

Y-axis marks can be set out using the Reference Line function or by using the tape and string lines in relation to reliable gridline reference points or as-built shutter positions.

24.25 WHAT IS THE PROCEDURE FOR FLOATING BOLTS?

When using the floating method to set out bolts, as soon as the concrete is poured, the bolt-box is pressed down into the concrete at the correct position (Fig. 24-18). This is a higher-risk method than hanging the bolts, as there is a small window of opportunity (less than one hour once the concrete has been poured) to set and check the position of each bolt-box. Advance preparation is a crucial element of this method. See Fig. 24-19.

METHOD 1: SET UP THE TOTAL STATION ON THE CENTRELINE

1. Install protruding nails on the X-axis centreline of the template, at the near and far edges.

2. Install Nail 1 on the centreline of the X-axis at least 10m back from the closest bolt group (using one of the methods described in section **24.23**).

3. Install Nail 2 on the centreline of the X-axis at the far end of the line of bolt groups. This may be in a peg, H-profile, blinding or the top edge of the shuttering.

4. Set up the total station over Nail 1.

5. Line up the single vertical x-hair on Nail 2.

6. Once the concrete has been poured to the correct level (see section **24.49** for level control of RC slabs), have the joiners move the template left or right until both nails on the X-axis of the template are in line with the vertical x-hair.

7. The Y-axis of the template can be set by placing marks on the shutters, pegs or H-profile in advance of the pour, or by using the Reference Line function as described in method 2.

8. Re-check the X-axis

9. Re-check the template position in both axes within 5 minutes of the concrete being poured to the correct level.

METHOD 2: REFERENCE LINE FUNCTION:

1. Set up a baseline as described in section **24.23**.

2. Set up and level your total station at a convenient point where you have visibility of all the bolt-groups to be set out.

3. Input the baseline into the Reference Line function by sighting to the mini-prism at Point 1 and Point 2 of the baseline.

4. If you don't have a reliable baseline set up for step 3, set up the position and orientation of the instrument. If you use the resection function for this, check extensively in relation to bolts set out from previous set-ups.

5. Define the line you wish to set out, in relation to the baseline.

6. Once the concrete has been poured to the correct level (see section **24.49** for level control of RC slabs), have the joiners move the template left or right, holding the prism at the front and back edges of the template, until they are both on the X-axis.

7. Have them move the template up and down the Y-axis and hold the prism on each edge of the template until both are on the correct line.

8. Re-check the X-axis.

9. Re-check the template position in both axes within 5 minutes of the concrete being poured to the correct level.

Fig. 24-19: Floating Bolts Section

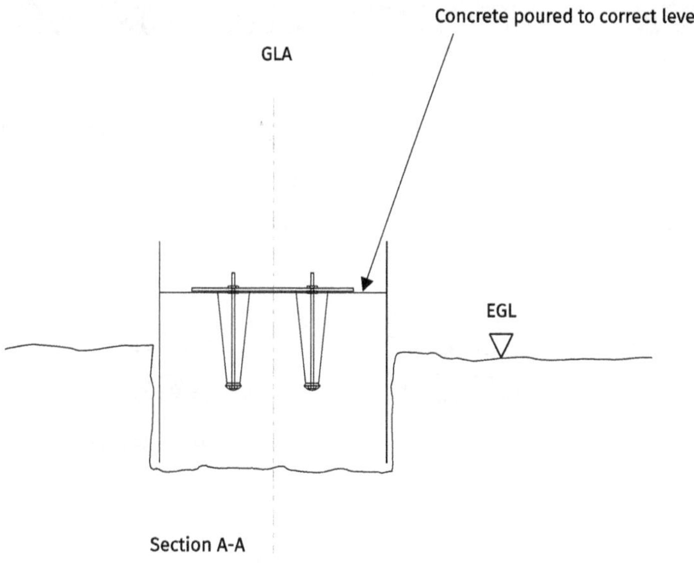

Fig. 24-20: Floating Bolts Plan

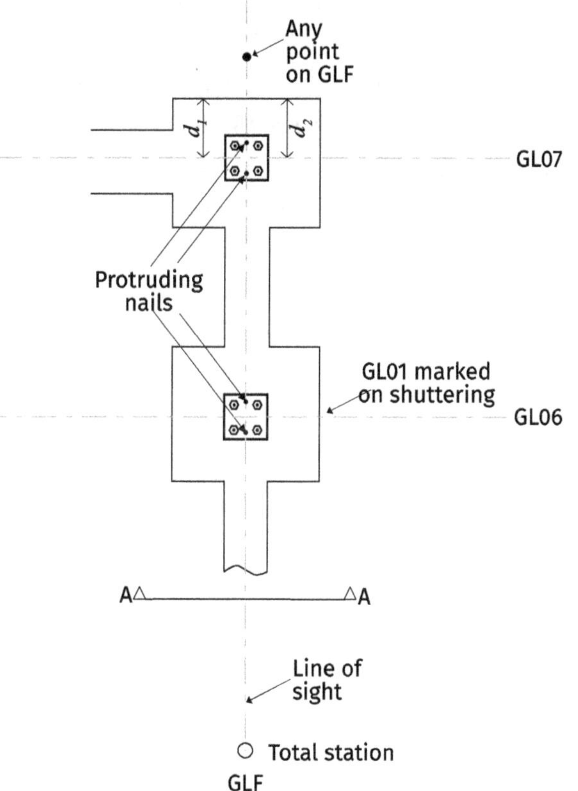

24.26 WHEN SHOULD THE FLOATING METHOD BE USED AND WHEN SHOULD THE HANGING METHOD?

'Floating the bolts' is used when there is no reinforcement steel, shutters or H-profile to attach the bolt-box to.

'Hanging the bolts' is used when there is a steel cage, or shuttering, or when floating is too high-risk.

24.27 WHAT IF THE STEEL REINFORCEMENT IS IN THE WAY OF THE BOLT CONES?

If there are steel reinforcement bars obstructing the bolt cones, the steel will need to be cut away or adjusted to create suitable gaps. Always seek written agreement from the designer.

24.28 WHEN SETTING OUT COLUMN CENTRES, WHICH IS MORE CRITICAL: PRIMARY CONTROL POINTS OR INTERNAL ACCURACY?

Internal accuracy is usually most critical when setting out a steel-framed building. Take into account as-built and as-set-out column positions. Full consideration should be given to interfaces with other buildings, structures and tie-ins. Use your engineering judgment based on all the information you have available to you.

24.29 WHAT IF THE BOLT POSITIONS ARE SET OUT INCORRECTLY?

If you discover that the bolt positions have been set out incorrectly, don't be tempted to force the steelwork into position or to 'make it fit'. Even though this may seem like a good way to fix the problem, be aware that sideways forces can shear off the top of the bolts once the columns have been installed or worse, later down the line when the building is in operation. Either one of these scenarios will lead to significant inconvenience, expensive remedial works and complex contractual implications.

If bolts are set out incorrectly, there are several possible courses of action, depending on the situation. The solution should be:

→ if the columns have already been manufactured and delivered to site, depending on the extent of the error, it may be possible to have additional holes drilled into the base plate, or the base plate may be removed, and a new base plate welded on to accommodate the as-built bolt positions.

→ if the option above is not possible or suitable, new holes can be drilled into the slab at the correct locations, the internal surface of the holes prepared so the grout will bond onto the concrete and the bolts set in the correct position using high strength grout, to a tolerance of +/-3mm.

→ if the columns have not yet been manufactured, the base plates can be redesigned to accommodate the as-built positions of the bolts.

24.30 FOUNDATION PILES

Piles are long steel or reinforced concrete columns which are either driven or bored into the ground to create a stable foundation for structures. On top of the piles, at ground level is an in-situ or precast pile cap which acts as a beam or slab (ground beam) connecting the piles. The piles and the pile cap form the sub-structure. Fig. 24-21 shows some reinforced concrete piles before they have been installed.

The purpose of foundation piles is to transfer the load of the structure down to solid strata. They act as columns that support roofs and upper slabs of buildings, except they are underground. They support the pile cap, and the bearing capacity of the soil assists the piles to support the load.

Sheet piles (Fig. 24-22) are used to create an underground retaining wall, which form the vertical faces of deep of excavations. Interlocking piles are driven to the required depth, and then the soil on one side of the sheet pile wall is excavated, with the piles supporting the soil behind it (Fig. 24-23).

Types of pile include:

→ driven concrete piles

→ driven steel piles

→ bored concrete piles also known as continuous flight auger (CFA) piles

→ sheet piles

→ contiguous piles

Fig. 24-21: Reinforced concrete piles prior to installation

Fig: 24-22: Sheet piles

Fig. 24-23: Piles supporting soil for excavation

24.31 WHAT IS INVOLVED IN SETTING OUT PILES?

The stages involved in setting out reinforced concrete driven piles are:

1. **Set out the formation level for the footprint of the structure** – use profile boards or a laser level.

2. **Set out the level of the piling mat (sound stone surface for the piling rig to sit on) if relevant** – using profile boards or a laser level.

3. **Set out the position of the piles** - set out the position of the piles using either the stake-out or Reference Line function and string lines and tape measure. You can mark out each pile individually using the total station, but it may be quicker to mark out piles which are approx. 30m apart and mark the intermediate ones using

a 30m tape. As a check, use the tie distance/compute inverse function to check distances between adjacent piles and also the cumulative distance from the first pile in each line. The piles can now be installed (Fig. 24-24).

4. **Carry out an as-built survey of the levels of the piles** - the length of the pile should be recorded before being installed. Once installed, the piles will be sticking up out of the ground. The level or height of the top of the protruding pile should be measured. This can be done by using the level, total station (see section **11.61** on measuring RLs using the total station) or by measuring the distance from formation to the top of the pile with a tape measure. This will allow the depth of the piles to be calculated to ensure they have reached the required depth.

5. **Carry out an as-built survey of the positions of the piles** – for a range of reasons, piles don't always end up in their design position. For example, if a pile is damaged before it has achieved its desired depth, a full length one must be installed as close as possible to it. Piles can be obstructed by large underground rocks, again, a new pile must be repositioned as close as possible to the design position. If the minimum required spacing is not achieved, the Designer should be consulted.

6. **Mark the level of the top of blinding on the piles** – a level or laser level can be used to mark the top of blinding level. This can be used as a guide along with the laser level when pouring the blinding. The mark should be a horizontal line with a 'V', pointing to it from above, which should still be visible once the blinding is poured.

7. **Mark the cut-off level of piles** – the cut-off level of driven concrete piles is usually something like 30mm above the RL of the blinding. This means the pile will connect fully with the slab and eventually transfer the load to the piles, but will not be protruding into the slab far enough to obstruct the bottom layer of steel reinforcement. Piles act as columns which are supporting the pile cap. If the piles are cut too low and a sound connection with the pile cap is not made, it will be as if there is no pile there at all. If they are too low they won't act as an integral part of the structure, so the span distance between piles will be greater than the designed span. Forces will cause the slab to sag, causing cracking in the underside of the slab, allowing water ingress and potentially differential movement causing damage to the structure.

8. **A level or laser level can be used to mark the cut-off levels of the piles.** The mark should be a horizontal line with a 'V' pointing to it from below, which, for checking

purposes, should still be visible once the piles have been cut. The concrete is broken out using a breaker and the reinforcing steel is cut off at the right level (Fig. 24-25). Note that CFA piles and other poured concreted or grouted piles should be placed to the exact level of the bottom of the base slab.

9. **Carry out an as-built survey of the cut-off piles** – the cut-off level of each pile should be measured using a level. Any piles that are outside of the design tolerance should either be cut down further, extended, or the Designer consulted.

Note: Control points in the vicinity of piling works should be regularly checked as ground movement can occur due to vibration or ground heave caused by deep excavations.

Fig. 24-24: Installing the piles

Fig. 24-25: Piles cut off to the correct level

24.32 SHUTTER POSITIONS FOR REINFORCED CONCRETE MEMBERS

The method for marking out shutter positions for a base slab will depend largely on the requirements of the joiners carrying out the work. Typically, they will require a nail in the blinding at each corner and they will use string lines to set the intermediate positions. To set out the corner positions you can:

→ use the stake-out or Reference Line function in your total station

→ set out the corners using a tape measure, in relation to existing features or offset marks.

Once the shutters are in position, the joiners will usually put a nail in the top of the shutter at each end to check that the shuttering is in a straight line between the two points. If it is appropriate, the engineer will then use an independent method to check the corners and several intermediate positions. Some possible checking methods include:

→ using the Reference Line function in the total station, with the start and end points being each end of the shutter

→ Set out nails beyond the extents of the line of the shutter (using set-out or Reference Line functions). Set up the total station on one of the nails and sight to the other. As you sight from one end of the shutter to the other, the edge of the shutter should be in line with the vertical cross-hair.

24.33 AS-BUILT SURVEYS

As-built surveys are the record of the as-built levels and positions. They may be required to be taken at each stage of construction, or on completion. The specific requirements depend on the nature of the project and the contractual requirements. The surveys may take the form of:

→ a 2D drawing showing levels, positions and dimensions

→ a 3D model

→ a marked up version of the original design drawings showing actual levels, positions and dimensions

→ hand-drawn diagrams

Time can be saved in preparing the drawings for submission to the client if the final output is considered in advance of taking the survey. For example, if the correct coding and a systematic method of storing the different type of features are used when taking the survey, minimum work will be needed when the information is exported from the total station into the electronic version of the drawing. Survey software can be more time- and cost-efficient than generic design or drawing packages.

24.34 HIGHWAYS (ROADS, FOOTWAYS, ROUNDABOUTS AND CAR PARKS)

Highways are traditionally set out in terms of chainage along the road and offset from the centreline. Negative offsets are to the left of the centreline of the road looking in the direction of increased chainage, and positive offsets are to the right of the centreline. The level and total station can be used for setting out highways projects, however due to the widespread availability and improved accuracy of modern GNSS equipment, Ordnance Survey coordinate data is often provided by the Designer in electronic format.

If you are involved in setting out highways, it is prudent to have an understanding of the horizontal and vertical curve geometry used in design (see section **26.2**).

The key components which need to be set out for highway construction are:

→ the different layers of road construction (excavation of sub-grade, sub-base, base-course, wearing course)

→ drainage, sewers (Fig. 24-26), manholes (Fig. 24-27) and inspection chambers

→ kerbs, including kerb logs (Fig.24-28)

→ dropped kerbs

→ block paving

→ gulley pots

→ traffic lights

→ loops and sensors

→ surfacing

→ white lining

→ associated earthworks

The stages are as follows:

1. Mark out centrelines, kerbs, heel kerbs and appropriate offsets.
2. Mark the locations of drainage pipes, manholes and gully tails.
3. Mark out the positions of excavations.
4. Set the levels of excavations using profile boards, batter rails (using the level or GNSS) or laser level.
5. Set out drainage runs using profile boards or pipe laser.
6. Set up profile boards or a laser level for setting out the base layers of road construction.
7. For the surface course, mark levels on pins or pegs, so that a string line can be set up across the carriageway, and the level of the top of the surface course 'dipped' from the string line.

Fig. 24-26: Drainage/sewerage pipe

Fig. 24-27: Manhole cover

Fig. 24-28: Preparation for a kerb

24.35 WHAT IS A ROAD STRING?

A road string is a set of design data for a road. Road string data can be imported directly into the total station. Setting out information is given in relation to the centreline of the road. The information is presented in tabular form, giving offsets and levels of points through a cross-section of the road at specified chainages. Design information includes levels and offsets for:

→ centreline

→ channels

→ kerbs

→ back of footway

→ gullys

24.36 REINFORCEMENT LEVELS AND POSITIONS

The critical dimensions when setting out reinforcement are the distance from the face of the concrete to the nearest edge of the steel and the distance between the top layer and the bottom layer of steel. If the depth of slab or beam is reduced, its strength will be compromised.

There are two possible scenarios when setting out the positions and levels of reinforcement:

1. the steel is fixed in situ and is set out in relation to marks or shutters

2. the steel cage is fixed off-site or adjacent to its final location and then lifted into position by crane.

The stages of setting out reinforcement are:

1. Mark the corner positions of the slab or wall.

2. Mark up the blinding with the as-built levels (Section **24.12**).

3. Check the levels of the top layer of steel. Be aware that if your assistant is standing on the steel cage with the staff, then the steel itself may sag indicating an artificially low level. Once your assistant steps away, the steel will spring back up, resulting in less cover than you expected.

24.37 COVER TO REINFORCEMENT

The implications of not achieving the correct depth of concrete cover to the reinforcement are significant. As with all setting out, the risk of error can be minimised by planning ahead, checking steel levels and spacers to shutters before the concrete pour commences, and checking frequently during the pour. Always carry out the checks yourself and seek an independent check from a third party.

24.38 WHAT IS THE PROBLEM WITH NOT ENOUGH COVER?

Insufficient depth of cover from the concrete to the steel (Fig. 24-29) can lead to:

- → water ingress

- → spalling

- → reduced beam or slab depth leading to reduced strength of the element

Fig. 24-29: Steel level too high

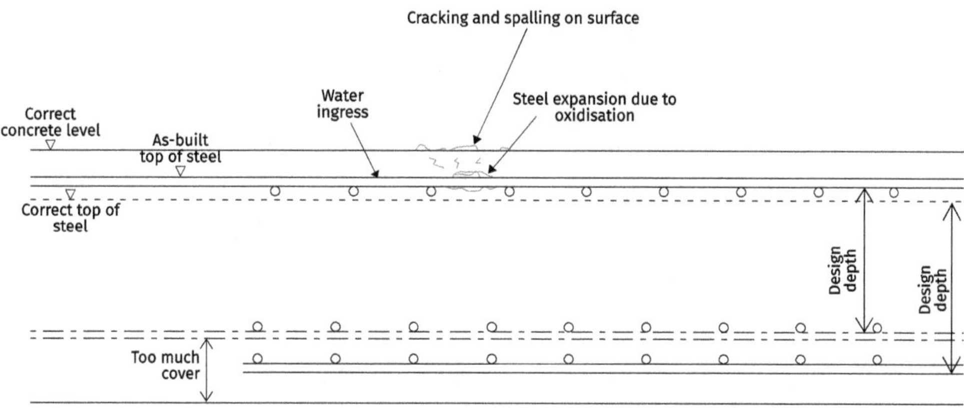

24.39 WHAT IS THE PROBLEM WITH TOO MUCH COVER?

Increased depth of cover (Fig. 24-30) can lead to:

- → spalling

- → punching shear

- → localised weakening due to reduced steel to concrete ratio

Fig. 24-30: Steel level too low

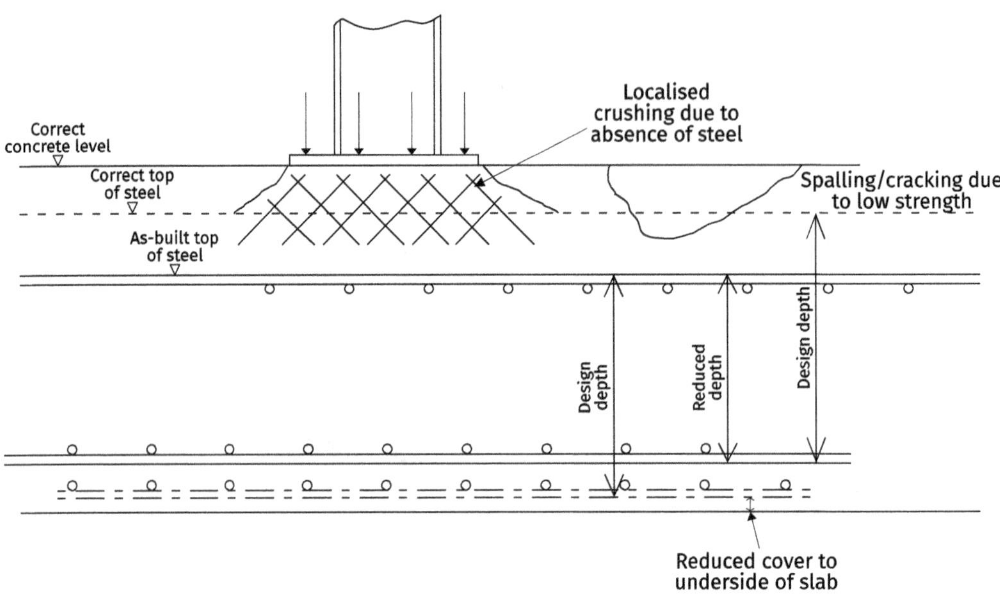

24.40 WHICH IS WORSE, NOT ENOUGH OR TOO MUCH COVER?

In most cases, insufficient cover would be worse than slightly too much cover.

24.41 WHAT HAPPENS IF IT IS WRONG?

It may seem that once the concrete is poured, nobody will be able to tell if there is the correct amount of cover. However, a cover meter can be used to measure the actual cover achieved. The remedial works involved in rectifying incorrect cover are significant, the costs astronomical, and can cause severe time delays.

In the event of incorrect depth of cover, the Contractor must submit a non-conformance report (NCR) to the Designer/Client. There are several possible outcomes:

→ the Designer/Client deems that no remedial work is required

→ re-design is required for subsequent stages of construction (at the Contractor's expense)

→ remedial work is required to rectify the defect

24.42 KICKERS

Where walls or columns connect to a slab, the bottom section (typically 50mm) of it is formed as an upstand as part of the slab pour (Fig.24-31). To form kickers, timber is secured into place, with the bottom of timber coincident with the top of slab (Fig. 24-32). Fig. 24-33 shows the steps for setting out a column kicker. Fig. 24-34 shows the steps for setting out a wall kicker. The steps for setting out kickers are as follows:

1. Mark the centreline of wall (or the centrelines of both axes for columns) on the top layer of steel. This can be done using the Reference Line function in the total station, or by using a tape measure in relation to shutters which have been verified to be in the correct position.

2. The steel-fixers can then position the starter bars in relation to the centreline marks.

3. The joiners will then fix the kickers in the correct position (see section **24.47**) in relation to the centreline marks. Note that steel-fixers and joiners may be willing, or prefer to, carry out their own setting out in relation to adjacent shutters once you have verified that they are in the correct position.

4. Once the kickers and starter bars are constructed, carry out an independent check on their relative and absolute positions, and the depth of cover to the steel.

Fig. 24-31: Column upstand as part of the slab pour

Fig. 24-32: Kicker section

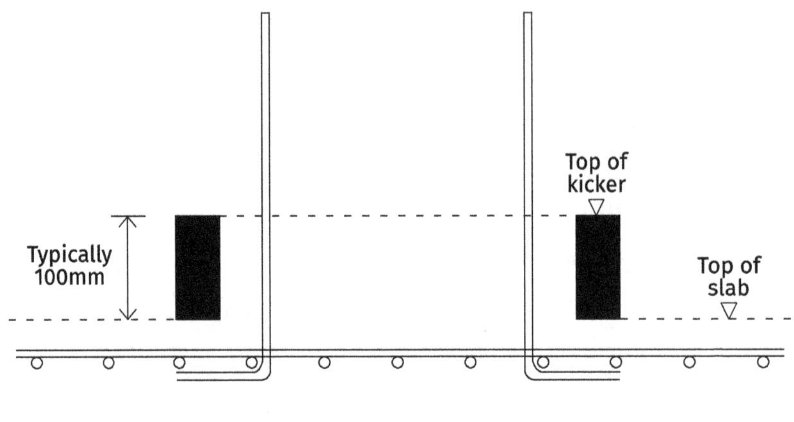

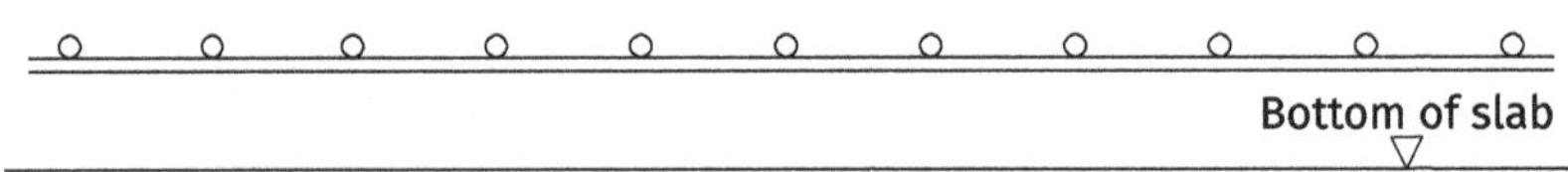

Fig. 24-33: Column Kicker Plan

Step 1

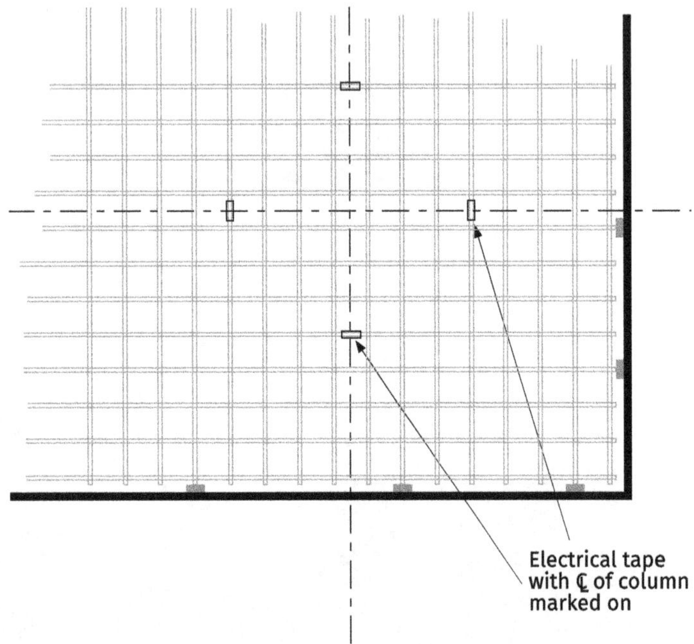

Fig 24-34: Wall Kicker Plan

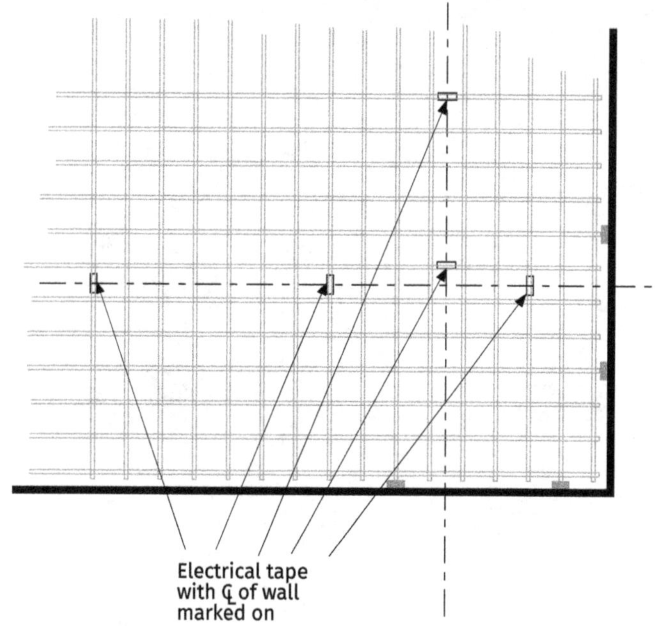

24.43   STARTER BARS

Starter bars are the bars projecting from a concrete slab or wall, which the reinforcement for the adjoining element ties into (Fig. 24-35). Starter bars are designed to transmit load between connecting elements. Without starter bars, the walls and columns would effectively be disjointed from each other and forces would not be transmitted as per the design. Starter bars can be cast into the original slab (Fig. 24-36) or wall or can be post-drilled. They may have threaded ends, which the adjoining bars screw onto (usually in the case of slabs connecting into walls) or may be plain reinforcement bars, which the adjoining bars are lapped onto using tying wire (usually in the case of walls connecting onto slabs).

The position of starter bars is critical as they pre-determine the position of the subsequent adjoining elements. The starter bars should be of sufficient length to achieve the required lap length given by the Designer. Starter bars should be set out along with the kicker positions (see section **24.47**).

Fig. 24-35: Starter bars

Fig. 24-36: Starter bars cast into original slab

24.44 SCABBLING

Most scabbling is carried out on top of wall kickers. Where a kicker is not used, the area to be scabbled can easily be marked out in relation to starter bars or adjacent shutters since high accuracy is not required. Where kickers are not used, marking the positions of connecting elements should be carried out after scabbling is complete, or offset marks should be used.

24.45 STOP-ENDS

A stop-end is an intermediate stopping point created when a slab or beam is too large to pour in one continuous pour (Fig. 24-37). When the joint between adjacent pours is not intended as an isolation joint, reinforcement bars continue through the stop-end (Fig. 24-38) so that the reinforcement in the next section can be lapped onto, allowing the structure to act as one integral component.

The stop-end can consists of either a shutter or sacrificial shuttering, typically 'expanded metal mesh', with holes for the longitudinal reinforcement bars to protrude through. The mesh sheet is clamped in place using vertical timber struts to resist the horizontal force of the wet concrete. The gaps in the mesh are sufficiently small that, provided the concrete mix is correct, the leakage is insignificant. The mesh creates a rough surface for the concrete on each side of the metal mesh to bond onto, or 'key into'. The position of a stop-end is usually set out as a distance from one end of the beam or slab. A tape measure or total station can be used for this.

Fig. 24-37: Stop ends

Fig. 24-38: Reinforcement bars continue through the stop-end

24.46 CAST-IN PIPES

The position of cast-in pipes is critical when tying-in to prefabricated pipes, as there is a very low tolerance in the connections (<3mm in some cases).

Cast-in pipes may be required in slabs or walls. In the case of slabs, the position can be marked on the bottom shutter.

For setting out cast-in pipes slabs, the steps are as follows:

1. Depending on the diameter of the pipe, mark 4-8 equal intervals around the perimeter of the pipe, at the inside and outside of the flange.

2. Once the falsework is in position, mark out the centre point and the internal perimeter of the pipe onto the falsework.

3. If the flange is an irregular shape, determine the correct orientation of the pipe and mark 4-8 equal intervals around the internal perimeter marked on the falsework, making sure the marks correspond with the marks made on the pipe.

4. Have the pipe secured in position and ensure the contact surface with the falsework is sealed around the outside of the pipe to prevent concrete leakage into the inside of the pipe or the threaded recesses for connections.

5. Mark the top level of concrete on the external perimeter of the pipe.

For setting out cast-in pipes in walls, the steps are as follows:

1. Prior to the fixing of the steel, or the erection of the shutters, mark out the positions of the centreline and edges of the pipe onto the concrete slab. This allows the steel fixers to position the recess correctly.

2. Mark four equal intervals around the perimeter of the pipe, at the inside and outside of the flange, at the top and bottom of the vertical centreline and the left and right of the horizontal centreline.

3. Once the first wall shutter is in position, mark out the vertical and horizontal centrelines on the internal face of the shutter. Transfer the marks down onto the kicker in case the marks on the shutter are lost. On the kicker, make a mark for the vertical centreline, and a level mark at a fixed distance below the horizontal centreline.

4. Line up the marks on the pipe with the horizontal and vertical centrelines. Have the pipe secured in position and ensure the contact surface with the shutter is sealed around the outside of the pipe to prevent concrete leakage into the inside of the pipe or the threaded recesses for connections.

5. Once the second shutter is secured in place and the pipe is hidden from view, mark the longitudinal position of the edge of the pipe on the top of the shutters and communicate this information to the concrete supervisor. This means that during placement of the concrete, the operatives can ensure sufficient vibration either side of the pipe to prevent voids and honeycombing.

24.47 RC WALLS

The steps to setting out RC walls are as follows:

1. Once the shutters for the base slab are in place, mark out the centrelines of any walls onto the shutters.

2. The marks made in Step 1 will subsequently be used by the steel-fixers to position the starter bars and by the joiners to position the wall kickers.

3. Once the starter bars and kickers are in place, check the position of the kickers (see section **24.42**). Ensure there is sufficient depth of cover.

4. Once the base slab has been poured, check the as-built positions of the kickers.

5. Mark the longitudinal position of any box-outs or cast-in items onto the kicker.

6. Once the first shutter is in position (Fig. 24-39), mark out the positions and levels of any cast-in items or box-outs onto the internal face of the shutter (see section **24.56**).

7. The wall steel will be constructed (Fig. 24-40) and the wall shutters clamped onto the kickers.

8. Once both sides of the wall shutters are fixed in position and you have access to the top of the wall, check that there is sufficient depth of cover from the face of the shutter to the steel.

9. Plumb the walls up using one of the methods described in sections **24.65 - 24.74**.

10. Determine the reduced level of the top of the shutter at approx. 2m intervals along it. At each position, determine the distance from the top of the shutter to the top of the concrete and either make a mark at the correct position, or indicate the required 'dip' at each point on the top of the shutter. Alternatively, determine the reduced level at each end of the shutter and use a string or chalk line to mark intermediate points.

11. Mark the positions of any cast-in items or box-outs on the top of the shutter.

12. As soon as the concrete has been poured, check the plumb of the wall before the concrete goes off, and have adjustments made as necessary. For very tall walls, it may also be necessary to check the plumb once half of the pour is complete.

Fig. 24-39: First shutter

Fig. 24-40: Wall steel constructed

24.48 SETTING OUT RC COLUMNS

The steps to setting out RC walls are as follows:

1. The kickers are formed as part of the slab. Check the positions of the kickers are correct.

2. Once the steel has been fixed (Fig. 24-41) spacer blocks should be attached to ensure that there is the correct depth of cover. The shutters are then erected (Fig. 24-42).

3. Prior to and during the concrete pour, plumb up the columns using the appropriate method from section **24.65 - 24.74**.

4. The columns should be checked for plumb on the shutters have removed (Fig. 24-43).

Fig. 24-41: Steel with spacers

Fig. 24-42: Shutters erected

Fig. 24-43: Columns at various stages of completion

24.49 SETTING OUT RC SLABS

The steps to setting out RC slabs are as follows, assuming the blinding is correct and has allowed for any voids or ducting:

1. Check the as-built levels of pile cut-off levels.

2. Mark the corners of the slab with a nail in the blinding or marks on falsework. The joiners will use these marks for erecting the shutters.

3. Mark up the as-built levels of the blinding onto the blinding with spray paint at approx. 2m intervals. The steel-fixers will use these as a guide for the packers required to the bottom level of steel.

4. Mark the positions of any ducting, apertures or recesses in the slab.

5. Check the straightness of shutters using a string line or a total station

6. Check the reduced level of the top layer of steel.

7. Accurately mark the positions of holding-down bolts, wall kickers, column kickers, starter bars, recesses, apertures, ducting and cast-in items onto the top layer of steel or falsework as appropriate. Use a tape measure in relation to verified shutter positions or the total station for this.

8. Before the concrete reaches full strength, put baseline and offset nails in the concrete (within 48 hours of pouring if possible).

24.50 ELEVATED RC SLABS

The steps for setting out elevated RC slabs (Fig. 24-44) are as follows:

1. The falsework is erected to fit snugly around the supporting walls and columns (Fig. 24-44).

2. Check the level of the soffit of the slab at regular intervals by taking inverted staff readings to the underside of the formwork (allowing for the thickness of the plywood).

3. Mark the edge shutter positions.

4. Follow steps 4 – 8 of section **24.49**.

Fig. 24-44: RC Slabs

Fig. 24-45: Falsework

24.51 CUBING UP

Calculating the volume of concrete required for a pour, or group of pours which will be carried out consecutively, is known as 'cubing up'.

Cubing up needs to be carried out both in advance of the pour taking place, to determine how much concrete should be ordered, and as the pour is nearing completion to determine the volume of the final part-load.

When cubing up to determine the quantity of concrete to be ordered, refer to the drawings, as-built formation or blinding levels and the as-built shuttering positions. It is prudent to overestimate rather than underestimate since in the worst-case scenario, overestimation may lead to wasted concrete, whereas the work involved in rectifying an unplanned cold joint in the concrete will be significantly more.

If the batching plant is within a 20-minute drive of the site, calculation of the final part-load is typically carried out immediately after the second last full load. The remaining volume should be approximated based on taped measurements and engineering judgment.

When cubing up, take into consideration:

→ the distance from the batching plant to the site

→ the demand on the batching plant

→ the loads may be slightly more or slightly less than 6m³ each

→ the volume of the pour (the larger the volume, the more scope for cumulative error)

→ the consequences of running out of concrete mid-pour and a cold-joint forming

→ the cost of disposal of wasted concrete (which may be zero if there is an area of the site where the concrete can be used)

→ The increased cost per m³ for a part load.

24.52 TRANSFERRING LEVELS BETWEEN FLOORS OF A BUILDING.

When setting up level control on different floors of buildings, first establish whether it is more important that the levels are correct in relation to your primary control points, in relation to lower floors or in relation to a datum such as finished floor level (FFL).

There are several methods for transferring levels vertically:

→ create a mark at a known level on a vertical surface such as a wall or column. Use a tape measure to transfer the datum vertically upwards. Be aware of cumulative errors

→ use the automatic level to transfer a TBM up staircases

→ use the tie distance/compute inverse function to measure a level change between two points

→ if your control points are reliable, set up the total station using a resection, and use the survey mode to measure the RLs of your new control points. Verify the level difference using an alternative method such as tie distance/compute inverse.

24.53 SETTING LEVELS FOR STEEL COLUMN BASES

There is little room (>5mm) for variation between the levels of column bases within a single structure. Since the column bases can be set higher than the as-built slab level, but not lower, it is necessary to set all the columns to the highest point on the slab. The process is as follows:

1. Give each column location a unique identifying code.

2. Use the automatic level to survey the slab level at the centre position of each bolt group.

3. Calculate the level of each point.

4. Identify the highest point.

5. For every column position, calculate the level difference between the highest point and the as-built slab level at that point.

6. Identify the required staff reading for the correct level.

7. Have your assistant place the required combination of packers at the centre of the bolt group. Packers are approx. 50mm x 50mm square and vary in thickness between 3mm and 10mm.

8. Check the staff reading to the top of the packers and adjust the number of packers if necessary.

24.54 BOX-OUTS AND RECESSES

A box-out is the arrangement of formwork which creates an aperture or recess in a reinforced concrete wall or slab. The surrounding reinforcement is formed in an arrangement which allows the required amount of concrete cover from the steel. Box-outs are most often square or rectangular but may be circular or irregular in shape.

24.55 WHAT CAN GO WRONG WITH BOX-OUTS AND CAST-IN ITEMS?

Possible defects with box-outs and cast-in items include:

→ **Incorrect setting out** – with box-outs and cast-in items, there is only one chance at getting the setting out right. Even small errors in position or level can have significant negative consequences for subsequent works. Setting out should always be checked from first principles in relation to original source information, by a fully competent independent person.

If a box-out is constructed in the wrong position, the incorrect void would need to be filled in by scabbling the smooth internal surface, drilling holes for anchor bars, grouting-in the anchor bars, installing reinforcement, securing shuttering and grouting the void. A new void would need to be created in the correct location by diamond saw cutting or drilling through the concrete and steel for the whole thickness of the wall, breaking the concrete back so that the reinforcement can be re-fixed accordingly, securing new shuttering and grouting to create the correct form.

For cast-in items, the remedial work would be similar, but rather than filling in the void, the cast-in item would need to be removed and may be damaged in the process. For specialist items, this could lead to delays in re-ordering the part.

→ **Movement of the shuttering or cast-in item** – this defect is particularly relevant in walls. The box-out shuttering or cast-in item must be secured extremely well, since during placement, concrete is likely to free-fall onto it from a significant height. Dense reinforcement may break the fall of the concrete and reduce the impact to some extent, but even so, it is essential to ensure that there is no scope for movement. The setting out and securing of the shuttering or cast-in item is critical, as there is no means of checking it until the formwork is removed, by which time the concrete has gone off and it is too late to correct any errors. Remedial work would require a combination of the remedial work for incorrect setting out and remedial work for voids and honeycombing around the box-out.

→ **Voids underneath the box-out or cast-in item** – this occurs when the concrete does not come fully into contact with the underside of the box-out or cast-in item. This causes a significant problem because a void is created and the reinforcement is exposed. Remedial work would involve breaking out any voids and honeycombing back to sound material, scabbling the exposed surface, installing new shuttering where necessary and filling with grout.

24.56 HOW DO I SET OUT A BOX-OUT IN A WALL?

The steps for setting out a box-out are as follows:

1. Face 1 of the shuttering is fixed in position.

2. Mark out the perimeter of the box-out on the inside face of the shuttering using a pencil. If pencil does not show up clearly enough, make the marks clearer using a chalk line, permanent marker or French chalk.

3. The steel reinforcement will be fixed in position based on the markings on the shutter.

4. Check the position of the reinforcement in relation to the box-out perimeter marks, paying particular attention to the required depth of cover.

5. The box-out shuttering will be fixed in position based on the markings on the shutter.

6. Check the position of the box-out shuttering in relation to the original marks you made on the shuttering, the reinforcement, critical as-set-out and as-built elements and/or critical control points (whichever are most relevant to the relative position of the box-out).

7. Ensure there is a strong seal between the box-out shuttering and the vertical wall formwork.

8. Face 2 of the wall formwork will be secured in position.

9. Clearly mark on the top edge of the wall formwork, the extents of the box-out and the depth to the bottom face of the shuttering. Communicate this to the concrete gang foreman verbally and with a diagram. This ensures that the concrete can be vibrated sufficiently at the edges of the box-out to ensure that the concrete flows to the underside of the horizontal surface, reducing the chance of an air lock leading to voids.

24.57 PLUMBING-UP WALLS AND COLUMNS

Plumbing-up a wall or column means adjusting it until it is truly vertical. For a wall, this means that the face of the wall must be plumb at both ends and at all intermediate points in between, and for columns, perpendicular faces must both be plumb.

There are many possible situations in which you may need to plumb-up walls or columns. Each situation will be unique, having its own constraints and considerations.

There are many variables including:

→ accessibility of the element itself

→ the space available to set up your equipment and install offset nails

→ visibility of control points

→ the height of the element you are plumbing-up

→ the required accuracy

→ the nature of the construction of the element e.g. steel, concrete, masonry or cladding

→ the equipment available to you

→ weather conditions

24.58 WHY DO WE NEED TO PLUMB-UP COLUMNS AND WALLS?

There are three reasons why we need to plumb-up walls and columns:

→ to ensure the position at the top of the element is correct - the principle of setting out vertical elements is that if the base is set out correctly and the element is plumb, then the top should also be correct. In theory, ensuring that an element is plumb will ensure the position at the top is correct. This cannot be taken as read due to the accumulation of base setting out and verticality errors. Where possible and where appropriate, the tops of the position of elements should also be checked.

→ so that the structural element acts as it has been designed to - being out of vertical alignment means the forces may not act as per the Designer's assumptions. Structural integrity may be compromised by the introduction of sideways forces. Span length may be affected and corners not square.

→ For aesthetic reasons – even if there is no impact on subsequent setting out or structural integrity, the presence of an unsightly 'dog-leg' shows bad workmanship and will reflect badly of the Contractor for the lifetime of the structure.

24.59 HOW DO WE ENSURE WALLS AND COLUMNS ARE PLUMB?

There are many different methods for plumbing, each with its own advantages and limitations. Detailed descriptions of these methods are provided below.

The basic principle is that we can check if an element is plumb by measuring the position of the top of the element in relation to the position of the bottom, checking that they are on the same vertical plane. Walls can only move on one plane so they only need to be plumbed on one axis. Columns can move in two different axes (think of them as X and Y) so need to be plumbed on both.

24.60 HOW DO I PLUMB-UP LONG SECTIONS OF WALL?

If a wall is long, it is necessary to check the plumb at both ends of the shuttering and at intermediate points. The straightness along the bottom should be checked, and not assumed. A string line should be set up above one of the internal faces of the shuttering or the plumb should be checked at intermediate points, particularly if the shuttering comprises multiple sections. Even if there are no obvious jumps in alignment, the wall may be bowed.

24.61 WHAT SHOULD BE CONSIDERED WHEN PLUMBING WALLS AND COLUMNS?

When plumbing walls and columns, effectively you are setting the position of the top of the element. The relative position may be critical, particularly if it acts as a connector or support for upper floors. For example, if two parallel walls support a precast unit or walkway, the tops of the walls must be the correct distance apart along the full length. In this case, the distance between the tops of the walls should be checked in conjunction with their plumb. If the bottoms of the walls have been set out incorrectly, you may need to seek guidance or make an engineering decision on whether plumb or relative position should be given priority, or whether redesign or remedial work is necessary.

24.62 AT WHAT POINT IN THE CONSTRUCTION PROCESS SHOULD PLUMBING BE CARRIED OUT?

For reinforced concrete elements, the joiners will use a spirit level for coarse plumbing of the element during the construction of the formwork. Accurate plumbing must then be carried out immediately before the concrete pour and immediately upon completion whilst the concrete is still green. In the case of particularly tall pours or where the risk of movement is high, it may be necessary to check the plumb at intervals throughout the pour.

For steel columns, again a spirit level can be used for coarse plumbing. Accurate plumbing must be carried out prior to the final bolt tightening and shimming.

24.63 HOW IS THE VERTICALITY OF REINFORCED CONCRETE ELEMENTS ADJUSTED?

Adjustable mechanical struts know as push-pulls allow the face of shuttering to be adjusted.

24.64 HOW IS THE VERTICALITY OF STEEL COLUMNS ADJUSTED?

Where is there is lack-of-fit, wedges, jacks and packers are employed to achieve verticality.

24.65 PLUMBING METHOD 1 - OFFSET AND STAFF

Suitable for:

→ walls and columns

Equipment:

→ total station or theodolite

→ levelling staff

Requirements:

→ approx. 10m clearance is needed beyond the extent of one end of the wall or column where the offset nail must be installed.

→ shuttering must be constant thickness.

→ access is needed to the bottom and the top of the wall or column, before during and after the concrete pour.

→ a clear line of sight is needed at a 1m offset from one end of the wall to the other.

→ your assistant must be able to view the staff from directly above to ensure that it is perpendicular to the shuttering.

Instructions:

1. Install offset nails 1m offset from the centreline of the wall. The offset nail at the instrument set-up end should be at least 5m back from the extent of the wall. The

sighting nail can be anywhere within a few meters of the extent of the other end of the wall.

2. Centre and accurately level the instrument over the set-up nail. Line up the single vertical cross-hair on the sighting nail (Fig. 24-46).

3. Have your assistant hold the staff horizontally against the shutter, perpendicular to the direction of the wall, as close to the bottom of the wall as possible and as close to the near end of the wall as possible (Fig. 24-47).

4. Record the staff reading at the single vertical cross-hair.

5. Have your assistant hold the staff horizontally against the shutter, perpendicular to the direction of the wall, as close to the top of the wall as possible.

6. Have the top of the wall or column adjusted until the staff reading matches the reading you took at the bottom of the wall.

7. If the wall shuttering has the potential to bow, or comprises more than one panel, carry out steps 8 and/or 9.

8. Repeat steps 3 – 6 for each individual panel (either the midpoint or each end of the panel depending on the panel length) or at the midpoint along the wall.

9. Set up a string line between the two extreme end points of the top of the wall at the internal face of the shuttering. Intermediate panels can be checked against the string line and adjusted accordingly.

Fig. 24-46 Plumbing a wall - plan

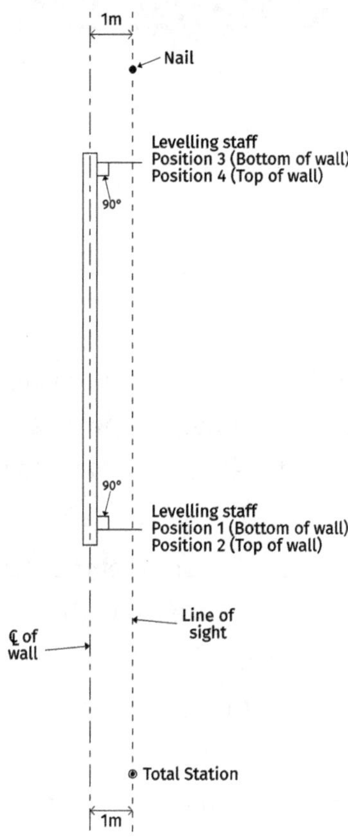

24-47: Plumbing a wall - section

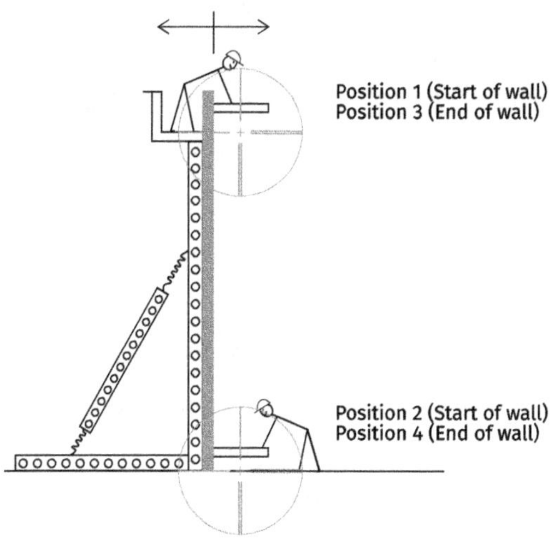

24.66 PLUMBING METHOD 2 – COORDINATES OF RETRO TARGETS

Suitable for:

→ columns

Equipment:

→ total station

→ retro targets

Requirements:

→ for RC columns, the shuttering must be a constant thickness

→ form multiple columns, you need to be able to set up at a position where you can get a good view of all of them (or as many as possible).

Advantages:

→ no need to set up over a point or to set the instrument position and orientation (no control points or offset nails are needed)

→ you don't need an assistant

→ you don't need access to the top of the column once the retro targets have been attached

→ both axes of the column are plumbed from a single set-up position.

Step-by-step instructions:

1. Position a retro target at the centreline of the column (either the flange or stem for RSJs), as close to the bottom and top as possible (Fig. 24-48). For steel columns, the retros can be attached whilst the columns are lying on the ground which means no access is needed by you or your assistant at all, once erected.

2. Set up the instrument at a convenient location. When plumbing multiple columns, optimise the position so you can see as many as possible.

3. Sight to the bottom target and record the coordinates either by hand or store in the total station memory.

4. Sight to the top target and measure the coordinates.

5. Have the top of the wall or column adjusted until the coordinates of the top target are equal to the coordinates of the bottom target (or are within acceptable limits). You can use the stake-out mode either to display the coordinates and check manually, or to set out the same coordinates of the bottom point.

Fig. 24-48: Plumbing columns - reflectorless

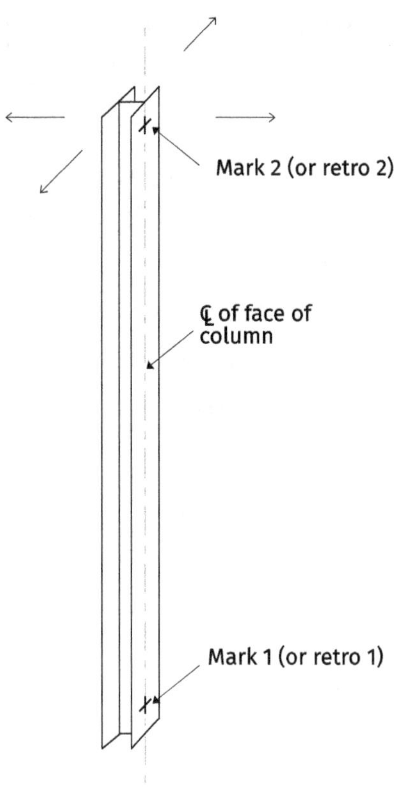

24.67 PLUMBING METHOD 3 - REFLECTORLESS VERSION OF PLUMBING METHOD 2

Suitable for:

→ columns

Equipment:

→ total station

Requirements:

→ for RC columns, the shuttering must be a constant thickness

→ form multiple columns, you need to be able to set up at a position where you can get a good view of all of them (or as many as possible).

Advantages:

→ no need to set up over a point or to set the instrument position and orientation (no control points or offset nails are needed)

→ you don't need an assistant

→ you don't need access to the top of the marks which have been made.

→ both axes of the column are plumbed from a single set-up position.

Instructions:

1. Visible marks, which you can sight to with your total station cross-hairs, must be made on the columns or walls. Position a visible mark at the centreline of the column (either the flange or stem for RSJs), as close to the bottom and top as possible. For steel columns, the mark can be made whilst the columns are lying on the ground which means no access is needed by you or your assistant at all, once erected.

2. Steps 2-5: as per plumbing method 2, but use the reflectorless function to sight to marks instead of retros.

24.68 PLUMBING METHOD 4 – USING THE 'REFERENCE LINE' FUNCTION WITH RETRO TARGETS (TOTAL STATION)

Suitable for:

→ walls

Equipment:

→ total station

→ retro targets

Requirements:

→ the wall shuttering must be a constant thickness

→ you need to be able to set up at a position where you can get a good view of the face of the shutter at both ends of the wall.

Advantages:

→ no need to set up over a point or to set the instrument position and orientation (no control points or offset nails are needed)

→ you don't need an assistant

→ you don't need access to the face of the wall once the retro targets have been attached

→ can be adapted for reflectorless if the surface of the shuttering is suitable (see method 5).

Step-by-step instructions:

1. Position retro targets as close to the bottom of the wall and as close to the top of the wall as possible, at each end (four in total). They do not need to be a fixed distance from any of the edges (Fig. 24-49).

2. Select the Reference Line function.

3. Enter the Reference Line by sighting to Point 1 (retro at one end of the wall) and Point 2 (the retro at the other end of the wall). You have now created a Reference Line. When the wall is plumb, the face of the shuttering at the top face of the wall will be 0mm offset from the Reference Line.

4. Sight to the retro at the top of the wall at one end. The screen will display the offset from the Reference Line. Have the top of the wall adjusted until the offset is 0mm.

5. Repeat for the other end of the wall. If the wall shuttering comprises more than one panel, carry out steps 6 or 7.

6. Repeat steps 4 – 5 for each individual panel (either the midpoint or each end of the panel depending on the panel length).

7. Set up a string line between the two extreme end points of the top of the wall at the internal face of the shuttering. Intermediate panels can be checked against the string line and adjusted accordingly.

Fig. 24-49: Plumbing a wall - reference line

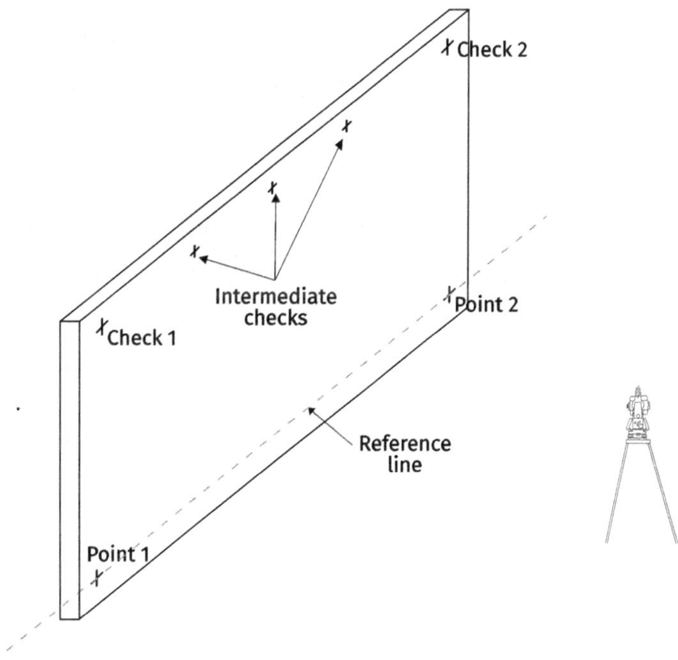

24.69 PLUMBING METHOD 5 – REFLECTORLESS VERSION OF PLUMBING METHOD 4

Suitable for:

→ walls

Equipment:

→ total station

Requirements:

→ the surface of the shuttering must be suitable for the reflectorless function

→ the wall shuttering must be a constant thickness

→ you need to be able to set up at a position where you can get a good view of the face of the shutter at both ends of the wall.

Advantages:

→ no need to set up over a point or to set the instrument position and orientation (no control points or offset nails are needed)

→ you don't need an assistant

→ you don't need access to the face of the wall at all.

Step-by-step instructions:

1. As per plumbing method 3. Step 1 is N/A (no marks are needed as the position and height of the reading relative to the edges of the shuttering is not critical).

2. Steps 2 – 9: as per plumbing method 3, but sight to the surface of the shuttering instead of marks. The exact position is not needed. To set the Reference Line it is ok to sight to anywhere in the region of the bottom face of the wall near end 1 and then the same again near end 2. When sighting to the top of the wall, again, anywhere close to the top of the wall at each end or any point along the wall will give you the offset distance at that point from the correct line.

24.70 PLUMBING METHOD 6 – STRING LINE

Suitable for:

→ walls and columns

Equipment:

→ string line and tape measure

Requirements:

> → non-windy conditions

> → the wall shuttering must be constant thickness

> → the face of the wall or column must be unobstructed from top to bottom.

Advantages:

> → no fancy equipment needed

> → quick to set up

> → you don't need an assistant

> → can be set up and left unattended for the duration of the pour for frequent checking

> → sufficiently accurate for pours up to 8m in height

> → no access to top of wall needed during concrete pour

> → measurements taken from ground level

> → the concrete foreman or other trustworthy person can check the plumb if you are not present.

For walls:

1. Attach a piece of timber securely to the top of the shuttering at one end of the wall (Fig. 24-50). The timber should be perpendicular to the face of the shuttering and protruding approx. 100mm.

2. Securely attach a string line which passes through a narrow 'V' cut into the end of the timber. Attach a heavy weight (e.g. a purpose-made plumb bob weight or a block of concrete) to the bottom of the string line approx. 400mm above ground level.

3. Use a ruler or tape measure to measure from the surface of the shuttering to the string line as close to the top of the shuttering as possible, i.e. the first point you come to where there are no obstructions between the string line and face of the shuttering.

4. Record the measurement.

5. Repeat step 2 as close to the bottom of the shutter as possible. Damp the movement of the string line as much as possible by steadying the weight. A bucket filled with oil or water may be used as a damper if necessary. Release the weight. When the

width of the swing is less than 5mm take a measurement from the face of the shutter to the string line. Note the extents of the swing of the string line. The mid-point of these two readings gives the correct measurement.

6. Have the top of the wall adjusted until the measurement at the bottom matches the measurement at the top.

7. Repeat steps 1 – 6 for the opposite end of the wall.

For columns:

Instead of each end of the wall, a piece of timber should be attached to any two perpendicular faces of the column as shown in Fig. 24-51. Carry out steps 1-6 for each face of the column. Once adjustments have been made, check both faces again.

Fig. 24-50: Plumbing a column - string line

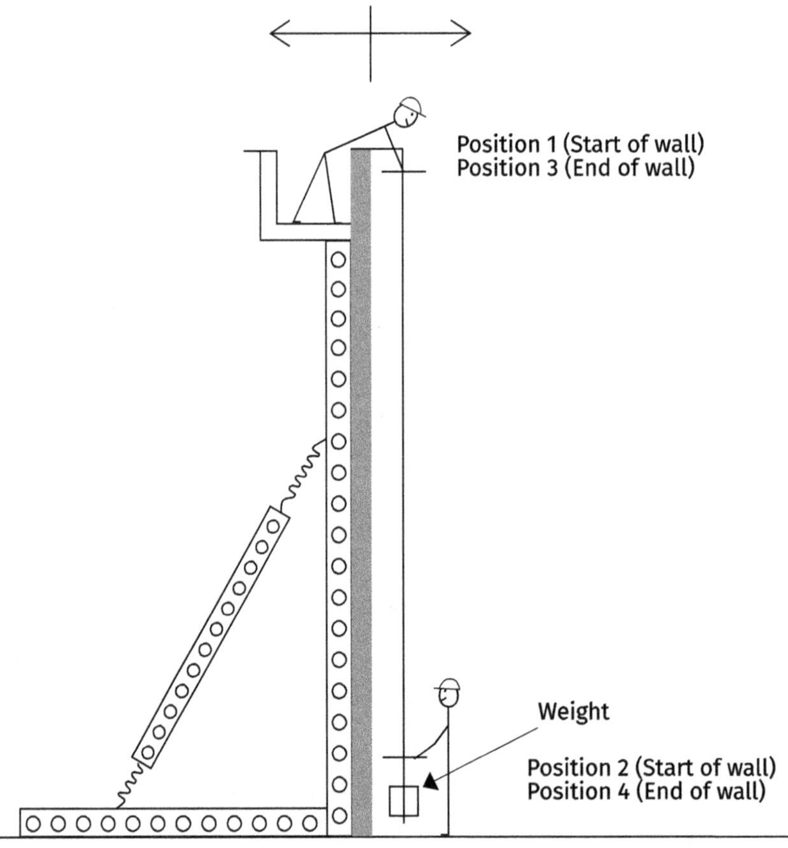

Fig. 24-51: Plumbing a column - stringlines

24.71 PLUMBING METHOD 7 – SET UP ON THE CENTRE LINE OF THE COLUMN

Suitable for:

→ columns

Equipment:

→ total station or theodolite

Requirements:

→ a relatively high amount of preparation is needed for each column, so this method is best-suited to cases when you only have one or two columns to plumb.

→ best suited to universal columns.

→ sufficient clearance is needed from the column you are plumbing to any obstructions, on two perpendicular faces of the column.

Advantages:

→ you don't need an assistant.

→ you don't need access to the top of the column once the mark has been made.

Instructions:

1. Mark the centreline of the flange and web at the top and bottom of the column. Highlight the marks with a sideways 'V' shape with the point touching the mark. Alternatively, attach a retro target. Do this before the steel is erected if possible.

2. Put a nail in the ground on the gridline of the column at a suitable distance from the column (Fig. 24-52).

3. Set up the total station (or theodolite) over the mark on the ground.

4. Line up the single vertical cross-hair on the bottom mark (Fig. 24-53).

5. Scroll up until the single vertical cross-hair is next to the top mark.

6. Have the column moved left or right until the centreline of the column is in line with the vertical cross-hair. Many total stations have a function which projects a red laser dot. This means that the steel-erector can move the top of the column into line without needing direction.

7. You have now plumbed the column along one axis only. Repeat steps 2 – 6 for the perpendicular axis (Fig. 24-54).

8. Return to your set-up point on the original axis to check that the plumb was not affected by adjustment of the perpendicular axis.

Fig. 24-52: Plumbing on centreline of column

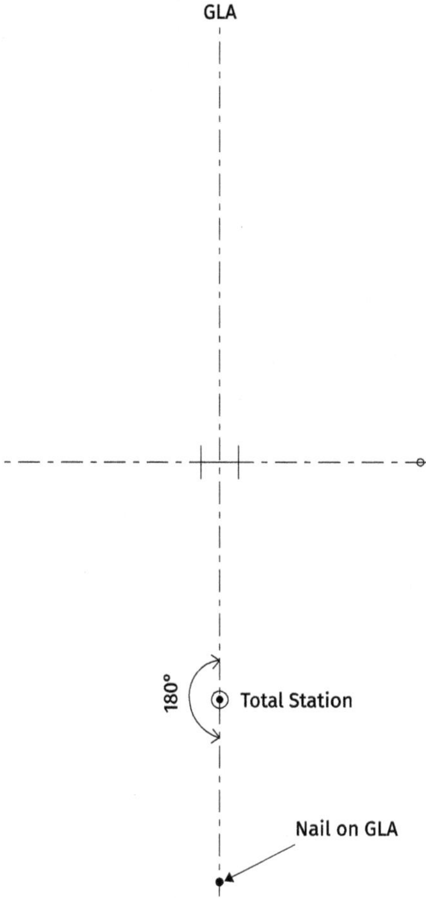

Fig. 24-53: Plumbing column on centreline - elevation

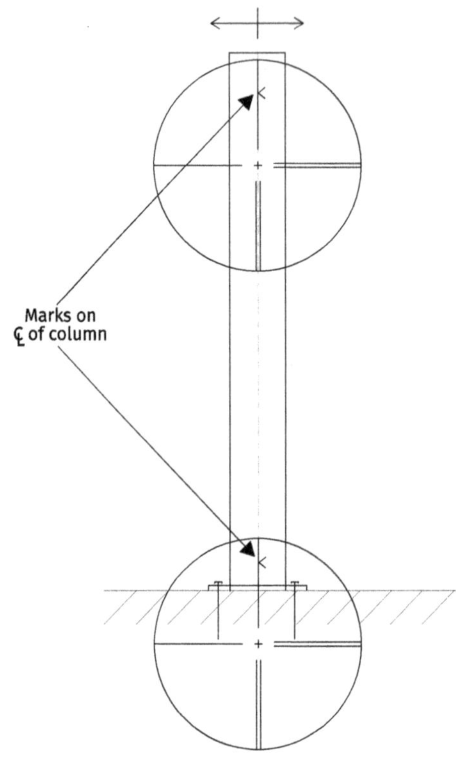

Fig. 24-54: Plumbing column on centreline - plan

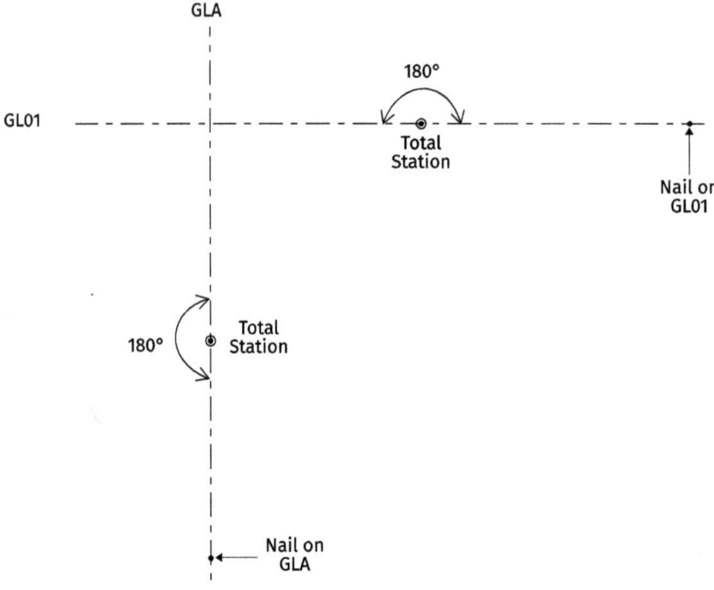

24.72 PLUMBING METHOD 8 – SET UP IN LINE WITH THE OUTSIDE EDGE OF THE COLUMN.

Suitable for:

→ walls

Equipment:

→ total station or theodolite

Requirements:

→ a relatively high amount of preparation is needed for each column, so this method is best suited to cases when you only have one or two columns to plumb

→ best suited to universal columns

→ sufficient clearance is needed from the column you are plumbing to any obstructions, on two perpendicular faces of the column.

Advantages:

→ you don't need access to the column

→ no mark is needed at the centreline of the column

→ you don't need an assistant.

Instructions:

1. Using a tape measure, measure the width of the column (the distance between the outside edges of the flanges) and divide the answer by 2 to find half the column width. Do this for both axes of the column.

2. Set out a mark on the gridline of the column.

3. Using a tape measure, make an offset mark at half the width of the column from the gridline (Fig. 24-55).

4. Set up over the offset mark.

5. Sight the outside edge of the column at the bottom using the cross (Fig. 24-56).

6. Scroll up to the top of the column.

7. Have the top of the column moved left or right until the outside edge of the column is in line with the cross.

8. You have now plumbed the column along one axis only. Repeat steps 2 – 7 for the perpendicular axis.

9. Return to your set-up point on the original axis to check that the plumb was not affected by adjustment of the perpendicular axis.

Fig. 24-55: Plumbing column on edge - plan

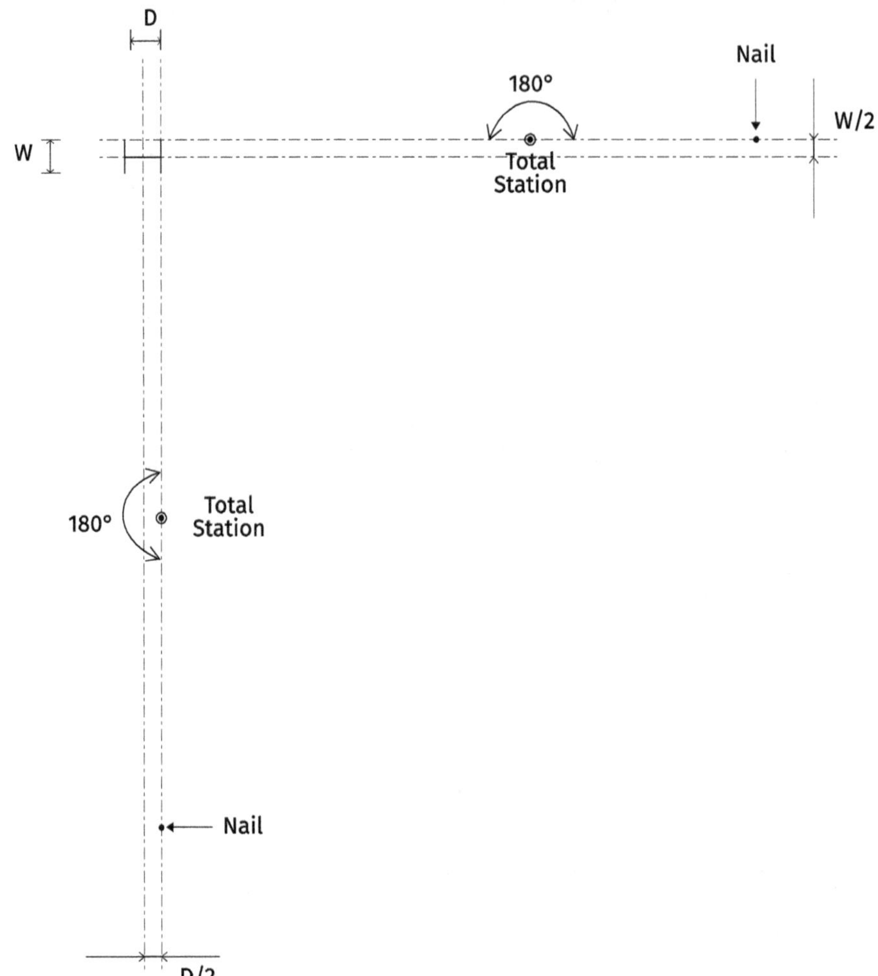

Fig. 24-56: Plumbing column on edge - elevation

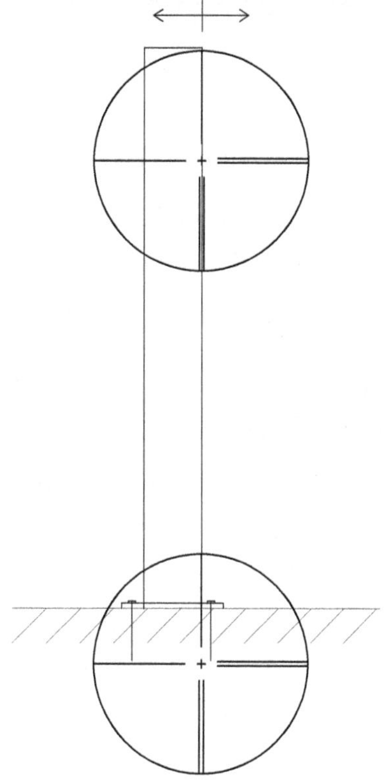

24.73 PLUMBING METHOD 9 – SPIRIT LEVEL

Suitable for:

> → walls

Equipment:

> → good quality 1.5m long spirit level

Requirements:

> → at least 1.5m vertical face of the column must be flat and accessible so this method is best suited to universal columns

> → due to the fact that the error increases with height, this method should not be used for elements greater than 3m high.

Advantages:

> → no fancy equipment is needed

→ very quick

→ you only need access to one point on the column

→ you don't need an assistant.

Instructions:

1. Check the adjustment of your spirit level (see section **6.9**).

2. Place the spirit level vertically on the column with the bubble in line with the direction of the line of adjustment. Have the column adjusted until the bubble is centred.

3. Repeat for the perpendicular face of the column.

24.74 PLUMBING METHOD 10 – HOW NOT TO PLUMB UP A COLUMN

There is a misconception that it is possible to set up the theodolite or total station at approx. 45 degrees to a column and plumb both axes in a single action by sighting the outside edge of the column at the bottom and scrolling up and checking at the top.

Because the edge of the column is in line with the vertical cross-hair, it may appear that both axes are plumb. In fact, this method does NOT mean that either of the faces are plumb. This method is incorrect. For plumbing columns, one of the methods described above should be used.

24.75 TUNNELLING PROJECTS

Tunnelling projects are more complex than above-ground projects due to the fact that once underground, the control points can exist only within the alignment of the tunnel itself. The underground control points are established as the tunnel progresses, using an open traverse. To achieve the highest possible accuracy, base plates are fixed into the tunnel walls to ensure that there is no variation in the positioning of the total station. Rounds of angles are recorded many times, and the averages taken.

Above-ground control will be set up using at least two methods so they can be checked against each other:

→ **GNSS** – the equipment may be set up gathering data for up to 8 hours to ensure the maximum possible accuracy of the co-ordinated points

→ **An above-ground traverse** – this will establish the closing error between the start and finish primary control points

Digital levels and 1" total stations are used, as well as other specialist equipment.

Tunnelling is considered a specialist setting out and activity and further details are not within the scope of this book.

24.76 COPLANING

Coplaning is a technique traditionally used to set up the instrument between two points so that other points can be set out between the two points. This is still a valid method if the only instrument available is a basic digital theodolite, however the 'reference line' function in modern total stations is much quicker and more accurate for this purpose.

The method is still used for transferring above-ground control down into a tunnel shaft. The tunnel alignment is marked on the inside face of the shaft at the surface. The marks are then transferred vertically down into the bottom of the shaft using either a plumb wire or surveying equipment. Once the alignment is marked on two opposite internal faces of the shaft, the total station is then set up and its position adjusted laterally until there is a 180-degree difference in the horizontal angle between the two alignment marks. This means that the total station is set up exactly on the centreline of the alignment and the alignment marks can be referenced in order to project the line as the tunnel progresses. Extreme care is needed and a great deal of checking required. Note that transiting the telescope is not suitable, and instead should be rotated horizontally through 180 degrees and checked on both faces.

24.77 HIGH-RISE BUILDINGS

The setting out of high-rise buildings would warrant a whole book in itself. This section gives an overview of the considerations.

For tall buildings which are not in the high-rise category, conventionally an auto-plumb would have been used in conjunction with a Perspex target which would be positioned in a purpose-built aperture or lift shaft opening at the corresponding position in each floor slab.

For total stations, a 90-degree eye piece for sighting steep vertical angles is available.

The setting out at ground floor is critical as this becomes the control for the rest of the structure. Subsequent floors are determined by the as-built positions of the ground floor. Walls and columns are plumbed as the project progresses. Setting out on each floor is carried out in relation to the predetermined positions of the walls and columns.

A structure of over forty floors is classed as high-rise. The methods used for setting out high-rise structures will depend on the height and nature of the building and the land available at ground level. Wind sway and floor slab deflections must be taken into consideration.

CHAPTER 25

CONCRETE DEFECTS

25.1 CONCRETE DEFECTS

The wider issue of concrete defects is a subject in itself and is not within the scope of this book. For in-depth practical information on how to avoid concrete defects, please see (**Concrete Practice; Guidance on the practical aspects of concreting GCG8,** *Concrete Society publication).* However, the relationship between setting out and concrete defects is relevant here.

25.2 SPALLING

Spalling is a defect which appears years after the concrete has been placed. When spalling occurs, the concrete between the surface and the rebar flakes or crumbles away, in serious cases until the rebar is exposed. Spalling occurs because water in the concrete capillaries expands as it freezes and shrinks as it thaws, putting repeated stresses on the material structure of the concrete. Over time, water reaches the rebar. When steel has contact with water, it rusts (oxidises) which causes it to expand, pressing outwards on the concrete and causing it to crack, crumble or delaminate. The effects of spalling are exacerbated by insufficient concrete depth-to-steel reinforcement. At the setting out stage, the risk of later spalling can be minimised by ensuring that the steel reinforcement and the top of the slab level are set to the correct levels and that the correct spacers are in place on vertical surfaces and the underside (soffit) of slabs.

25.3 HONEYCOMBING

Honeycombing is a defect which shows up as voids, usually on vertical surfaces of concrete. It occurs when the concrete does not properly reach the faces of the shuttering during place-ment. The defect can be shallow or deep and in serious cases, the rebar may be exposed. Remedial works must be carried out. There are several possible causes of honeycombing:

→ the concrete is not workable enough

→ there is not enough fine material in the concrete mix

→ the aggregate size is too large, preventing it flowing into the gaps between the rebar and the shuttering

→ the concrete is not vibrated sufficiently during placement

→ the water content of the mix is too high, causing grout to leak out of formwork at joints, leaving voids in the concrete which the grout should have filled

→ there is insufficient cover to the rebar, meaning the concrete cannot easily flow into the gap between the shutter and the rebar

→ obstructions such cast-in pipes or formwork for box-outs prevent the poker-vibrator reaching concrete which is vertically below them

Honeycombing on the surface of concrete means there is reduced cover to rebar, and the gaps in concrete allow water ingress, reducing the durability of the concrete.

More seriously, the presence of honeycombing could indicate that there are voids within the concrete which mean that the concrete has not bonded fully with the steel reinforcement and may not transfer loads in the way it is designed to.

To reduce the risk of honeycombing at the setting out stage:

→ ensure the correct distance between the shuttering and the rebar

→ mark the horizontal position of any cast-in items or box-outs within the pour at the top surface of the shuttering, indicating the depth. This will enable the concrete operatives to carry out extra vibration directly either side of and below the obstruction.

25.4 VOIDS

Voids can occur when an airlock prevents the concrete flowing to the underside of horizontal obstructions such as cast-in items or formwork for box-outs. The defect may be visible at the surface of the concrete or may be hidden.

Hidden voids significantly reduce the strength and durability of reinforced concrete in a localised area. Voids which appear at the surface can be repaired using grout.

As with honeycombing, to minimise the risk of voids, mark the horizontal position of any cast-in items or box-outs within the pour at the top surface of the shuttering, indicating the depth. This will enable the concrete operatives to carry out extra vibration directly either side of and below the obstruction.

25.5 MISSING REINFORCEMENT

Before a concrete pour takes place, the arrangement of the rebar should be thoroughly checked against the drawings. If the spacing between bars is too great, the structural strength is compromised. If the spacing is too small, the concrete may not flow properly into the gaps, causing voids and preventing the concrete bonding to the steel. Where bars are missing, the strength is reduced in a localised area and can lead to lack of structural integrity as well as localised issues such as punching shear.

CHAPTER 26

SETTING OUT CURVES

26.1 CURVE GEOMETRY

The widespread use of total stations and CAD makes it possible to carry out almost any setting out task, including setting out roads and curved elements such as building façades without having any understanding of curve geometry. However, understanding the maths will help you understand the design and will reduce the chances of error on your part and increase the chances of you spotting errors in the design drawings.

26.2 CURVE GEOMETRY IN ROAD DESIGN

Road alignments are made up of a combination of straight and curved sections laid end to end. There are two components to a road alignment:

1. The horizontal alignment

2. The vertical alignment

The horizontal alignment consists of a series of straight and curved sections joined together (Fig. 26-1). Curved sections are usually an arc of a circle. As well as straight and curved sections being joined, curved sections of different radii may also be joined together. In the vertical alignment, parabolic curves are used instead of circular curves.

For both horizontal and vertical alignments, straight sections meet curved sections at a tangent. Horizontal and vertical curve radii and length are designed for road user comfort and safety, allowing for centrifugal forces and visibility.

Fig. 26-1: Curve alignment

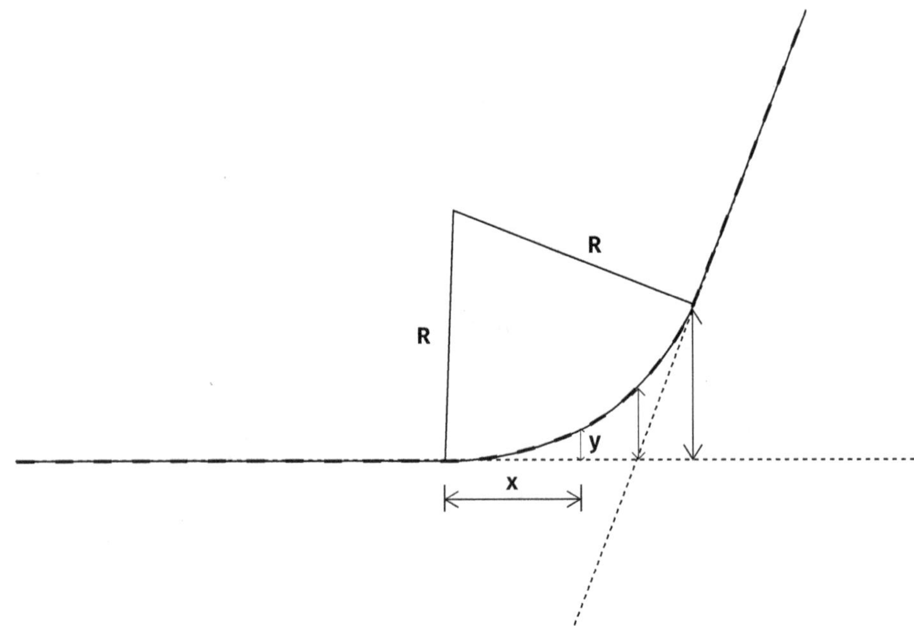

26.3 CIRCULAR CURVE TERMINOLOGY

Fig. 26-2   Circular curves

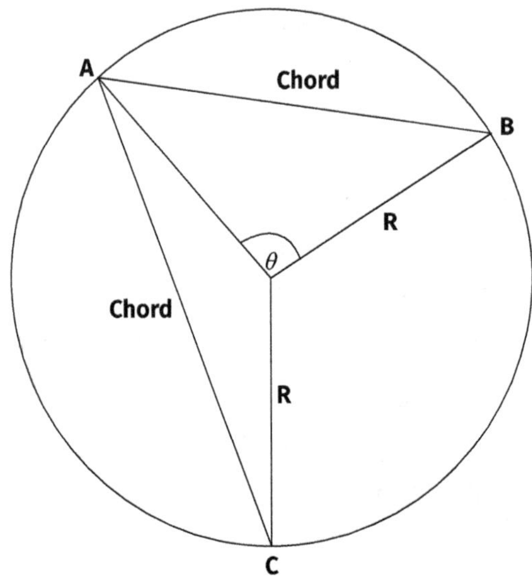

Arc – the section of the circle between two points on the circumference

Chord – the straight line which joins two points on the circumference

Circumference – the distance around the circle

Intersection angle (θ) – the angle subtended at the origin of the circle, by the lines joining the two end points of a chord and the origin

Intersection point (I) – the point at which the tangent lines intersect

Parabola – a curve which has a constant rate of change of gradient

Perimeter – the outside boundary of a circle

Pi (π) - the constant which is the ratio of the diameter to the circumference

Radius (R) – the shortest line from the origin of the circle to the circumference

Tangent lines – the straight lines which touch the circumference at either end of the chord, at an instantaneous point at right angles to the radius

Tangent point – the instantaneous point where the tangent line touches the circumference.

26.4 FORMULAE FOR CIRCULAR CURVES

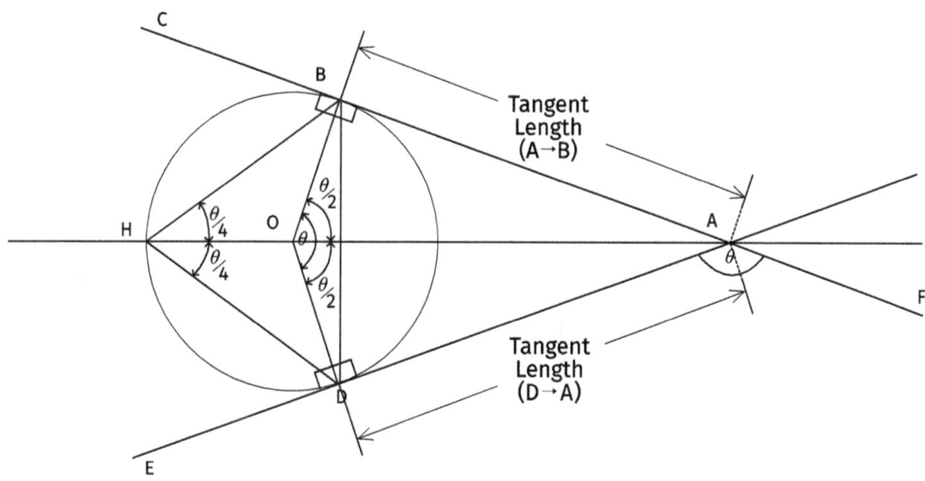

Length of arc = $2\pi R\left(\dfrac{\theta}{360}\right)$

Tangent length = $R\tan\left(\dfrac{\theta}{2}\right)$

Chord length = $2R\sin\left(\dfrac{\theta}{2}\right)$

Major offset = $R\left(1-\cos\left(\dfrac{\theta}{2}\right)\right)$

Intersection angle = WCB(CA) – WCB (AD)

CIRCULAR CURVE GEOMETRY WORKED EXAMPLE

Two straight roads meet at junction I. The junction is to be replaced by a curve or radius 300m

Information:

Point	Eastings (m)	Northings (m)
A	50.000	65.000
I	909.230	216.505
B	1473.046	11.293

Calculate:

a) lengths AI and IB

b) the intersection angle (deviation angle)

c) the tangent length

d) the length of the curve

e) the chainage at each tangent point assuming that the chainage at A = 275.000

f) the chord length

g) the major offset

1) Length of AI = $\sqrt{(\Delta E^2 + \Delta N^2)}$

ΔE = 909.230- 50.000 = 859.230m

ΔN = 216.505-65.000 = 151.505m

AI = $\sqrt{(859.230^2 + 151.505^2)}$ = 872.485m

Length of IB = $\sqrt{(\Delta E^2 + \Delta N^2)}$

= 1473.046 - 909.230 = 563.816m

= 216.505 - 11.293 = 205.212m

IB = $\sqrt{(563.816^2 + 205.212^2)}$ = 600.000m

2) Intersection angle = WCB (IB) − WCB (AI)

a) WCB(AI) = $\tan^{-1}\left(\frac{859.230}{151.505}\right)$ = 80°00'00"

b) WCB(IC) = 90 + $\tan^{-1}\left(\frac{205.212}{563.816}\right)$ = 90° + 20°00'00" = 110°00'00"

c) Intersection angle = 110°00'00" - 80°00'00" = 30°00'00"

c) Tangent length = $R\tan\left(\frac{\theta}{2}\right)$ = $300\tan\left(\frac{30}{2}\right)$ = 80.385m

d) Length of arc = $2\pi R\left(\frac{\theta}{360}\right)$ = $2\pi 300\left(\frac{30}{360}\right)$ = 157.080m

e) Chainage at TP1 = Length AJ − tangent length = 872.485 − 80.385 = 792.100m

d) Chainage at TP2 = Chainage at TP1 + Length of arc = 792.100 + 157.080 = 949.180m

f) Chord length = $2R\sin\left(\frac{\theta}{2}\right)$ = $2\times300\sin\left(\frac{30}{2}\right)$ = 155.291m

g) Major offset = $R1 - \cos\left(\frac{\theta}{2}\right)$ = $300\left(1-\cos\left(\frac{30}{2}\right)\right)$ = 10.222m

EXERCISE:

1. Two straight roads meet at junction I. The junction is to be replaced by a curve of radius 200m

Information:

Point	Eastings (m)	Northings (m)
A	60.000	79.000
I	827.381	246.911
B	1177.861	168.198

   a. lengths AI and IB

   b. the intersection angle (deviation angle)

   c. the tangent length

   d. the length of the curve

   e. the chainage at each tangent point assuming that the chainage at A = 275.000

   f. the chord length

   g. the major offset

26.5 DIFFERENT METHODS OF SETTING OUT HORIZONTAL CURVES

Most decent modern total stations offer a function for setting out curves, allowing for the different connotations of information available. The point information can be input as coordinates, or by sighting to a prism held on the point. The options are:

→ enter the start and end points and the radius

→ enter the centre of the circle and the start point of the curve

→ enter any 3 points on the curve and the total station will resolve the radius and position of the curve.

There are numerous ways for setting out curves using a total station or theodolite and tape measure, which you are unlikely to ever need, however they are included in this book since they are still included as part of some academic modules.

26.6 CALCULATING THE COORDINATES AT POINTS ON A CURVE

Once the required coordinates have been calculated, the Stakeout function can be used as with any other coordinates. The steps for calculating the coordinates of points on a curve are as follows:

1. Calculate the WCB of the two straight sections AJ and JC

2. Calculate the intersection angle, θ, where $\theta =$ WCB(JC) – WCB (AJ)

3. Calculate the WCB(AO)

4. Calculate the partial coordinates of AO

5. Calculate the coordinates of O

6. Calculate the angle, alpha, as a ratio of the arc length to the circumference

7. Calculate the WCB of OP_1

8. Calculate the partial coordinates of OP_1

9. Apply the partial coordinates of OP_1 to O to find the coordinates of P_1

10. Repeat steps 6 – 9 for all points required.

WORKED EXAMPLE FOR CALCULATING THE COORDINATES ON A CURVE

Chainage at A = 0m

R = 300

Point	E	N
A	1003.520	871.605
B	1104.907	887.206
C	1292.245	855.114
D	1393.901	800.632

For the horizontal alignment shown above, calculate:

a) The chainage at B and C

b) The coordinates at chainage 120m (Ch.120)

c) The coordinates at Ch.200

1) To calculate the chainage at B, calculate the length AB and add it onto the chainage at A.

The length AB = $\sqrt{(\Delta E_{AB}^2 + \Delta N_{AB}^2)}$

Where:

$\Delta E_{AB} = E_B - E_A$ = 1104.907 − 1003.520 = 101.387m

$\Delta N_{AB} = N_B - N_A$ = 887.206 − 871.605 = 15.601m

So AB = $\sqrt{(101.387^2 + 15.601^2)}$ = 102.580m

Therefore the chainage at B = 0 + 102.580 = 102.580m

To calculate the chainage at C, calculate the arc length and add it onto the chainage at B.

To calculate the arc length, we first need to calculate the intersection angle, θ, by finding the difference between the whole circle bearing from A - B, WCB (AB), and the whole circle bearing from C – D, WCB(CD).

WCB(AB) = $\tan^{-1} (\Delta E_{AB} / \Delta N_{AB})$ = $\tan^{-1}$ (101.387/15.601) = 81°15′ 08″

WCB(CD) = $90 + \tan^{-1} (\Delta E_{CD} / \Delta N_{CD})$

Where:

$\Delta E_{CD} = E_D - E_C = 1393.901 - 1292.245 = 101.656m$

$\Delta N_{AB} = N_D - N_C = 800.632 - 855.114 = -54.482m$

So WCB(CD) = 90 + tan-1 (101.656/ 54.482) = 118°11'19"

and Intersection angle = WCB(CA) – WCB (AD) = 118°11'19" - 81°15' 08" = 36°56'11"

and arc length $2\pi R\left(\frac{0}{360}\right) = 2\pi R\left(\frac{36°56'11"}{360}\right)$ 193.398m

Therefore the chainage at C = Chainage at B + arc length = 102.580 + 193.398 = 295.978m

2) To calculate the coordinates at Ch.120, we need to calculate the coordinates at O by first calculating the WCB of the line BO. We can then calculate the difference in eastings and northings from B to O.

Then we will calculate WCB(OB) then the WCB (O→Ch.120) by calculating α. We can then calculate the difference in eastings and northings between O and Ch.120

The WCB(BO)= WCB(AB) + 90 = 171°15'08"

$\Delta E_{(BO)}$ = R x sin (WCB) = R x sin (171°15'08") = 45.626m

$\Delta N_{(BO)}$ = R x cos (WCB) = R x cos (171°15'08") = -296.510m

which gives us the coordinates of O as:

$E_O = E_B + \Delta E_{(BO)}$ = 1104.907 + 45.626 = 1150.533m

$N_O = N_B + \Delta N_{(BO)}$ = 887.206 + (-296.510) = 590.696m

The WCB(OB) = 180 + 171°15'08" = 351°15'08"

To find α:

$\alpha/360$ = Arc/circumference

So rearrange to get:

α = 360 x Arc/ circumference

where the circumference = $2\pi R = 2 \times \pi \times 300$ = 1884.956m

and the arc length = 120 - 102.580 = 17.420m

So α = 360 x 17.420/1884.956 = 3°19'37"

Therefore the WCB(O→Ch.120) = 351°15'08" + 3°19'37" = 354°34'45"

and

$\Delta E_{(O \to Ch.120)}$ = R x sin (WCB) = R x sin (354°34'45") = -28.328m

$\Delta N_{(O \to Ch.120)}$ = R x cos (WCB) = R x cos (354°34'45") = 298.660m

So

$$E_{Ch.120} = E_0 + \Delta E_{(O \to Ch.120)} = 1150.533 + (-28.328) = 1122.205m$$

$$N_{Ch.120} = N0 + \Delta N_{(O \to Ch.120)} = 590.696 + 298.660 = 889.356m$$

Therefore, the coordinates at Ch. 120 = (1122.205mE, 889.356mN)

3) To calculate the coordinates at Ch.200, we already have most of the information we need from the previous step so there are just a couple more steps:

The arc length = 200 - 102.580 = 97.420m

and

$$\alpha = 360 \times 97.420/1884.956 = 18°36'21''$$

$$WCB(O \to Ch.200) = 351°15'08'' + 18°36'21'' = 369°05'29''$$

There are only 360° in a full circle so subtract 360° to give WCB(O→ Ch.200) = 9°05'29''

So:

$$\Delta E_{(O \to Ch.200)} = R \times \sin (WCB) = R \times \sin (9°05'29'')= 51.362m$$

$$\Delta N_{(O \to Ch.200)} = R \times \cos (WCB) = R \times \cos (9°05'29'') = 295.570m$$

which gives:

$$E_{Ch.200} = E_0 + \Delta E_{(O \to Ch.200)} = 1150.533 + 51.362 = 1201.895m$$

$$N_{Ch.200} = N_0 + \Delta N_{(O \to Ch.200)} = 590.696 + 298.660 = 886.266m$$

Therefore, the coordinates at Ch. 200 = (1201.895m, 886.266m mN)

EXERCISE:

1. For the horizontal alignment B – C calculate:

   a. The chainage at B and C

   b. The coordinates at chainage 120m (Ch.120)

   c. The coordinates at Ch.200

The Chainage at A = 0m, R = 250

Point	E	N
A	3003.520	1271.605
B	3104.907	1287.206
C	3292.245	1255.114
D	3393.901	1200.632

26.7 SETTING OUT A CURVE BY LOCATING THE CENTRE

The steps for setting out a curve by locating the centre (Fig. 26-3) are as follows:

1. Set out tangent lines and project them using the total station or a string line to locate the IP.

2. Calculate the tangent length TP→IP.

3. Measure back the tangent length from IP to set out the TPs.

4. Hammer a steel pin into the ground at each TP.

5. To set out the origin, O, hook a tape over each tangent point and spray an arc. Where the arcs intersect, hammer a steel pin into the ground. Alternatively, use step 6.

6. Set up the total station over the TP and sight onto the tangent line. Rotate 90 degrees and make a mark at distance, R, from the TP. Repeat from the other TP to check the mark. Hammer a steel pin into the ground at O.

7. Check chord length on ground = theoretical chord length (or some other geometrical check).

8. Hook the tape over the steel pin at O.

9. Any point on the curve can be marked at distance R.

26.8 SETTING OUT A CURVE BY OFFSETS FROM THE TANGENT

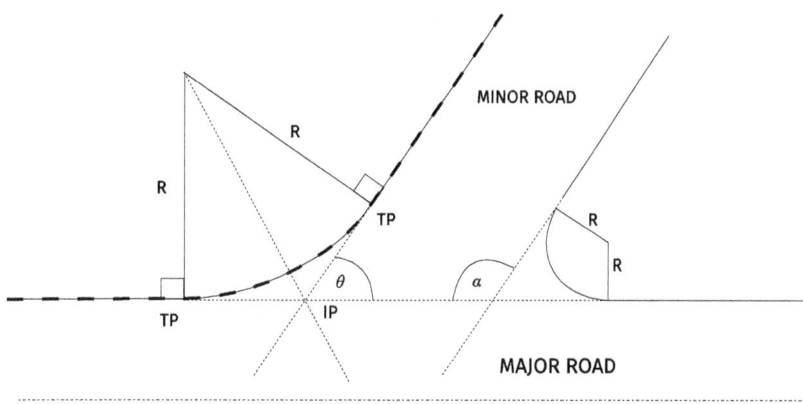

See diagram above

Formula:

$$Y = R - \sqrt{(R^2 - x^2)}$$

Where:

Y = offset

X = distance from TP1

To set out a curve using offsets from the tangent, the steps are:

1. Set out TP1.

2. Set out the tangent line to any point beyond the corresponding point at TP2.

3. Set up the total station at a known distance, x_1, along the tangent line from TP1.

4. Sight along the tangent line.

5. Rotate 90 degrees.

6. Set out the corresponding distance, y_1, using a tape measure or distance measurement function.

7. Move the total station to set up at X_2 from TP1 along the tangent line.

8. Sight along the tangent line.

9. Rotate 90 degrees.

10. Set out the corresponding distance, y_2.

26.9 SETTING OUT A CURVE BY OFFSETS FROM THE LONG CHORD

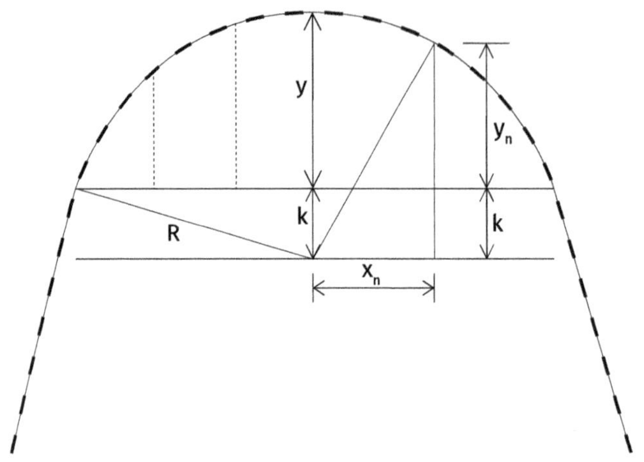

The major offset, $y = (R - k)$

Where

$$k = \sqrt{(R_2 - X_2)}$$

$$y_n = (R^2 - X_n^2) - k$$

and

y_n = offset

X_n = the distance from the midpoint of the chord

26.10 SETTING OUT A CURVE BY DEFLECTION ANGLES USING THE TOTAL STATION

The steps for setting out a curve by deflection angles using a total station(Fig. 26-3) are:

1. Set up on T_1 and sight to the IP.

2. Set the horizontal angle to 0.

3. Turn the horizontal angle until the angle α_1 is displayed on the screen.

4. Directing your assistant left to right until they are in line with the vertical cross-hair and using either a tape measure or the distance measurement function in your total station, hammer a steel pin into the ground at distance T1→B.

5. Turn the horizontal angle until the angle $(\alpha_1 + \alpha_2)$ is displayed on the screen.

6. Directing your assistant left to right until they are in line with the vertical cross-hair and using either a tape measure or the distance measurement function in your total station, hammer a steel pin into the ground at distance T1→C.

7. Turn the horizontal angle until the angle $(\alpha_1 + \alpha_2 + \alpha_3)$ is displayed on the screen.

8. Directing your assistant left to right until they are in line with the vertical cross-hair and using either a tape measure or the distance measurement function in your total station, hammer a steel pin into the ground at distance T1→D.

9. Continue like this until all points are set out.

Fig. 26-3:

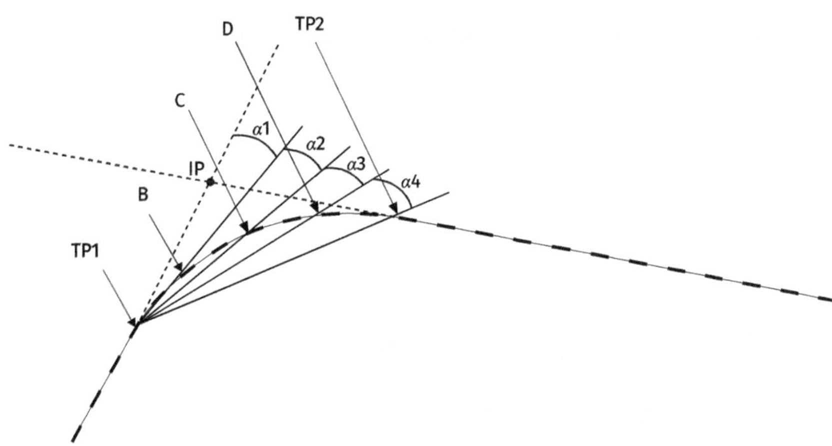

26.11 SETTING OUT A CURVE BY DEFLECTION ANGLES USING A TAPE AND DIGITAL THEODOLITE

See Fig. 26-3

This method involves calculating deflection angles and tangential angles where:

→ the deflection angle, α, is the angle turned from the previous point on the curve.

→ the tangential angle is the total angle turned from the tangent line, i.e. the cumulative sum of the deflection angles.

The steps for setting out a curve by deflection angles using a tape and digital theodolite are:

1. Hammer a steel pin into the ground at T1.

2. Set the total station over T1 and sight to the IP, or in fact any point on the tangent line.

3. Set the horizontal angle to 0.

4. Turn the horizontal circle until the angle α_1 shows on the screen.

5. Hook the zero end of a tape measure over the steel pin at T1.

6. Have your assistant pull the tape taut and locate the chord length, AB, on the tape measure with the tip of their thumb.

7. Sight through the theodolite and direct your assistant left or right until the tip of their thumb is in line with the vertical cross-hair.

8. Have them hammer in a steel pin at this point, B.

9. Turn the horizontal circle until the angle $(\alpha_1 + \alpha_2)$ shows on the screen.

10. Have your assistant hook the zero end of the tape measure over the pin at B.

11. Have your assistant pull the tape taut and locate the chord length, BC, on the tape measure with the tip of their thumb.

12. Sight through the theodolite and direct your assistant left or right until the tip of their thumb intersects with the vertical cross-hair.

13. Have them hammer in a steel pin at this point, C.

14. Continue like this until all points are set out.

WORKED EXAMPLE FOR CALCULATING DEFLECTION ANGLES

Two straight sections of road are connected by a curve of radius 200m. The intersection angle, θ, is 36°. The chainage at A is 354m.

1) Draw a diagram of the arrangement

2) Calculate the deflection angles and tangential angles for every 20m, starting at the first multiple of 20.

First calculate the arc length, and hence the chainage at B:

$$\text{arc length} = 2\pi R\left(\frac{\theta}{360}\right) = 2\times\pi\times250\left(\frac{36}{360}\right) = 157.080\text{m}$$

The chainage at B = chainage at A + arc length = 354.000 + 157.080 = 511.080m

Since the chord length is less than 1/20th of the radius, we can assume that the chord length and the arc length are equal, and use the formula:

$$\alpha = \left(\frac{c}{R} \times 1718.9\right) \text{minutes}$$

Where c = chord length

We must calculate the initial and final chord lengths:

The first multiple of 20 after A is 360m, so the initial chord length = 360 − 354 = 6m

The last multiple of 20 before B is 500m, so the final chord length = 11.080m

We can now tabulate the results and use the information for setting out the curve:

Chord	Chainage	Length	Deflection angle	Tangential angle
A	354	6	00°41'13"	00°41'13"
1	360	20	02°30'31"	03°11'44"
2	380	20	02°30'31"	05°42'15"
3	400	20	02°30'31"	08°12'46"
4	420	20	02°30'31"	10°43'17"
5	440	20	02°30'31"	13°13'48"
6	460	20	02°30'31"	15°44'19"
7	480	20	02°30'31"	18°16'50"
8	500	20	02°30'31"	20°45'21"
9 (B)	511.078	11.078	00°16'10"	21°01'31"
			Σ = 21°01'31"	

26.12 VERTICAL CURVES

The vertical alignment consists of straight sections joined by parabolic curves. Vertical curves can be crest (summit) curves or sag (valley) curves. Vertical curves are parabolic since they have a uniform rate of change of gradient which compensates for the effects of centrifugal forces in gravity. Theoretically, if a crest curve was circular instead of parabolic, a vehicle travelling at speed would follow the trajectory of the curve and would leave the surface of the road and land. This would be dangerous and also uncomfortable for road users. Even though in practice, the effect of using circular curves instead of parabolic would have negligible effect, parabolic curves are still used.

This chapter covers the basics of vertical curve geometry and explains how to calculate the reduced levels at intervals along a vertical curve alignment. For a more in-depth explanation of vertical curve design including k values, minimum curve length and minimum and maximum gradients, refer to *Surveying for Engineers* by Uren and Price and *Surveying for Construction* by Irvine and MacLennan.

26.13 METHODS FOR SETTING OUT VERTICAL CURVES

Vertical alignments are usually set out using road pins or profile boards (chapter **22**). For small projects, the positions of the pins or boards may be set out using a tape measure, however for medium and larger schemes, a total station or GNSS will most likely be available. The levels of the pins or boards may be set using the total station, GNSS or automatic level. The information provided in the design will vary from project to project and in some cases it may be necessary to manually calculate the reduced levels. This section explains the theory and maths involved in calculating the levels of vertical curves.

26.14 VERTICAL CURVE CALCULATIONS

For the purposes of the calculations in this chapter:

→ the chainage increases from left to right

→ the gradient of the left hand straight (approaching the curve) is G_1

→ the gradient of the right hand straight (exiting the curve) is G_2

→ gradients are expressed as a percentage (see section **28.33**)

→ a straight section which increases in level has a positive gradient

→ a straight section which decreases in level has a negative gradient

→ the tangent point at the beginning of the curve is C

→ the tangent point at the end of the curve is D

→ the length of the curve is L

→ the reduced level is RL

→ vertical curves can be assumed to be 'flat' since the curve length is less than 1/10 of the approximated radius

→ the tangent lengths of the straight sections approaching the curve and exiting it are equal

→ the intersection point, I, is midway between C and D in plan

→ the arc length of the curve can be assumed to be the same as the straight length between the two tangent points, C and D

→ the plan distance between the two tangent points, C and D, can be assumed to be equal to the slope distance.

Formulae:

$$RL_x = RL_C + bx + ax^2$$

Where:

$$a = (G_2 - G_1)/\ 200L$$
$$b = G_1/100$$

WORKED EXAMPLE

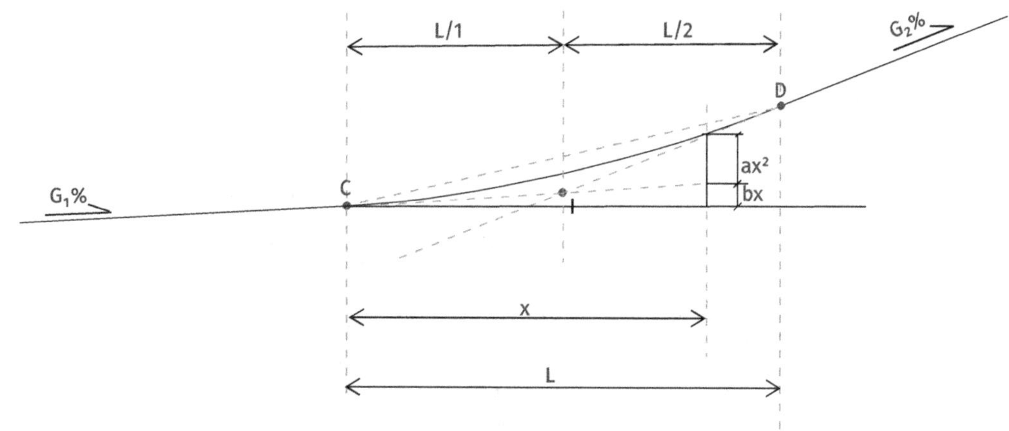

For the vertical curve in the image above, the RL at the initial tangent point is 60.52m and the chainage is 51.35m. The straight section entering the curve has a gradient of 1:40 and the straight section exiting the curve has a gradient of -1:25.

Calculate:

1. The chainage at the final tangent point, D.

2. The reduced levels at 20m intervals, starting at the next multiple of 20m, and at the final tangent point, D.

Answers:

1. Since the curve can be assumed to be 'flat', then the arc length CD = the curve length of 120m. Therefore the chainage at D = 51.35 + 120 = 171.35m

2. The next multiple of 20 after 51.35 is 60m. Therefore we need to calculate the RLs at Ch.60m, Ch.80m, etc. up to Ch. 160m and 171.35m.

$$RLx = RL_x = RL_C + bx + ax^2$$

Where:

$$a = \frac{(G_2 - G_1)}{200L}$$

$$b = \frac{G_1}{100}$$

and

$$G1 = \frac{1}{40} \times 100 = 2.5\%$$

$$G2 = \frac{-1}{25} \times 100 = -4\%$$

So,

$$a = \frac{(-4 - 2.5)}{(200 \times 120)} = 2.7 \times 10^{-3} = 0.000271$$

$$b = 2.5/100 = 0.025$$

Ch (m)	X (m)	RLC + bx	ax^2	RLC + bx + ax^2
51.35	0	60.52	0	60.52
60	8.65	60.74	-0.02	60.72
80	28.65	61.24	-0.22	61.02
100	48.65	61.74	-0.64	61.10
120	68.65	62.24	-1.28	60.96
140	88.65	62.74	-2.13	60.61
160	108.65	63.24	-3.20	60.04
171.35	120	63.52	-3.90	59.62

CHAPTER 27

CONTOURING

27.1 WHAT IS CONTOURING?

Contouring is a technique which enables us to create a visual representation of a 3D profile on a flat page.

Representing an irregular surface in this way allows us to get an overview of the profile of the ground, observe gradient patterns, calculate surface areas and volumes and develop design. We can use contour lines to analyse a section through the model.

27.2 WHAT ARE CONTOUR LINES?

Contour lines are smooth lines on a map or plan, joining points of equal height. The closer together the contour lines, the steeper the slope. The further apart the lines, the more gradual the slope.

27.3 HOW DO I CREATE CONTOUR LINES?

To create contour lines, you will need a survey of the area showing spot heights (reduced levels plotted to scale in their relative positions). The survey may have been carried out using a total station or an automatic level. The closer together the spot heights, the greater the level of detail. Contour lines can be plotted using software or by hand. Plotting contour lines uses the principle of similar triangles, the triangle being formed by a level change and plan distance. Features such as fence lines should be ignored in this exercise, as they do represent the shape of the ground.

To plot contour lines by hand:

1. Look for the highest and lowest point and subtract them from each other to determine the range of heights.

2. Decide on the height interval between contour lines. This will depend on the range of heights and the level of detail you require.

3. Decide which contour line you want to plot first.

4. Select any two adjacent spot heights whose levels are either side of a contour line (Fig. 27-1).

5. Calculate the level difference between the two points by subtracting one from the other.

6. Measure the distance between the points with a ruler.

7. Show the level change and plan distance on a diagram.

8. Calculate the level change between the highest point and the level of the contour line.

9. Calculate the ratio of the level change between the highest point and the contour line to the full level change.

10. Apply this ratio to the plan distance.

11. In a straight line between the two points, plot the position of the contour line at the correct distance from the highest point.

12. Repeat steps 4 – 11 for all sets of points which fall either side of the contour line.

13. Repeat steps 4 – 11 for each interval.

14. Connect the points plotted for each interval to form contour lines (Fig. 27-2).

Fig. 27-1: Plot contour lines - Step 1

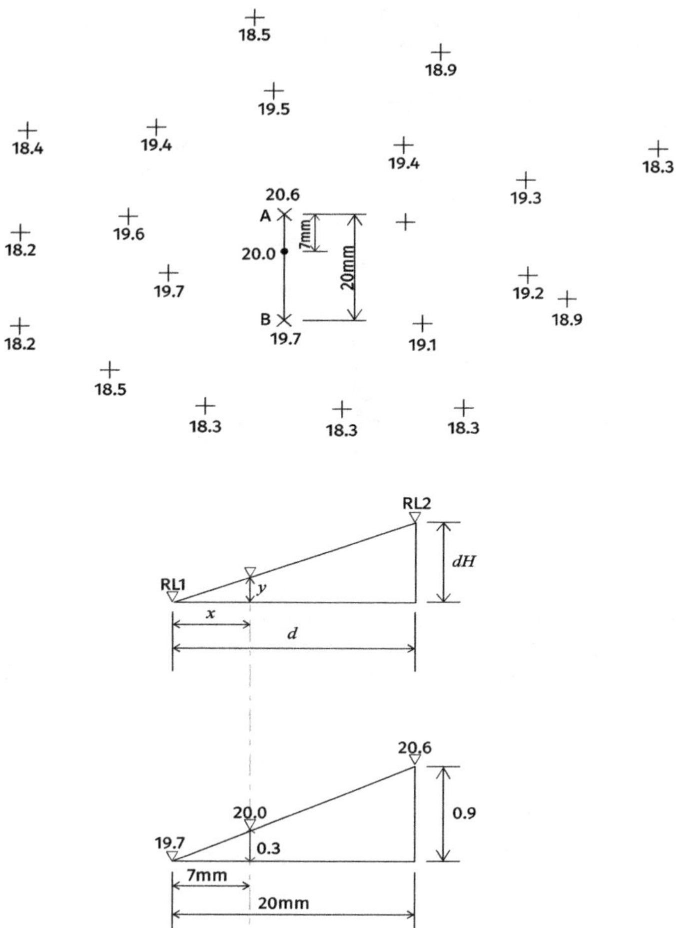

Fig. 27-2: Plot contour lines - Step 2

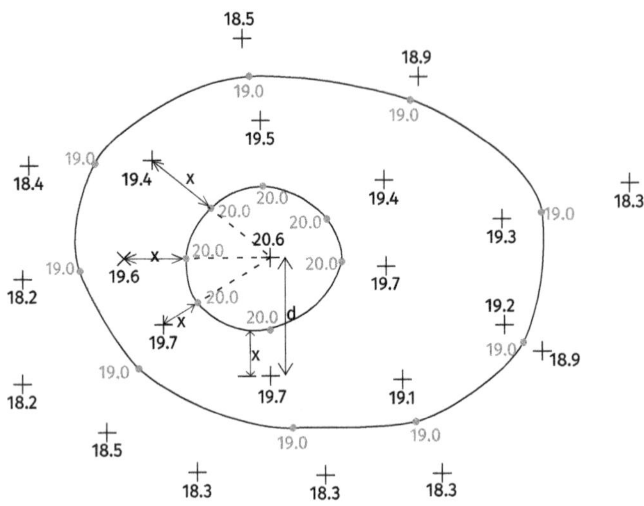

27.4 HOW DO I USE CONTOUR LINES TO CREATE A SECTION?

To create a section by hand (Fig. 27-3):

1. Draw the section line through the plan.

2. Set up a graph where the x-axis represents the distance along the section and the y-axis represents the contour intervals.

3. At each point that the section line crosses a contour line either draw a construction line vertically down the page or measure its distance along the section line to determine its position.

4. Plot the position along the x-axis vs the contour interval on the y-axis.

5. Join the points with a smooth line.

Fig. 27-3: Contour lines

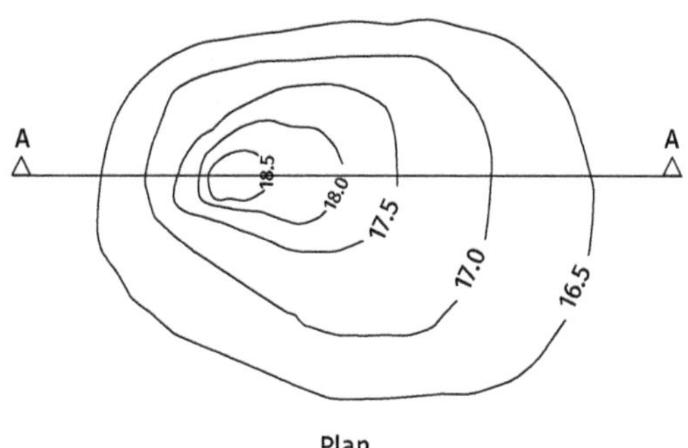

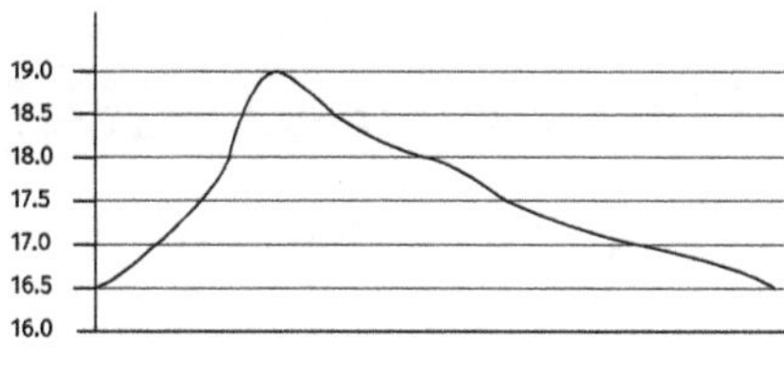

Section A-A

27.5 HOW DO I USE CONTOUR LINES TO CALCULATE A GRADIENT?

The gradient of a slope is measured tangentially to the contour lines. To calculate the gradient at any given point:

1. Draw a perpendicular line between two contour lines (or as close to perpendicular as possible).

2. Measure the distance between the two points with a ruler.

3. Convert the measured distance into a horizontal distance between the two points (see section 4.8).

4. The gradient is the horizontal distance divided by the level change, expressed as 1: x.

CHAPTER 28

MATHS

28.1 PYTHAGORAS'S THEOREM (CALCULATING THE DIAGONAL)

Pythagoras's Theorem is the name of the mathematical process used for calculating the hypotenuse of a right-angled triangle, or in other words, the length of the diagonal of a square or rectangle.

The process derives from the fact that the square of the hypotenuse adds up to the sum of the squares of the other two sides of the triangle as represented in the diagram below.

The formula is:

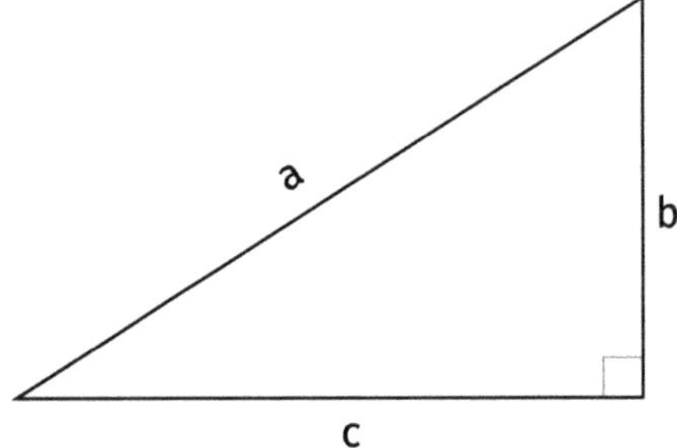

$$a^2 = b^2 + c^2$$

To find the length of the hypotenuse, the equation can be arranged to get:

$$a = \sqrt{(b^2 + c^2)}$$

Where:

a = the length of the hypotenuse

b and c = the lengths of the other two sides of the triangle

Note: for triangles which do not contain a right-angle, see sections **28.23 and 28.24** for the sine and cosine rules.

488

28.2 WHY WOULD I WANT TO CALCULATE THE DIAGONAL OF A SQUARE OR RECTANGLE?

In the context of setting out, there are many scenarios where you might need to calculate the hypotenuse of a right angled-triangle. Some common scenarios include:

→ calculating the theoretical distance of the hypotenuse so you can compare it to the as-set-out distance between two points

→ calculating the distance between two points which you know the coordinates of, by calculating the partial coordinates (difference in eastings and difference in northings)

→ calculating the slope distance between two points when you know the level difference and the plan distance.

WORKED EXAMPLE

Q: What is the length of the hypotenuse?

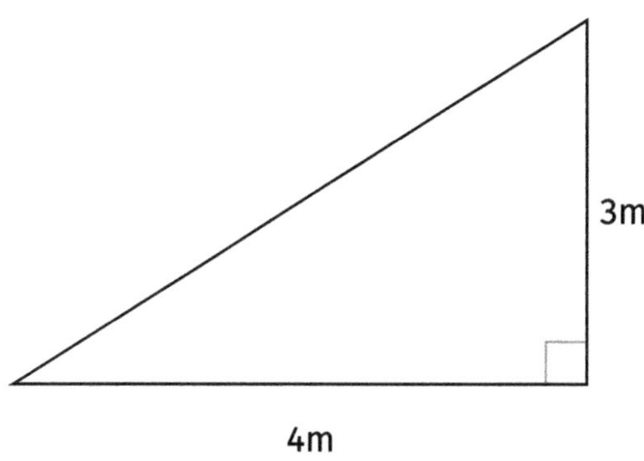

A: In this case,

$b = 3m$

$c = 4m$

So if

$a = \sqrt{(b^2+c^2)}$

Then

$a = \sqrt{(3^2+4^2)}$

So

$a = \sqrt{25} = 5m$

EXERCISE:

1. If two sides of a building are 11.5m and 8.1m, what is the length of the diagonal?

28.3 HOW DO I FIND THE LENGTH OF A SIDE IF I ALREADY KNOW THE DIAGONAL?

In some cases, you know the length of the diagonal and one of the sides, and want to find the length of the unknown side. In this case you can rearrange the formula to get:

$$c = \sqrt{(a^2 - b^2)}$$

WORKED EXAMPLE

Q: What is the length of side c?

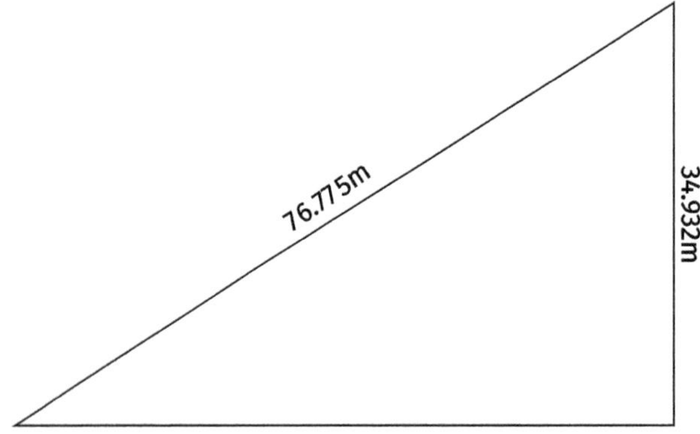

A: In this case,

a = 76.775m

b = 34.932m

So if

$$c = \sqrt{(a^2 - b^2)}$$

Then

$$c = \sqrt{(76.775^2 - 34.932^2)}$$

So

c = 68.641m

EXERCISE:

2. If the the level change between two points is 2.340m and the distance measured with a tape measure is 25.435, what is the plan distance?

28.4 ANGLE GEOMETRY

Relative angles and distances are the mathematical basis for all surveying and setting out activities.

There are a few simple rules of angle geometry that you should be conversant with and they are described in further detail in this section.

28.5 THE OPPOSITE ANGLES RULE

The opposite angles rule, is that when two lines intersect, the opposite angles are equal. In the diagram below, 'A' and 'B' show equal, opposite angles.

Fig. 28-1: The opposite and corresponding angles rule

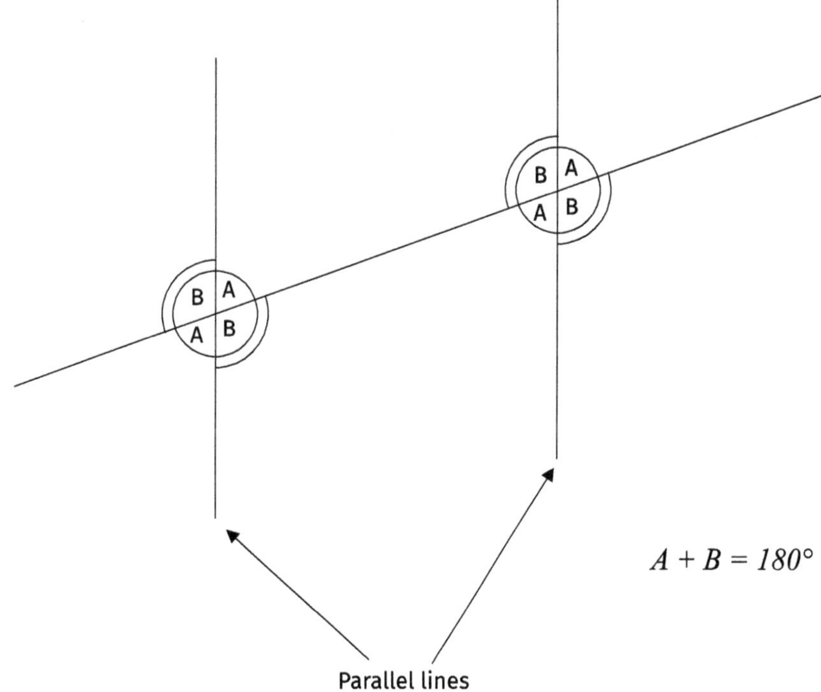

28.6 THE CORRESPONDING ANGLES RULE

When two parallel lines are intersected by a third line, the corresponding angles are equal. In the diagram below, all the angles labelled 'a' are equal and all the angles labelled 'b' are equal.

EXERCISE:

3. In the arrangement below, what is angle Q?

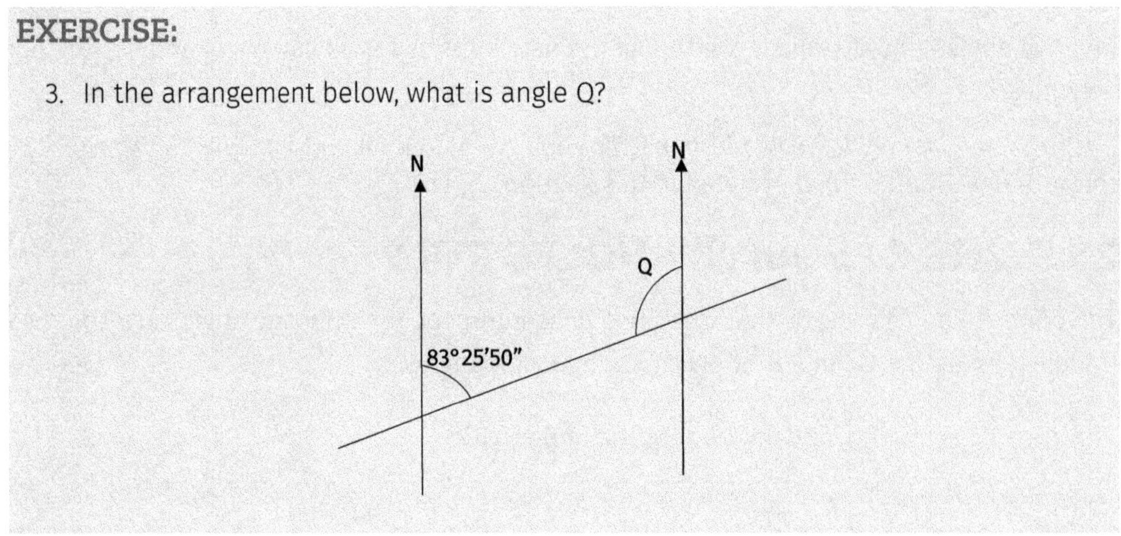

28.8 ADDING AND SUBTRACTING ANGLES

There are 360 degrees in a full circle. In the diagram below, a + b + c + d + e = 360∬

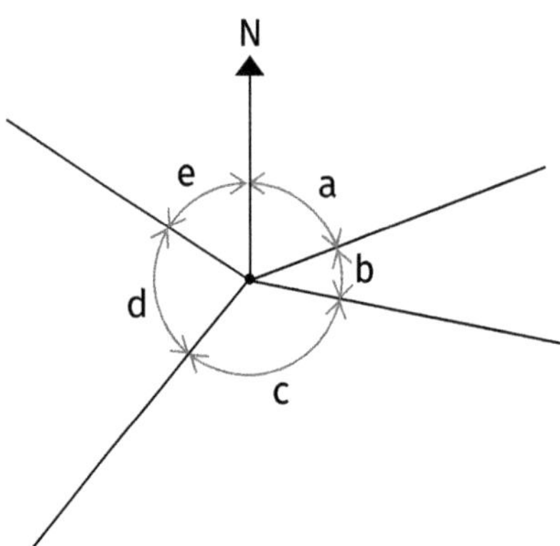

There are 180 degrees in half a circle. In the diagram below, a + b + c + d + e = 180°

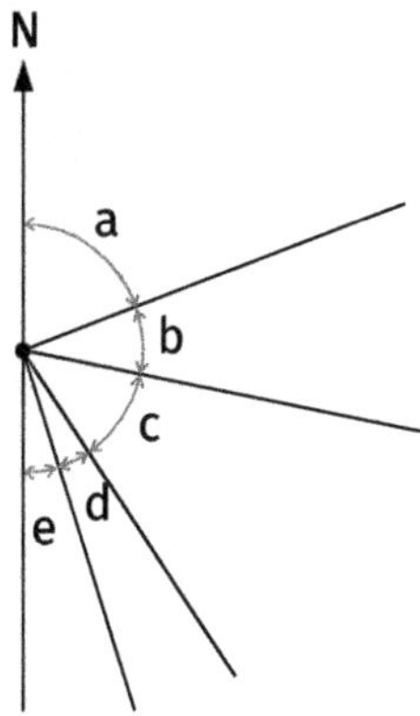

EXERCISE:

4. In the arrangement below, what is angle G?

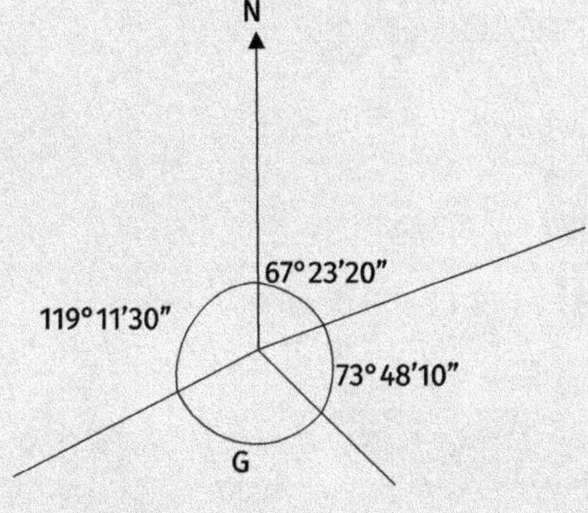

28.9 DEGREES, MINUTES AND SECONDS

'Degrees, minutes and seconds' is an alternative way of expressing decimal angles. Total stations display angle readings using this convention, so it is important to know how to work with them. It is best practice to work with degrees, minutes and seconds in your calculations, rather than with decimal angles, for several reasons:

→ it minimises the risk of error arising from converting back and forth between different formats

→ it is a more convenient format and easier to remember, so there is less chance of writing down the number incorrectly, than there is with decimal numbers

→ the format is different to that used for distances, so there is less chance of accidentally inputting an angle into your calculations instead of a distance, and vice versa.

28.10 WHAT ARE DEGREES MINUTES AND SECONDS (DMS)?

A full circle is split into 360 degrees. Each degree is split into 60 minutes. Each minute is split into 60 seconds.

28.11 HOW DO I CONVERT DEGREES, MINUTES AND SECONDS INTO A DECIMAL ANGLE?

To convert an angle expressed in DMS into a decimal angle, the formula is:

$$\text{Decimal angle} = \text{degrees} + \frac{\text{minutes}}{60} + \frac{\text{seconds}}{3600}$$

WORKED EXAMPLE:

Q: Convert 67° 53'17" into a decimal angle

A: Decimal angle = $67 + \frac{53}{60} + \frac{17}{3600}$ = 67.88806 degrees

EXERCISE:

5. Convert 329° 37'29" in to a decimal angle

28.12 HOW DO I CONVERT A DECIMAL ANGLE INTO DEGREES MINUTES AND SECONDS?

To convert a decimal angle into DMS, the process is:

1. Subtract the whole number and multiply the remaining decimal by 60. This gives the number of minutes.

2. Subtract the whole number from the answer and multiply the remaining decimal by 60. This gives the number of seconds. Round to the nearest whole number.

WORKED EXAMPLE:

Q: Convert 159.2837486 into DMS

A: Step 1:

(159.2837469 – 159) x 60 = 17.0248'

Step 2:

(17.0248 – 17) x 60 = 1.48" (round up to 2")

Therefore, the answer is 159 17' 2"

EXERCISE:

6. Convert 83.283048287 in to DMS

28.13 HOW DO I WORK WITH DEGREES, MINUTES AND SECONDS IN MY CALCULATOR?

There is a button in most scientific calculators for converting decimal degrees to degrees, minutes and seconds and vice versa. It may be labelled DMS or ° " '

To input an angle in degrees, minutes and seconds:

1. Input the degrees then press the button

2. Input the minutes then press the button

3. Input the seconds then press the button

To convert to decimal degrees:

1. Input the degrees, minutes and seconds

2. Press =

3. Pressing the DMS button will toggle between the decimal angle and DMS

To convert to degrees, minutes and seconds:

1. Input the decimal degrees

2. Pressing the DMS button will toggle between the decimal angle and DMS

28.14 WHAT ARE RADIANS?

A radian is a unit of measurement for angles. There are 2π radians in a full circle, where:

π = 3.141592654

1 radian = $\frac{180}{\pi}$ degrees = 57.295779 degrees

28.15 WHEN DO I NEED TO USE RADIANS IN SETTING OUT?

For setting out, it is unlikely that you will need to convert degrees to radians in the course of your everyday work. It is worth noting however, that spreadsheets measure all angles in radians, so if you create a spreadsheet where you need to input angles, you will need to convert them to radians before carrying out any trigonometrical functions. If you are using a spreadsheet to calculate angles, the answer will be given in radians, so you will need to convert the answer to degrees.

28.16 CALCULATION OF AREAS

The most suitable method for calculating an area will depend on the format of the information, the nature of the task and the accuracy required. Some possible methods include:

→ component parts - Breaking an irregular area down into component parts by inspection, calculating the areas of the individual regular shapes and summing them to find the total area

→ grid method - Breaking an irregular area down into a grid and calculating the area based on the number of full squares, number of half-filled squares, number of squares more than half- filled, number of squares less than half-filled

→ approximating the area of an irregular shape to a regular shape

→ using survey software or CAD to define the perimeter of the shape, and using the 'measure area' function to determine the area contained within the perimeter

→ using the total station to survey the perimeter and subsequently calculate the area using one of the above methods

→ using the 'measure area' function on the total station

Formulae for basic shapes:

Area of a rectangle = base x height

Area of a triangle = ½ (base x height)

Area of a parallelogram = base x height

Area of a trapezium = ½ (base 1 + base 2) x height

Area of a circle = π x radius²

WORKED EXAMPLE FOR CALCULATING AN AREA BY COMPONENT PARTS:

You have been asked to calculate the area of a car park so that the correct amount of tarmac can be ordered. Calculate the areas of the following shape:

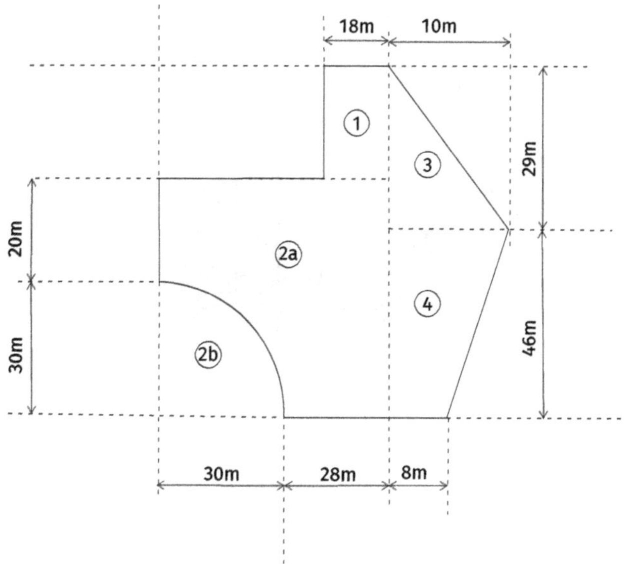

Step 1: Break the shape down into component parts and calculate the area of each part

Part 1: base x height = 18 x 25 = 450m^2

Part 2a: base x height = 58 x 50 = 2900m^2

Part 2b: ¼ x π x radius2 = ¼ x π x 302 = 706.85m^2

Part 3 = ½ (base x height) = ½ x 10 x 29 = 145m^2

Part 4 = ½ (base 1 + base 2) x height = ½ (10 + 8) x 46 = 414m^2

Step 2: Calculate the total area:

Total Area   = Part 1 + Part 2a – Part 2b + Part 4 + Part 4

= 450 + 2900 – 706.85 +145 + 414 = 2792.15m2

WORKED EXAMPLE FOR CALCULATING AN AREA USING THE GRID METHOD

You are given a plan showing an area of forest. Use the grid method to calculate the area if the grid is at 20m spacing :

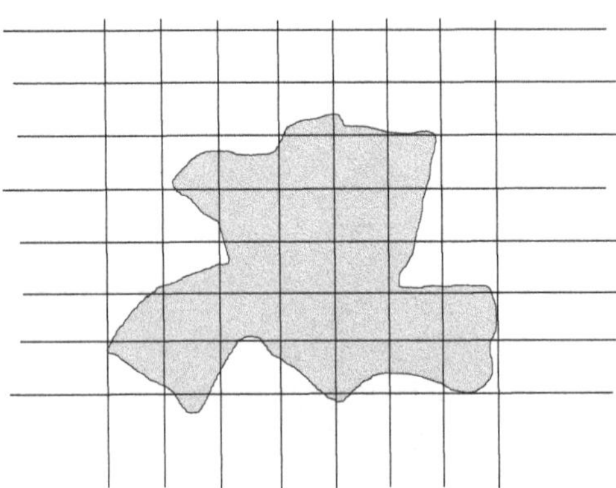

Draw on by hand, a 20m grid:

Step 1: Calculate the number of squares:

Number of full squares = 9

Number of half squares = 4

Number of squares greater than half = 12

Ignore squares which are less than half

Total number of squares = 9 + 4 + 12 = 25

Step 2: Calculate the area of each square:

Area of each square = 20 x 20 = 400m^2

Step 3: Calculate the total area:

Total area = number of squares x area of each square = 25 x 400 = 10000m^2

28.17 CALCULATION OF VOLUMES

As with areas, the most suitable method for calculating a volume will depend of the format of the information, the nature of the task and the accuracy required.

Some possible methods include:

→ component parts - breaking an irregular volume down into component parts by inspection, calculating the volumes of the individual regular shapes and summing them to find the total volume

→ approximating the area of an irregular volume to a regular volume

→ using survey software or CAD to define the boundaries of a 3D model, and using the volume measurement function to determine the volume contained within the boundary

→ using the total station to survey the boundary and surface and subsequently calculate the area using one of the above methods

→ using the volume measurement function on the total station

Formulae for basic 3D shapes:

Volume of a cuboid = base x height x length
Volume of a triangular prism = ½ (base x height x length)
Volume of a cylinder = height x π x radius²
Volume of a pyramid = length x width x height
Volume of a cone = ⅓ (π x radius² x height)
Volume of a sphere = ⁴⁄₃ (π x radius³)

WORKED EXAMPLE FOR CALCULATING THE VOLUME OF A COMPOSITE SHAPE

You have been asked to calculate the total volume of concrete required for a length of retaining wall so that the costs can be calculated. Calculate the volume required for this 20m-long retaining wall, allowing 5% for the volume of the steel reinforcement:

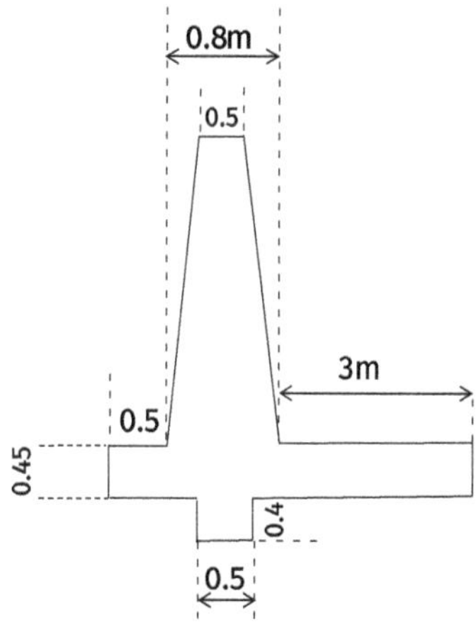

Step 1: Calculate the area of the cross-section by breaking it down into component parts:

Part 1: ½ (base 1 + base 2) x height = ½ (0.8 + 0.5) x 6 = 17.4m²

Part 2: base x height = (0.5 + 0.8 + 3) x 0.45 = 1.935m²

Part 3: base x height = 0.5 x 0.4 = 0.2m²

Step 2: Calculate the total cross-sectional area:

Part 1 + Part 2 + Part 3 = 17.4 + 1.935 + 0.2 = 19.535m²

Part 3: Calculate the gross volume:

Volume = area x length = 19.535 x 20 = 390.7m³

Part 4: Calculate the nett volume:

Volume of reinforcement = 5% x 390.7 = 1.95m³

Nett volume = 390.7 – 1.95 = 388.75m³

28.18 SIMILAR TRIANGLES

Triangles are similar if:

→ All the angles are the same

→ The ratios of corresponding sides are the same

In the similar triangles below, $\frac{a}{A} = \frac{b}{B} = \frac{c}{C}$

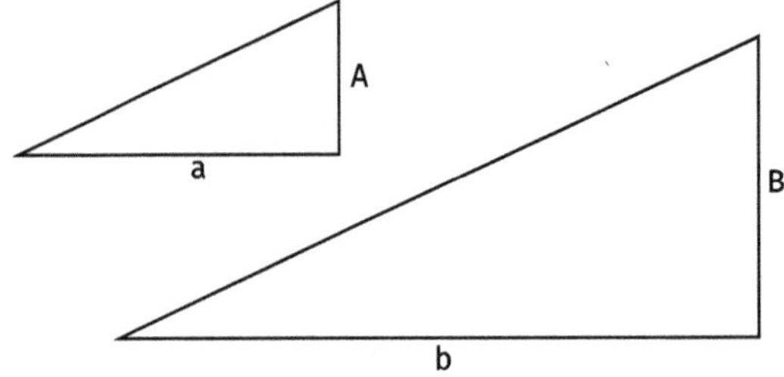

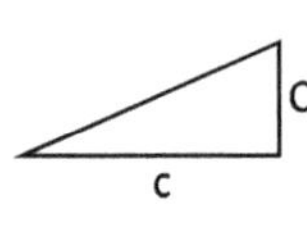

WORKED EXAMPLE FOR SIMILAR TRIANGLES

A concrete gravity sewer pipe connecting two manholes, MH1 and MH2, has a length of 78m and an overall level change of -0.78m. What is the level change at 20m from MH1?

Answer:

$$\frac{a}{A} = \frac{b}{B} = \frac{c}{C}$$

Where A = 0.78m, a = ?, B = 78m and b = 20m

Rearrange the equation to get $= \frac{Ab}{B} = \frac{0.78 \times 20}{78} = 0.2m$

EXERCISE:

7. A concrete gravity sewer pipe connecting two manholes, MH1 and MH2, has a length of 43m and an overall level change of -0.538m. What is the level change at 17m from MH1?

28.19 TRIGONOMETRY

Trigonometry is the method used for calculating lengths or internal angles of right-angled triangles, when some of the information about the triangle is known. Although trigonometry has numerous practical applications, it is most commonly used by site engineers for calculating whole circle bearings and partial coordinates between two points.

This section is not intended as a complete lesson in trigonometry, but an overview of the basic formulae. If you wish to gain an in-depth understanding of the history, background theory and derivations of the formulae, there is a wealth of excellent material available on the internet.

Trigonometry formulae:

Sin	Cos	Tan
$\sin\theta = \dfrac{\text{Opposite}}{\text{Hypotenuse}}$	$\cos\theta = \dfrac{\text{Adjacent}}{\text{Hypotenuse}}$	$\tan\theta = \dfrac{\text{Opposite}}{\text{Adjacent}}$
$\theta = \sin^{-1}\left(\dfrac{\text{Opposite}}{\text{Hypotenuse}}\right)$	$\theta = \cos^{-1}\left(\dfrac{\text{Adjacent}}{\text{Hypotenuse}}\right)$	$\theta = \tan^{-1}\left(\dfrac{\text{Opposite}}{\text{Adjacent}}\right)$
Opposite =Hypotenuse x $\sin\theta$	Adjacent = Hypotenuse x $\cos\theta$	Opposite = Adjacent x $\tan\theta$
Hypotenuse = $\dfrac{\text{Opposite}}{\sin\theta}$	Hypotenuse = $\dfrac{\text{Adjacent}}{\cos\theta}$	Adjacent = $\dfrac{\text{Opposite}}{\tan\theta}$

28.20 HOW DO I KNOW WHICH IS THE 'OPPOSITE', 'ADJACENT' AND 'HYPOTENUSE'?

The 'opposite' and 'adjacent' are defined based on their position relative to θ.

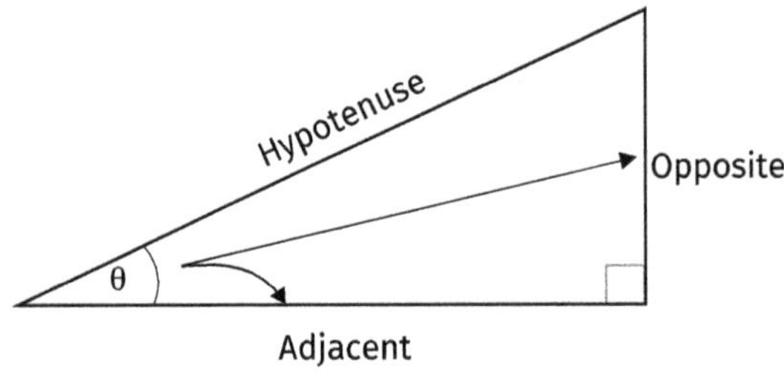

A simple way to work it out is:

→ the 'hypotenuse' is the side of the triangle directly opposite the 90° angle. It is the diagonal of the right-angle triangle.

→ the 'opposite' is the side of the triangle which is not touching θ at all.

→ the 'adjacent' is the side of the triangle which is touching θ.

28.21 WHAT DO SIN⁻¹, COS⁻¹ AND TAN⁻¹ MEAN?

Sin^{-1} means 'inverse sin'. You can think of it as the reverse of sin, and is used when you know two of the sides of the triangle and you want to use these to calculate the angle.

To enter it into your calculator, it is either:

→ 'Shift' then 'sin', or:

→ 'Inv' then 'sin'

This should display 'sin^{-1}' on your screen. Most calculators will automatically open the brackets so you can then enter the 'opposite' divided by 'hypotenuse' then 'close brackets' then 'equals'.

The same principles apply for cos^{-1} and tan^{-1} , replacing the numbers in brackets with the numbers from the relevant formulae.

Once the angle is displayed on your screen, press the 'Degrees, minutes, seconds' button to convert your answer.

28.22 HOW DO I USE THE TRIGONOMETRY FORMULAE?

To use the formulae, follow these steps:

1. Draw a diagram showing your right-angled triangle, marking on it the information that you know.
2. Identify on the diagram the angle or length you would like to calculate.
3. Based on 1. and 2. above, select the appropriate formula.
4. Carry out your calculation.

WORKED EXAMPLE FOR USING THE TRIGONOMETRY FORMULAE (1)

In the right-angled triangle below, calculate θ:

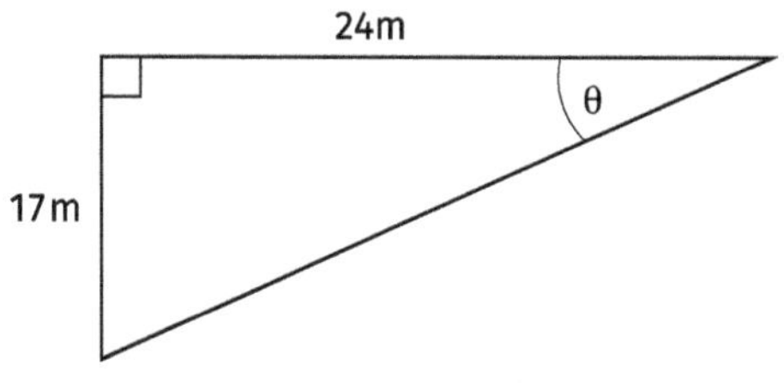

The opposite and the adjacent are known. Therefore, we can use them to find θ.

So the equation we need is:

$$\tan \theta = \frac{\text{Opposite}}{\text{Adjacent}}$$

Rearrange it to get θ on its own:

$$\theta = \tan^{-1}\left(\frac{\text{Opposite}}{\text{Adjacent}}\right)$$

Put the numbers into the equation:

$$\theta = \tan^{-1}\left(\frac{17}{24}\right) = 35° \ 18' \ 40"$$

WORKED EXAMPLE FOR USING THE TRIGONOMETRY FORMULAE (2)

In the right-angled triangle below, calculate θ:

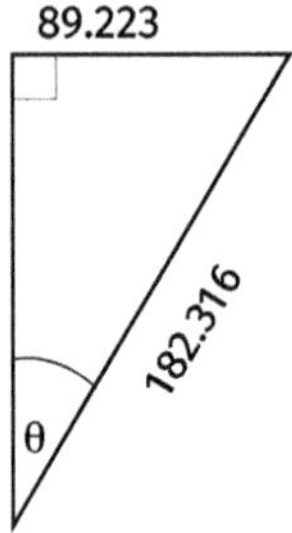

The opposite and the hypotenuse are known. Therefore, we can use them to find θ.

So the equation we need is:

$$\sin \theta = \frac{\text{Opposite}}{\text{Hypotenuse}}$$

Rearrange it to get θ on its own:

$$\theta = \sin^{-1}\left(\frac{\text{Opposite}}{\text{Hyptenuse}}\right)$$

Put the numbers into the equation:

$$\theta = \sin^{-1}\left(\frac{89.223}{182.316}\right) = 29° \ 18' \ 01"$$

WORKED EXAMPLE FOR USING THE TRIGONOMETRY FORMULAE (3)

In the right-angled triangle below, calculate θ:

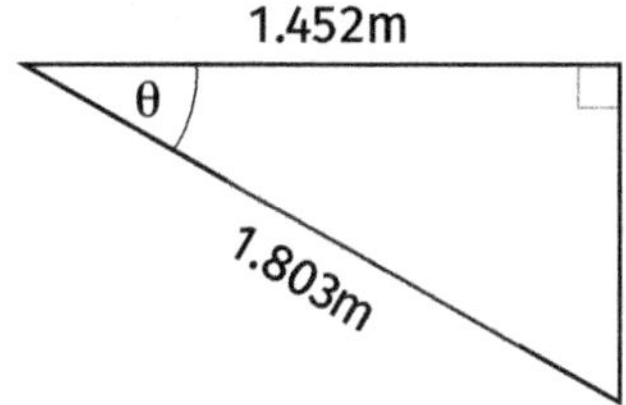

The adjacent and the hypotenuse are known. Therefore, we can use them to find θ:

So the equation we need is:

$$\cos \theta = \frac{\text{Adjacent}}{\text{Hypotenuse}}$$

Rearrange it to get θ on its own:

$$\theta = \cos^{-1}\left(\frac{\text{Adjacent}}{\text{Hypotenuse}}\right)$$

Put the numbers into the equation:

$$\theta = \cos^{-1}\left(\frac{1.452}{1.803}\right) = 36°\ 21'\ 30''$$

WORKED EXAMPLE FOR USING THE TRIGONOMETRY FORMULAE (4)

In the right-angled triangle below, calculate x:

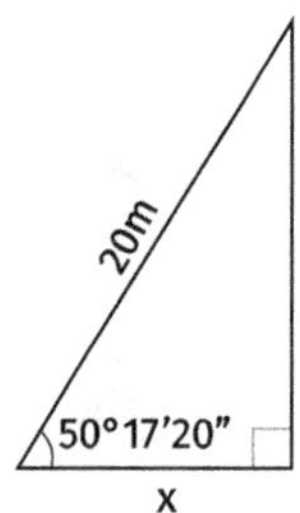

x is the adjacent. The angle, θ, and the hypotenuse are known. Therefore, we can use them to find x.

So the equation we need is:

$$\theta = \cos{-1}\left(\frac{\text{Adjacent}}{\text{Hypotenuse}}\right)$$

Rearrange it to get the adjacent on its own:

Adjacent = Hypotenuse x cos θ

Put the numbers into the equation:

Adjacent, x = 20×cos (50° 17 '20") = 12.778m

WORKED EXAMPLE FOR USING THE TRIGONOMETRY FORMULAE (5)

In the right- angled triangle below, calculate x:

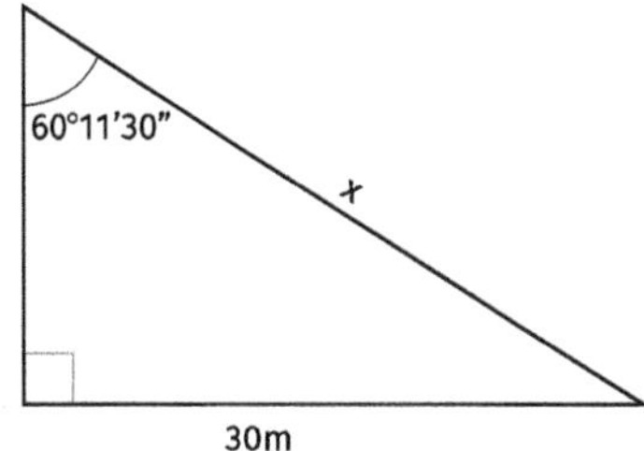

x is the hypotenuse. The angle, θ, and the opposite are known. Therefore, we can use them to find x.

So the equation we need is:

$$\sin\theta = \frac{\text{Opposite}}{\text{Hypotenuse}}$$

Rearrange it to get the hypotenuse on its own:

$$\text{Hypotenuse} = \frac{30.000}{\sin(60°11'30")} = 34.574m$$

WORKED EXAMPLE FOR USING THE TRIGONOMETRY FORMULAE (6)

In the right-angled triangle below, calculate x:

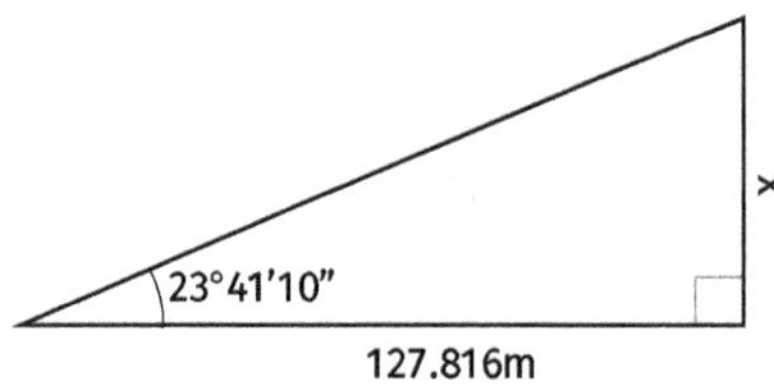

x is the opposite. The angle, θ, and the adjacent are known. Therefore, we can use them to find x.

So the equation we need is:

$$\tan\theta = \left(\frac{\text{Opposite}}{\text{Adjacent}}\right)$$

Rearrange it to get the opposite on its own:

$$\text{Opposite} = \text{Adjacent} \times \tan\theta$$

Put the numbers into the equation:

$$\text{Opposite}, x = 128.816 \times \tan(23°\ 41\ '10") = 56.070m$$

28.23 SINE RULE

The sine rule, or law of sines, is a set of formulae for calculating angles and distances in a triangle. Unlike trigonometry, the triangle does not have to be a right-angled triangle.

The sine rule states that:

$$\frac{a}{\sin A} = \frac{b}{\sin B} = \frac{c}{\sin C}$$

Where:

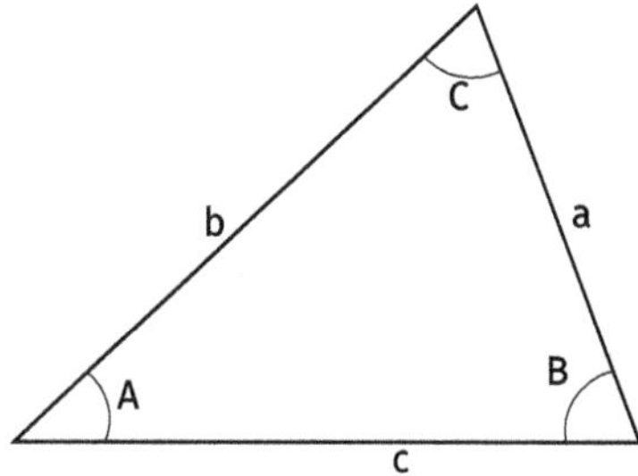

WORKED EXAMPLE FOR THE SINE RULE (1)

In the triangle below, calculate c:

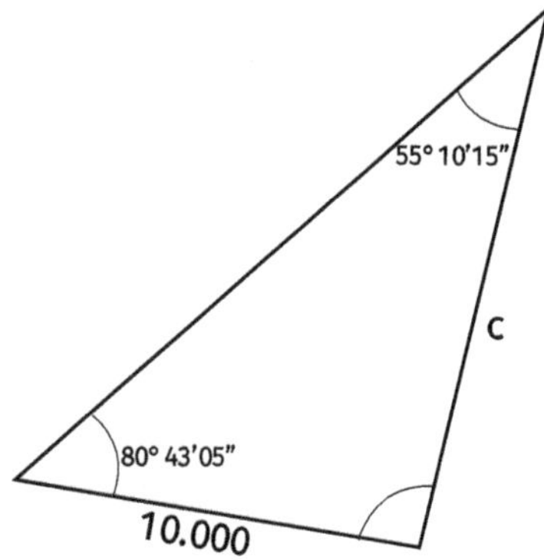

a, A and C are known, so we can use:

$$\frac{a}{\sin A} = \frac{c}{\sin C}$$

Where:

a = 10.000m

A = 55°10'15"

C = 80°43'05"

Rearrange to get c on its own:

$$c = \frac{a}{\sin A} \times \sin C$$

So,

$$c = \frac{10.000}{\sin 55°10'15} \times s(80°43'05'') = 12.023m$$

WORKED EXAMPLE FOR THE SINE RULE (2)

In the triangle below, calculate B.

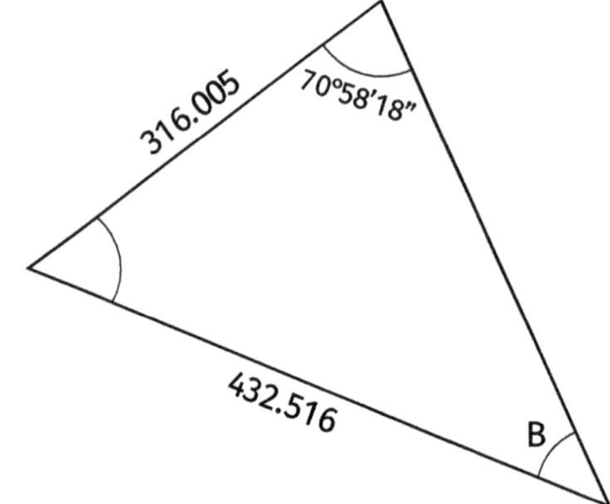

a, A and b are known, so we can use:

$$\frac{a}{\sin A} = \frac{b}{\sin B}$$

Where:

    a = 432.516

    A = 70°58'18"

    b = 3

Rearrange to get B on its own:

$$B = \sin^{-1}\left(b \times \frac{\sin A}{a}\right)$$

$$B = \sin^{-1}\left(316.005 \times \frac{\sin(70°58'18'')}{432.516}\right) = 43°41'46''$$

28.24 COSINE RULE

The cosine rule, or law of cosines, is a formula for calculating angles and distances in a triangle. Unlike trigonometry, the triangle does not have to be a right-angled triangle.

$$c^2 = a^2 + b^2 - 2ab.\cos C$$

Or

$$\cos C = \frac{a^2 + b^2 - c^2}{2ab}$$

which can rearranged to get:

$$a^2 = b^2 + c^2 - 2bc.\cos A$$

$$b^2 = a^2 + c^2 - 2ac.\cos B$$

$$\cos A = \frac{b^2 + c^2 - a^2}{2bc}$$

$$\cos B = \frac{a^2 + c^2 - b^2}{2ac}$$

WORKED EXAMPLE FOR THE COSINE RULE (1)

In the triangle below, calculate a:

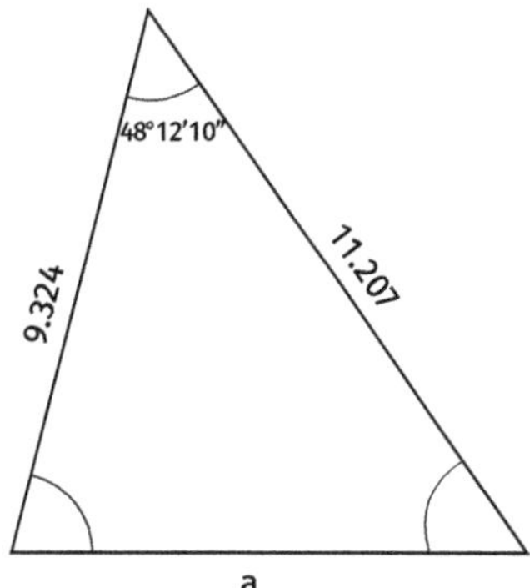

The formula we need is:

$$a^2 = b^2 + c^2 - 2bc.\cos A$$

Where:

A = 48°12'10"

b = 11.207m

c = 9.324m

Rearrange the equation to get a:

$$a = \sqrt{(b^2 + c^2 - 2bc\,\cos A)}$$

Put the numbers into the equation:

$$a = \sqrt{(11.207^2 + 9.324^2 - 2(11.207)(9.324)\cos 48°12'10")} = 8.558m$$

WORKED EXAMPLE FOR THE COSINE RULE (2)

In the triangle below, calculate B:

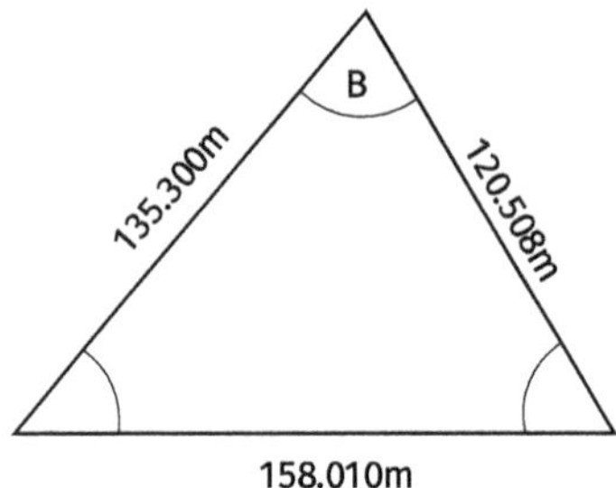

The formula we need is:

$$\cos B = \frac{a^2 + c^2 - b^2}{2ac}$$

Where:

   a = 158.010m

   b = 135.300m

   c = 120.508m

Rearrange the equation to get B:

$$B = \cos^{-1}\left(\frac{a^2 + c^2 - b^2}{2ac}\right)$$

Put the numbers into the equation:

$$B = \cos^{-1}\left(\frac{158.010^2 + 120.508^2 - 135.300^2}{2(158.010)(120.508)}\right) = 56°12'14"$$

28.25 COORDINATE GEOMETRY

Coordinates are a way of defining the position of a point in relation to a set of perpendicular axes. Coordinates can have two dimensions (2D) or three (3D)

Coordinates can be expressed in several different formats. In setting out and surveying, they are usually expressed as cartesian (sometimes known as rectangular) coordinates or polar coordinates.

This section does not cover the geographic coordinate system in which coordinates are expressed latitudinal and longitudinal angles in relation to the north and south pole.

28.26 CARTESIAN COORDINATES (ALSO KNOWN AS RECTANGULAR COORDINATES)

Cartesian coordinates define the point as the distance from the origin along each axis, separated by commas. For two dimensional coordinates, the horizontal (or X) axis represents the 'eastings' and the vertical (or Y) axis represents the 'northings' of a point. For three dimensional coordinates, the height of elevation is also given. Cartesian coordinates are the most convenient system to use in surveying and setting out.

For construction purposes, coordinates are expressed in following way:

The x component is x mE (metres Eastings)

The y component is y mN (metres Northings)

The coordinates of (x,y) are (xmE, xmN)

If x = 10.673m and y = 8.113m, then the coordinates are (10.673mE, 8.113mN)

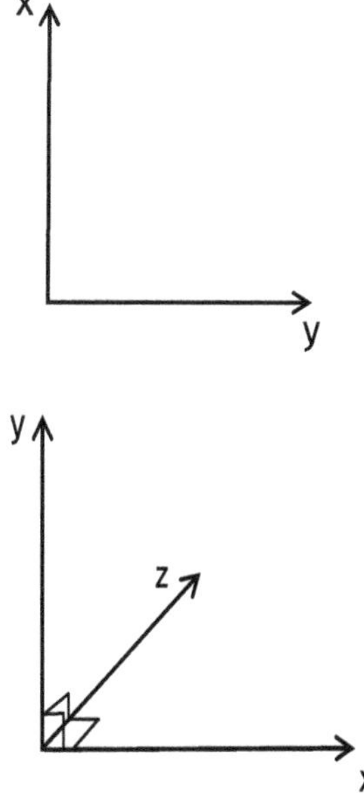

28.27 POLAR COORDINATES

Polar coordinates define a point relative the origin of a set of axes as an angle relative to the x axis and distance from the origin to the point. It is an alternative way of expressing rectangular coordinates but is rarely used in practice.

Polar coordinates are expressed as $(r, \theta°)$ where r is the hypotenuse and θ is the angle relative to the positive x-axis measured anticlockwise.

To convert rectangular to polar:

$$q = \tan^{-1}\left(\frac{N}{E}\right)$$
$$r = \sqrt{(E^2 + N^2)}$$

To convert polar to rectangular:

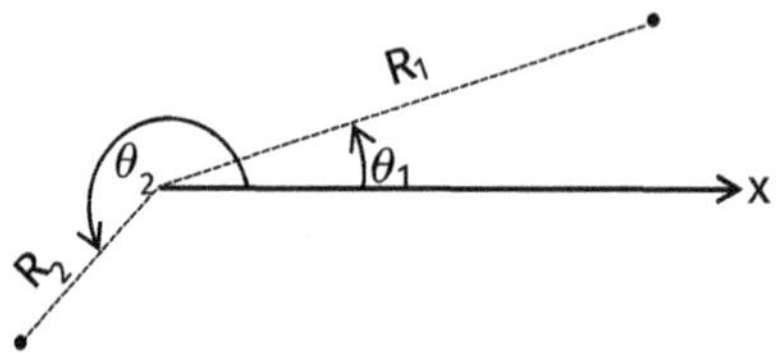

$$E = r \times \cos\theta$$
$$N = r \times \sin\theta$$

There is a function, Pol, in the Casio calculator for converting rectangular to polar. Be aware that if θ is negative, this indicates the clockwise angle, so you need to subtract this from 360° to convert it to the clockwise angle. To use the 'Pol' function:

1. Press 'shift' (top left blue button)
2. Press '+' to access the 'Pol' function. Brackets will open automatically
3. Enter in the E value
4. Enter a comma by pressing 'shift' then ')' which is just above the '9' key
5. Enter the N value
6. You can close the brackets by pressing ')' but the function will still work if you omit this step.
7. Press '='
8. The answer will be displayed as 'r = value, θ = value'
9. If θ is not visible on the screen, use the 'right' and 'left' arrows on the round arrow key to scroll across.

WORKED EXAMPLE FOR CONVERTING RECTANGULAR TO POLAR COORDINATES

Express the rectangular coordinates of P as polar coordinates, where P = (50.000mE, 61.300mN)

Answer

Step 1: draw a diagram of the arrangement

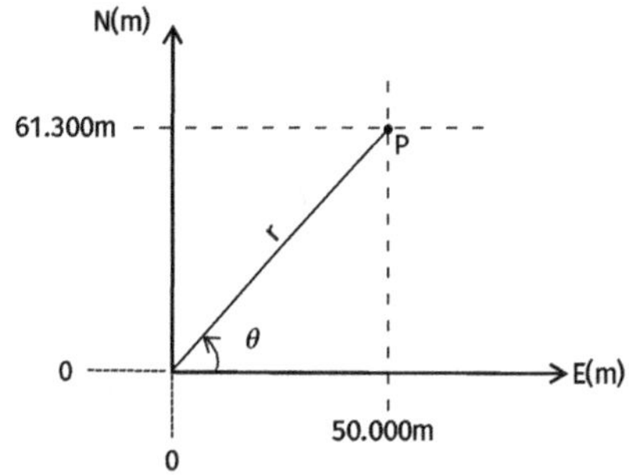

$r = \sqrt{(E^2 + N^2)} = \sqrt{(50.000^2 + 61.300^2)} = 79.106m$

$\theta = \tan^{-1}(\frac{N}{E}) = \tan^{-1}(\frac{50.000}{61.300}) = 39°12'10''$

Therefore P = (79.106m, 39°12'10")

WORKED EXAMPLE FOR CONVERTING POLAR TO RECTANGULAR COORDINATES

Express the rectangular coordinates of Q as rectangular coordinates, where Q = (100.000m, 30°)

Answer

Step 1: draw a diagram of the arrangement

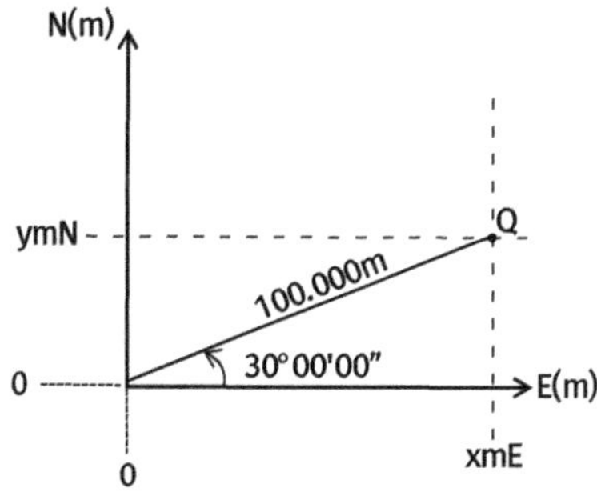

$E = r \times \cos\theta = 100 \times \cos30 = 86.603m$

$N = r \times \sin\theta = 100 \times \sin30 = 50.000m$

Therefore Q = (86.603mE, 50.000mN)

28.28 PARTIAL COORDINATES

Partial coordinates refer to the relative difference in eastings and northings between two points. The partial coordinates can be used to calculate the whole circle bearing and length of a line, or if the whole circle bearing and distance are known, they can be used to calculate the partial coordinates. This can then be used to calculate the coordinates of one point, relative to another within the same system or grid.

Partial coordinates are calculated by finding the difference in eastings and the difference in northings of two points.

For the points A and B:

$$\Delta E_{A\text{-}B} = E_B - E_A$$

$$\Delta N_{A\text{-}B} = N_B - N_A$$

WORKED EXAMPLE FOR CALCULATING PARTIAL COORDINATES (1)

Find the partial coordinates of the line P-Q if:

P = (45.219mE, 59.143mN)

Q = (31.673mE, 122.500mN)

Answer:

Draw a sketch showing the approx. location of Q relative to P:

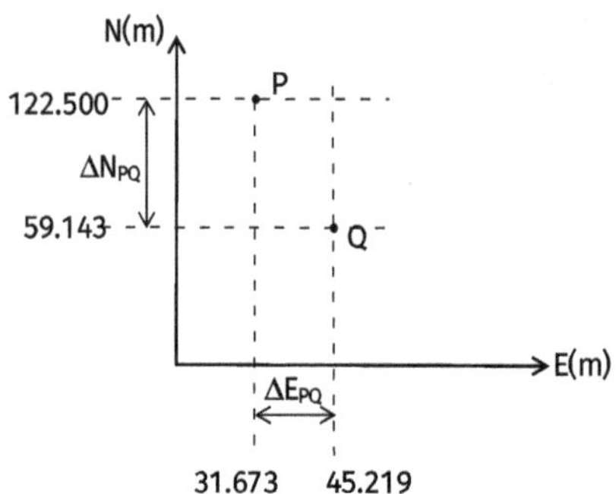

$$\Delta E_{PQ} = E_Q - E_P = 31.673 - 45.219 = -13.546m$$

$$\Delta N_{PQ} = N_Q - N_P = 122.500 - 59.143 = 63.357m$$

28.29 WHOLE CIRCLE BEARINGS

The definition of the whole circle bearing of a line is the clockwise angle from the line drawn true north of the first point, round to the line that joins the two points.

We can use the partial coordinates to calculate the whole circle bearing of a line. The line will be in one of four quadrants:

$$WCB = 90 - \tan^{-1}\left(\frac{\Delta E}{\Delta N}\right) \quad WCB = 180 - \tan^{-1}\left(\frac{\Delta E}{\Delta N}\right) \quad WCB = 180 + \tan^{-1}\left(\frac{\Delta E}{\Delta N}\right) \quad WCB = 360 - \tan^{-1}\left(\frac{\Delta E}{\Delta N}\right)$$

WORKED EXAMPLE USING PARTIAL COORDINATES TO CALCULATE WHOLE CIRCLE BEARING (1)

Calculate the WCB of the line AB if:

$\Delta E_{AB} = 17.600m$

$\Delta N_{AB} = 13.200m$

Step 1: Draw a diagram showing the arrangement

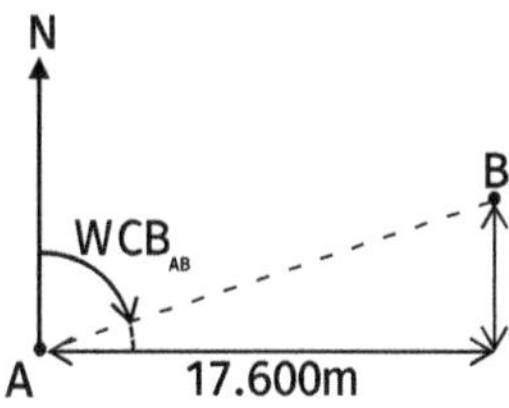

Step 2: Calculate the WCBAB

$$WCB_{AB} = 90 - \tan^{-1}\left(\frac{\Delta E}{\Delta N}\right)$$

$$= 90 - \tan^{-1}\left(\frac{17.600}{13.200}\right) = 36°52'12''$$

Quadrant 1   **Quadrant 2**

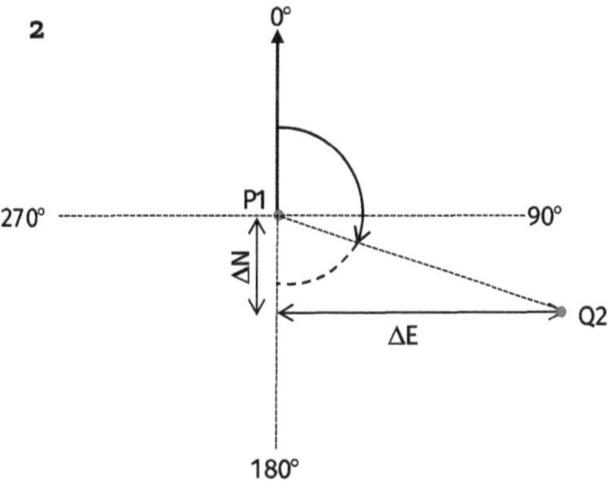

$$WCB = \tan^{-1}\left(\frac{\Delta E}{\Delta N}\right) \qquad\qquad WCB = 180 - \tan^{-1}\left(\frac{\Delta E}{\Delta N}\right)$$

Quadrant 3

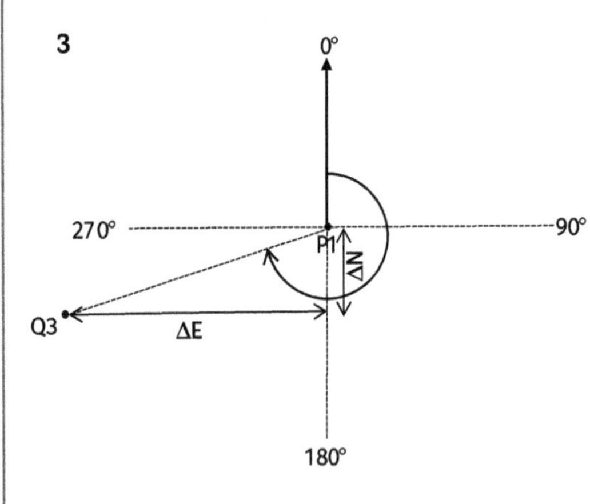

$$WCB = 180 + \tan^{-1}\left(\frac{\Delta E}{\Delta N}\right)$$

Quadrant 4

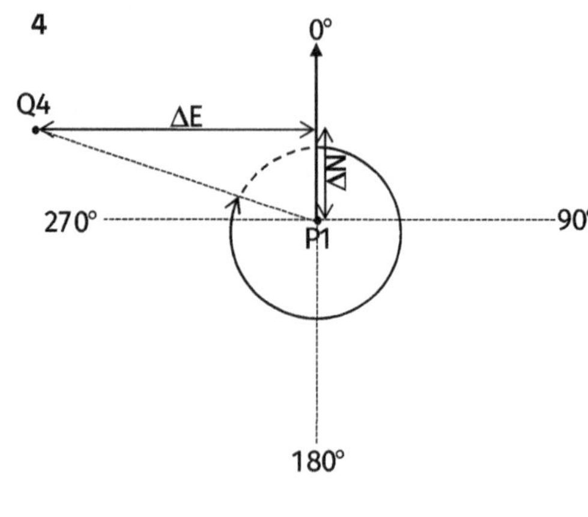

$$WCB = 360 - \tan^{-1}\left(\frac{\Delta E}{\Delta N}\right)$$

WORKED EXAMPLE USING PARTIAL COORDINATES TO CALCULATE WHOLE CIRCLE BEARING (2)

Calculate the WCB of the line CD if:

ΔE_{CD} = 1392.190m

ΔN_{CD} = 726.951m

Step 1: Draw a diagram showing the arrangement

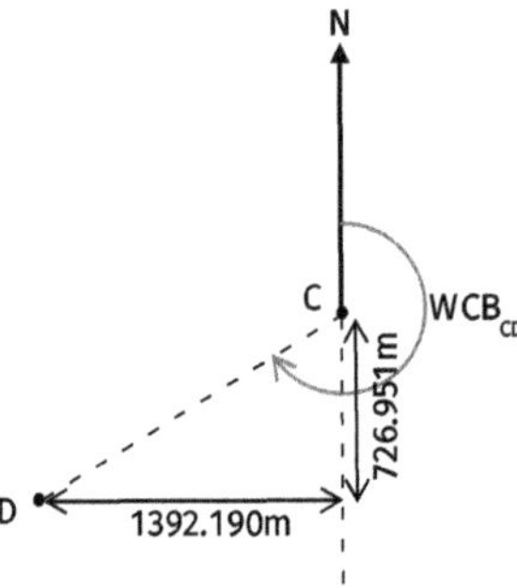

Step 2: Calculate the WCBCD

$$WCB_{CD} = 180 + \tan^{-1}\left(\frac{\Delta E}{\Delta N}\right)$$

$$= 180 + \tan^{-1}\left(\frac{1392.190}{726.951}\right) = 242°25'41''$$

28.30 LOCAL OR GLOBAL COORDINATES

A coordinate system may be local or global. A local coordinate system is created using an arbitrary origin and arbitrary 'north'. It may be established at the initial survey or design stage, or for construction purposes once the works have commenced. A local coordinate system can exist in isolation or tie into a global coordinate system such as WGS84 as used in GNSS positioning. The global coordinate system may also be arbitrarily defined, i.e. there is a global coordinate system for a whole project, but individual sections or buildings may have their own local coordinate system which is more convenient to use.

28.31 GRID ROTATIONS AND SCALE FACTORS

A grid transformation is the process of mathematically converting the coordinates of points from one coordinate system to another.

A transformation can involve rotation, shift (displacement) and scale factors:

→ **Rotation** - a rotation occurs when the grid system is pivoted about an axis (Fig. 28-2). It is expressed as an angle.

→ **Shift** – the shift is defined as the displacement and is split into the displacement along the X axis and the displacement along the Y axis.

→ **Scale factor** – this is the scale factor by which dimensions in the original grid are multiplied.

The hand calculations for performing a grid transformation involve in-depth maths such as calculus and matrices which is beyond the scope of this book. However, the operation can be performed easily in any survey software package and there are various conversion programmes and spreadsheets available on the internet.

Fig. 28-2: Co-ordinates on gridlines

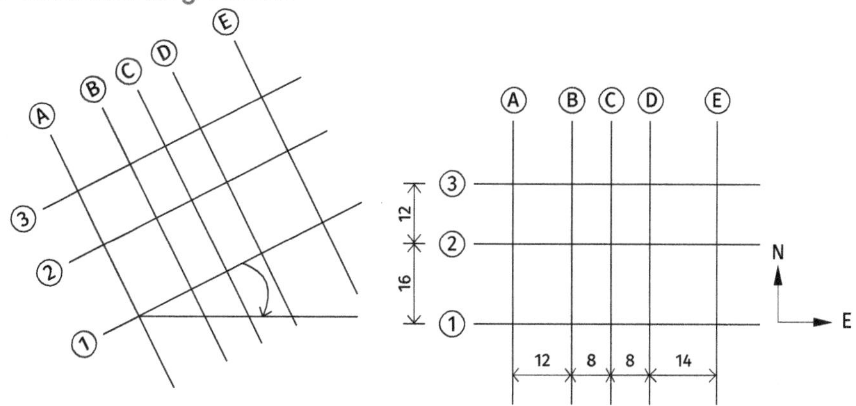

28.32 CALCULATOR TROUBLE SHOOTING

The most common calculator readily available in stationary shops and most supermarkets is the Casio fx-83GT PLUS and similar models. This section relates to this type of calculator.

When performing calculations, there are a few common issues that can cause problems.

Typing errors - using the calculator app on a mobile phone increases the chance of error. This is because on many phone apps, once you have entered the number and the operation, the previous number disappears from the screen, making it less likely that you will spot a typing error. It is also more likely that you will mis-type a number on the touch screen than on a calculator. Finally, on a Casio calculator, when you have performed a series of operations, you can press the 'up' and 'down' arrows to navigate through the previous steps of the calculations, and the 'left' and 'right' arrows to edit the data you have input. This means that if you have made an error, you can amend the relevant number without having to redo the whole calculation which may introduce a different error.

If you do choose to use your phone app, it is recommended that you perform each calculation twice and ensure that you are getting the same answer each time by writing it down.

Syntax error when working with degrees, minutes and seconds – it is likely that you have forgotten to press the DMS button after you have entered the seconds.

Syntax error when using trigonometry – if you get a syntax error when using trigonometry, it could be that you have entered the numbers in the wrong order. For example, when using sin, if you confused the opposite with the hypotenuse, the calculator will display 'syntax error' because it will recognise that the opposite cannot be longer than the hypotenuse.

The answer is a fraction – in the Casio calculator, the default display is as a fraction. For example, if you enter 5÷10, the answer will display as ½. The same will apply for larger and more complicated fractions. However, most of the time in setting out, we need the answer to display as a decimal. There are two solutions to this problem.

Solution 1: When the answer displays as a fraction, press the 'S⇔D' button. This will convert the answer to a decimal. The problem with this is that you have to press the button every time a fraction is displayed.

Solution 2: Change the mode in your calculator:

1. 'shift' (top left blue button)

2. 'mode' (blue button to the left of the 'on' button)

3. Type '1' which is for 'MthIO'

4. In the 'Result Format?' screen, type '2' for 'LineO'

5. Now all your answers should display as decimals and not fractions.

A 'strange answer' – when the answer displayed doesn't look right, it could be that the answer is so small that the calculator has automatically displayed it as 'standard form'. Standard form is a convention for writing very large or very small numbers without having to write large numbers of zeros. There is plenty of information on the internet about standard form if you wish to look into this in more detail, but here is a 'quick fix':

If you enter 0.009 – 0.001 into your Casio, it will first display as a fraction i.e. $\frac{1}{125}$. You then press the 'S⇔D' button which will convert it to 8 x 10^{-3}. In standard form, 10^{-3} means the decimal place should be moved three places to the left. So 8 x 10^{-3} would become 0.008. To convert this using your calculator, press 'shift' then 'ENG' (which is found just about the '8' key). Then press 'shift' then 'ENG' again. Repeat this until the answer displayed is x 10^{0}. In maths, any number to the power of zero equals one, so this is your answer i.e. 0.008 x 10^{0} = 0.008 x 1 = 0.008.

Answer isn't right – it could be that you have put the brackets in the wrong place or combined several operations. For example in the sum 56 x (3 + 10) = 728, however if you enter 56 x 3 + 10, the answer will display as 178. If you suspect that you may be using your calculator incorrectly in this way, make sure you are pressing '=' after each part of the sum. If you are still having problems, try writing the sum out on a piece of paper showing the brackets in the correct places and enter the sum into the calculator as you have written it down.

28.33 GRADIENT/ GRADE/ SLOPE/ FALL

There are many situations in construction where it is necessary to work with sloping features for example, pipe laying, road construction and earthworks.

There are four different methods of quantifying a slope:

a) a ratio (1:G)

b) a percentage (P%)

c) level change per horizontal distance (mm/m)

d) the angle relative to the horizontal plane ($\theta°$)

The gradient as ratio – calculate the ratio of the vertical to the horizontal i.e 1: (horizontal distance ÷ level change) = 1:G

The gradient as a percentage - calculate the level change as a percentage of the horizontal distance i.e. (level change ÷ horizontal distance) x 100 = P%

The gradient as a level change per horizontal distance – calculate the level change(mm) for every m of horizontal distance i.e. (level change ÷ horizontal distance) x 1000 = Ymm/m

The gradient as an angle relative to the horizontal – calculate the angle using trigonometry where the level change is the 'opposite' and the horizontal distance is the 'adjacent' i.e tan⁻¹ (level change ÷ horizontal distance) = θ°

WORKED EXAMPLE FOR EXPRESSING GRADIENTS IN DIFFERENT FORMATS:

A pipe has a level change of 0.713mm over a horizontal distance of 57m. Express this as:

a) a ratio

b) a percentage

c) level change per horizontal distance

d) the angle relative to the horizontal

Step 1: Draw a diagram of the arrangement

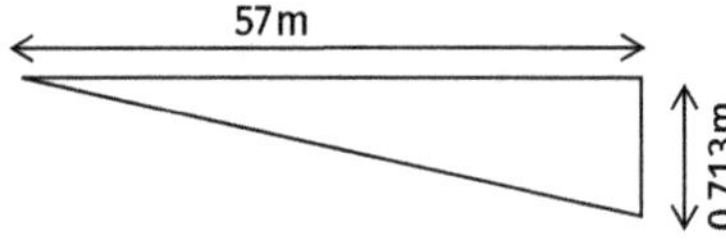

Answers:

a) The ratio = 1: (horizontal distance ÷ level change)

$$= 1: \left(\frac{57}{0.713}\right) = 1{:}79.94 \text{ (round up to 1:80)} = 1{:}80$$

b) The percentage = (level change ÷ horizontal distance) x 100 = P%

$$= \left(\frac{1}{80}\right) \times 100 = 1.25\%$$

c) The level change per horizontal dist. = (level change ÷ horizontal dist.) x 1000

$$= \left(\frac{1}{80}\right) \times 1000$$

$$= 12.5\text{mm/m}$$

d) The angle relative to the horizontal = tan⁻¹ (level change ÷ horizontal distance)

$$= \tan^{-1}\left(\frac{1}{80}\right)$$

$$= 00°42'58''$$

28.34 CALCULATOR SHORTCUT FOR PYTHAGORAS

There is a function in Casio calculators which allows you to calculate the hypotenuse of a triangle, r, and the angle of the line relative to the horizontal, θ. The function is called 'Pol' and is used to convert rectangular coordinates into polar coordinates. It is not necessarily any quicker than entering the numbers as per the formula, but worth knowing in case you need to use it as a check or if you prefer this method. See section **28.27** for instructions.

28.35 SETTING OUT BY BEARING AND DISTANCE

Although, in practice, this method of setting out is obsolete, it is included here to demonstrate the fundamental principles of setting out a new point in relation to control points. When the 'stakeout' mode in the total station guides you to a point, its onboard software is using these principles.

The basic principle of setting out by bearing and distance is that by setting up directly over a point of known coordinates (the station), and sighting to another point of known coordinates (the backsight), you are defining the position (E,N) of the instrument and its orientation in relation to your grid north. The position of the point you are setting out is then defined as a relative horizontal angle from the reference object (RO) and a distance from the total station.

The steps are as follows:

Part 1 – Set up your total station and check that the set-up is correct.

1. Carry out the calculations to determine the check angle, setting out angle and setting out distance. (See worked example on section **28.36**)

2. Set up the total station over your 'set-up point' i.e. a control point of known coordinates.

3. Sight to your backsight. Remember that the horizontal angle is critical so line up the vertical x-hair accurately on the point. No prism is needed here, it is purely an angle you are measuring.

4. Manually set the horizontal circle to 00° 00' 00" using the relevant function in your total station.

5. Sight to your check point. No prism needed, it is just the single vertical x-hair you are lining up.

6. Compare the measured check angle to your theoretical check angle (calculated in step 1).

7. If the error is acceptable, continue to step 8. If it is unacceptable, repeat steps 2 – 6 and if necessary, check your calculations in step 1.

Part 2 – Set out your point(s)

8. Manually adjust the horizontal circle until the angle displayed matches your setting out angle (calculated in step 1).

9. Get your assistant to pace out the setting out distance (calculated in step 1) in roughly the right direction.

10. Use hand signals to direct your assistant left or right until the prism is in line with your vertical x-hair.

11. Scroll up until the cross at the centre of the x-hairs is on the prism. Measure the horizontal distance.

12. Note the difference between the measured distance and the setting out distance.

13. Direct your assistant forwards or backwards as necessary by the difference calculated in step 12.

14. Repeat steps 10 – 13 until the point you are setting out is within an acceptable error. To achieve the highest accuracy possible, ensure the bottom of the detail pole is lined up with the vertical x-hair.

28.36 BEARING AND DISTANCE CALCULATIONS

To calculate the check angle, setting out angle and setting out distance, follow these steps:

1. Identify your station, backsight, checkpoint and setting out points.

2. On a sketch, plot the approximate location of the points relative to each other by inspecting the eastings and northings of each point.

3. Draw a north arrow from the station.

4. Draw a line from the station to each of the other points.

5. Using partial coordinates, calculate the WCB from the station to each of the other points.

6. Calculate the check angle as the difference between the WCB to the check point and the WCB to the backsight.

7. Calculate the setting out angle as the difference between the WCB to the setting out point and the WCB to the backsight.

8. Calculate the setting out distance using Pythagoras, i.e. the hypotenuse of the difference in eastings and the difference in northings between the station and the setting out point.

WORKED EXAMPLE FOR SETTING OUT BY BEARING AND DISTANCE

Carry out the calculations for setting out the point, P, using the bearing and distance method using the following points:

	Point	Eastings	Northings
Station	S1	1140.350	919.005
Backsight	S4	1328.900	832.315
Check point	S2	526.105	863.490
Setting out point	P	1074.385	572.245

Answer:

Step 1: Draw a sketch - plot the points relative to each other. Draw a north arrow from the station. Draw lines from the station to the backsight, check point and setting out point.

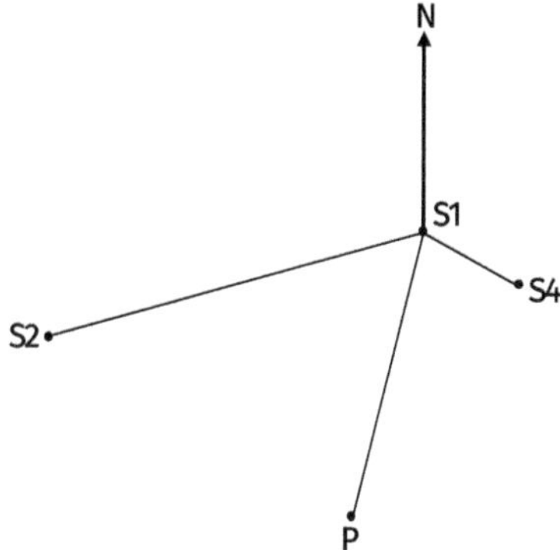

Step 2: Calculate the WCB of each line:

WCB_{S1-S4}

$\Delta E_{S1-S4} = E_{S4} - E_{S1} = 1328.900 - 1140.350 = 188.550m$

$\Delta N_{S1-S4} = N_{S4} - N_{S1} = 832.315 - 919.005 = -86.690m$

$WCB_{S1-S4} = 90 + \tan^{-1}\left(\frac{\Delta N}{\Delta E}\right) = = 90 + \tan^{-1}\left(\frac{86.690}{188.550}\right) = 114°41'29''$

WCB$_{S1-S2}$

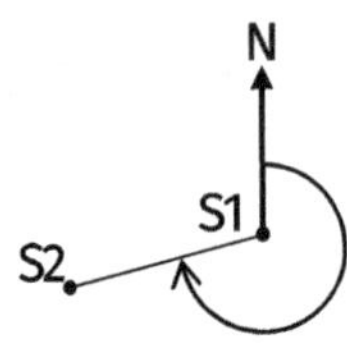

$$\Delta E_{S1-S2} = E_{S2} - E_{S1} = 526.105 - 1140.350 = -614.245m$$

$$\Delta N_{S1-S2} = N_{S2} - N_{S1} = 863.490 - 919.005 = -55.515m$$

$$WCB_{S1-S2} = 270 - \tan^{-1}\left(\frac{\Delta N}{\Delta E}\right) = = 90 + \tan^{-1}\left(\frac{55.515}{614.245}\right) = 264°50'08''$$

WCB$_{S1-P}$

$$\Delta E_{S1-P} = E_P - E_{S1} = 1074.385 - 1140.350 = -65.965m$$

$$\Delta N_{S1-P} = N_P - N_{S1} = 572.245 - 919.005 = -346.760m$$

$$WCB_{S1-P} = 270 - \tan^{-1}\left(\frac{\Delta N}{\Delta E}\right) = = 90 + \tan^{-1}\left(\frac{346.760}{65.965}\right) = 190°46'15''$$

Step 3: Calculate the angles:

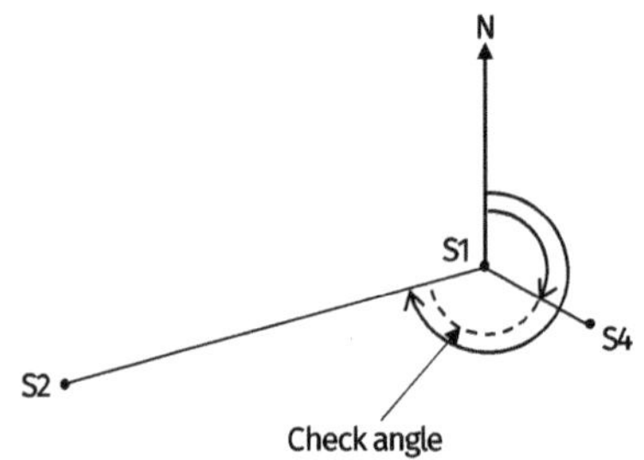

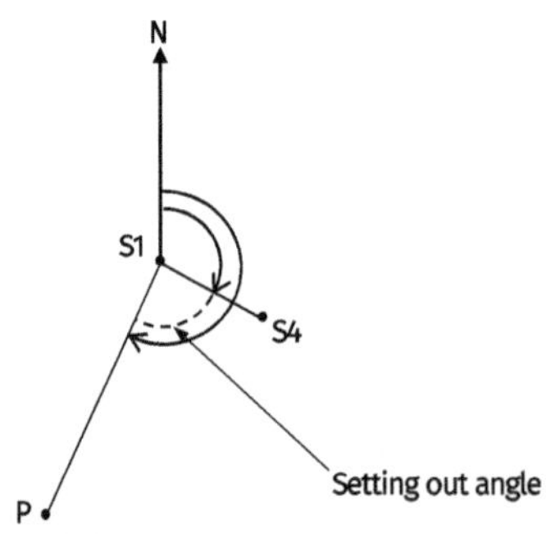

Check angle = $WCB_{S1\text{-}S2}$ - $WCB_{S1\text{-}S4}$ = 264°50'08" - 114°41'29" = 150°08'39"

Setting out angle = WCB_{S1} -P - $WCB_{S1\text{-}S4}$ = 190°46'15" - 114°41'29" = 76°04'46"

Step 4: Calculate the setting out distance

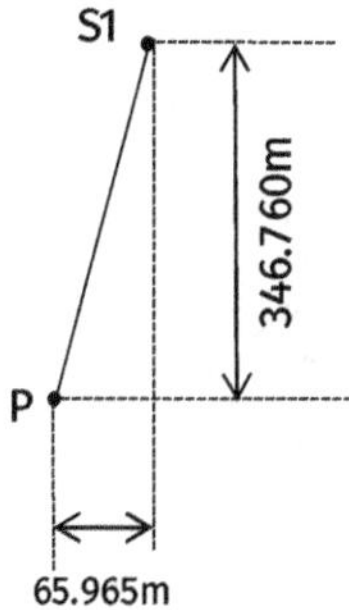

Setting out distance = $\sqrt{(\Delta E_{S1\text{-}P}{}^2 + \Delta N_{S1\text{-}P}{}^2)}$ = $\sqrt{65.965^2 + 346.760^2}$ = 352.979m

To set out point P:

a) Set up the total station over S1 (station)

b) Sight to S4 (backsight)

c) Set the horizontal circle to zero

d) Sight to S2 (check point)

e) The horizontal angle reading should be within an acceptable error of 150°08'39"

f) Turn the horizontal angle manually until it reads 76°04'46"

g) Direct your assistant until the bottom tip of the prism pole is in line with your vertical x-hair

h) Scroll up or down until the cross is on the prism and measure the distance

i) Direct your assistant until the prism is in the correct line and at the correct distance.

WORKED EXAMPLE SURVEYING A POINT USING BEARING AND DISTANCE

You have set out the point in the worked example above. From the same set-up position, you need to determine the as-built position of the corner of a concrete slab.

Work out the coordinates of the corner of the slab, C, if:

the horizontal angle from the backsight to C = $HA_{S4\text{-}C}$ = 16°52'20"

the plan distance, $L_{S4\text{-}C}$ = 268.165m

Answer

Step 1: Draw a diagram of the arrangement

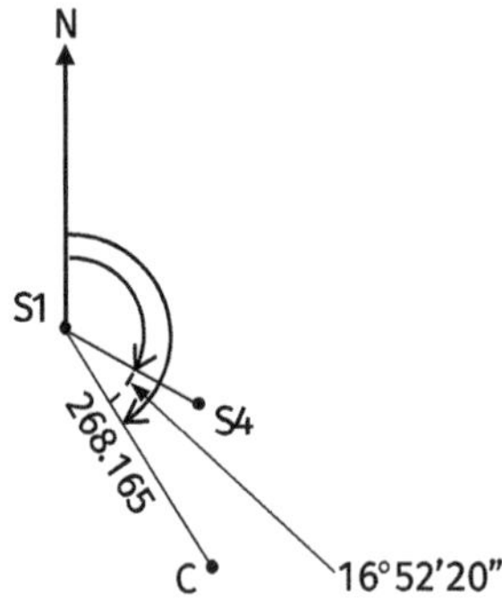

Step 2: Calculate the WCB_{S1-C}

$WCB_{S1-C} = WCB_{S1-S4} + HA_{S1-C} = 114°41'29" + 16°52'20" = 131°33'49"$

Step 3: Calculate the partial coordinates

$\Delta E_{S1-C} = L \sin WCB_{S1-C} = 268.165 \times \sin(131°33'49") = 200.646m$

$\Delta N_{S1-C} = L \sin WCB_{S1-C} = 268.165 \times \cos(131°33'49") = -177.914m$

Step 4: Calculate the eastings and northings of C

$E_C = E_{S1} + \Delta E_{S1-C} = 1140.350 + 200.646 = 1340.996mE$

$N_C = N_{S1} + \Delta N_{S1-C} = 919.005 + (-177.914) = 741.091mN$

Therefore C = (1340.996mE, 741.091mN)

28.37 FINDING THE COORDINATES OF A POINT ON A LINE.

If you know the beginning (A) and end (B) coordinates of a line you can calculate the coordinates of a point M, at any distance along that line. There are a range of situations where this might be necessary, but one example is, if you know the coordinates at the beginning and end of a straight section of road centreline, and you need to calculate the coordinates at a given chainage. The steps are as follows:

1. Draw a sketch of the arrangement
2. Determine the length of the line AM
3. Determine the length of the line AB
4. Calculate the partial coordinates from AB

5. Calculate the partial coordinates of AM as a ratio of the lengths AM:AB

6. Calculate the coordinates of M

WORKED EXAMPLE FOR FINDING THE COORDINATES OF A POINT ON A LINE

Find the coordinates of T, which is 10m along the line QR, where:

Q = (120.000mE, 150.000mN)

R = (155.000mE, 138.000mN)

Answer

Step 1: Draw a sketch of the arrangement

  • Q = (120.000mE, 150.000mN)

    • R = (155.000mE, 138.000mN)

Step 2: Determine the length of the line QR

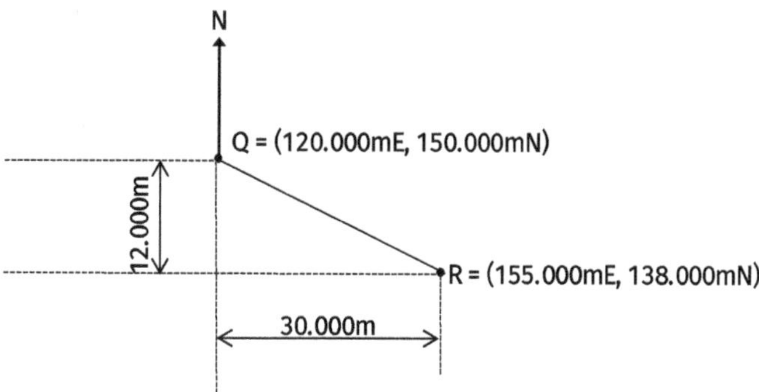

$\Delta E_{QR} = E_R - E_Q = 155.000 - 120.000 = 35.000m$

$\Delta N_{QR} = N_R - N_Q = 138.000 - 150.000 = -12.000m$

$QR = \sqrt{(\Delta E_{QR}^2 + \Delta N_{QR}^2)} = \sqrt{(35.000^2 + (-12.000)^2)} = 37.000m$

Step 3: Determine the length of the line QT

QT = 10.000m

Step 4: Calculate the partial coordinates from AB

$\Delta E_{QR} = 35.000m$

$\Delta N_{QR} = -12.000m$

Step 5: Calculate the partial coordinates of QT as a ratio of the lengths QT:QR

$$\Delta E_{QT} = \left(\frac{QT}{QR}\right) \times \Delta E_{QR} = \left(\frac{10}{37}\right) \times 35.000 = 9.459m$$

$$\Delta N_{QT} = \left(\frac{QT}{QR}\right) \times \Delta E_{QR} = \left(\frac{10}{37}\right) \times (-12.000) = -3.243m$$

Step 6: Calculate the coordinates of T

$$E_T = E_Q + \Delta E_{QT} = 120.000 + 9.459 = 129.459m$$

$$N_T = N_Q + \Delta N_{QT} = 155.000 + (-3.243) = 151.757m$$

Therefore T = (129.459mE, 152.757mN)

APPENDICES

Appendix A

LEICA TS06

USEFUL INFORMATION

Toggling the options menu along the bottom of the screen

When you are in a program, pressing F4 will scroll through different menu options along the bottom of the screen. If you cannot see what you are looking for, it is probably hidden. Press F4 until you find it.

Changing the target type, prism constant and reflector height

'EDM settings' is the instrument's terminology for target type, prism constant and reflector height.

Within each function, one of the 'F' keys along the bottom will be 'EDM'. Remember to set the correct EDM settings for the target type you are sighting to. If you are working with heights the reflector height must be input manually each time you change from one EDM type to another. If you are not working with heights, you can ignore the reflector height.

SCREEN ARRANGEMENT

Icons down the right hand side of the screen (when in a program)

Top right – TS battery level indicator

Second one down – Indicates the EDM mode the instrument is currently set to. It will be one of the following:

- → Reflectorless mode (A house with a laser beam bouncing off it)
- → Prism (Picture of a prism). Remember to check that it is the correct prism constant!
- → Tape (Picture of a retro target)

Third one down – What looks like a capital I is actually a roman numeral 1. This indicates that you are on face one. If you change faces the I will change to II, which indicates you are on face two.

Forth one down - Auto

Bottom – '345' means the keypad is set to type numbers. 'ABC' means it is set to type letters.

Small round buttons immediately to the right of the

Top button – Menu button (looks like a piece of paper with writing on it) – This button will switch between the different tabs found at the top of the screen.

Second button down – FNC – Contains quick links to handy functions e.g. Home screen, level/ plummet, toggle between prism and no-prism, laser pointer and screen illumination.

Third button down – User defined button (Person 1) – This key can be set by the user. Many people choose to set this to the Level & Plummet screen.

Fourth button down – User defined button (Person 2) – This key can be set by the user. Many people set this to the laser pointer.

Escape button – round white button at the very bottom of the panel. Escape will always take you back to the previous menu. Keep pressing until you reach the home screen.

MAIN MENU

'Q – Survey'

This will take you straight into quick survey mode without performing a station setup. Be aware that if a station set-up is not performed, all co-ordinates given will be arbitrary and not correct in relation to control points.

'Programs'

This is where you will find all the functions in including Station setup, Survey, Stakeout, Ref. Line, Ref. Arc, Ref. Plane, Tie- Distance, COGO and Area & Volume

'Manage'

Job - Create, edit, view or delete jobs

Fixpoints - Create, edit, view or delete Fixpoints

Meas. Data – View measured data. Every measurement taken is assigned a block number.

Codes – Create, edit, view or delete point codes

Formats – Manage formats

Del. Data – Delete data

'Transfer data'

This is where you can import and export data.

'Settings'

Work – Set trigger keys, set user keys (third and fourth round buttons), set the order of the fields on the screen, set map parameters (e.g. which types of points to show, labels etc), manage icons in the top icon bar.

Regional – Set units, horizontal circle direction, language, time, date etc

LEICA TS06

Data – order how data is stored and select where it is stored (internal memory or interface)

Screen – Set keyboard illumination and intensity of screen illumination, screensaver, touchscreen, switch off stakeout bleep

EDM – Set atmospheric pressure, prism type, switch laser pointer and guide light on/off.

Interface – Select the interface e.g. USB, Bluetooth etc

'Tools'

Adjust – Carry out instrument checks and adjust accordingly

Startup – Configure start up sequence

Info – Software, memory, system, maintenance information

Licence – Install licence keys for software

PIN – Set a PIN code for access to the instrument

Load FW – Load firmware

INSTRUMENT SET-UP

Set up over a known point ('Ori. With Coord')

1. >>Main menu
2. >>Programs
3. >>Survey (tab at the top)
4. >>Stn. Setup > "Set up a station and determine its coordinates and orientation"
5. >>F4 Cont.
6. Use the left/ right arrows to select 'Ori. With Coord.
7. >> Ori. With Coord. > "Station is known. Measure 1 or more known targets to compute orientation and height"
8. >>F4 Cont
9. The top of the screen says 'Enter Station Data".
10. >>F2 List and select the point you are set up over. Alternatively select F2 ENH and manually enter the coordinates.
11. >>F4 Cont.
12. If you are happy with the station name, press F3 Cont.
13. The top of the screen says 'Enter Target Data".
14. >>F2 List and select the point you are backsighting to.
15. The top of the screen says 'Sight target point!".

16. Line up on the target. Press F1 Meas.

17. You can now choose from the following:

18. F1 Measure more points (to measure more backsights)

19. F2 Measure in other face (to improve the accuracy of your set up)

20. F3 Access tolerances

21. F4 Compute (once you are satisfied that you have taken enough backsights)

22. >>F4 Compute

23. The top of the screen says 'Station setup result 1/2'. The coordinates displayed here are the coordinates you entered earlier.

24. >>F4 Set

25. >>F4 New

26. The screen will say 'Station and orientation set' then it will return to the survey menu.

Set up at an unknown point ('Resection')

1. >>Main menu

2. >>Programs

3. >>Survey (tab at the top)

4. >>Stn. Setup > "Set up a station and determine its coordinates and orientation"

5. >>F4 Cont.

6. Use the left/ right arrows to select 'Resection."

7. >> Resection. > "Station is unknown. Measure 2 or more known targets to compute station and orientation"

8. >>F4 Cont

9. The top of the screen says 'Enter Station Data". Give the station a name.

10. >>F3 Cont

11. The top of the screen says 'Enter Target Point".

12. >>F1 List and select the point you are backsighting to.

13. >>F4 Cont

14. The top of the screen says 'Sight target point!".

15. Line up on the target. Press F1 Meas.

16. You can now choose from the following:

    a. F1 Measure more points *(to measure more backsights)*

    b. F2 Measure in other face *(to improve the accuracy of your set up)*

    c. F3 Access tolerances

17. >>F1 Measure more points

18. The top of the screen says 'Enter Target Point".

19. >>F1 List and select the point you are backsighting to.

20. The top of the screen says 'Sight target point!".

21. Line up on the target. Press F1 Meas.

22. The accuracy of your set up. Accur. Posit. is displayed. If this is within acceptable limits press F4 Compute. If you would like to improve the accuracy of your set up press F1 Measure more points.

23. Repeat until you have an acceptable error.

24. >>F4 Compute

25. The eastings, northings and height of your total station position will be displayed on the screen.

26. >>F4 Set

27. The screen will say 'Station and orientation set' then it will return to the survey menu.

INSTRUCTIONS

Measure points/ take a survey ('Survey')

Set up the instrument

1. >>Main menu

2. >>Q-Survey will take you straight into the survey function and jump to Enter PtID below.

3. Alternatively:

4. >>Programs

5. >>Survey (tab at the top)

6. >>Survey > "Measure unlimited number of points. Pre-settings for job and stations can be done"

7. >>F4 Cont

8. Enter the PtID

9. Use the left/right arrows to select a Code from the code list. If there is no code list stored, and you would like to create codes as you go along press F2 Code then F1 New. However it is much easier to add codes in Manage>Codes or by importing them in Transfer>Import.

10. Sight to the point you would like to survey. Press F2 Dist to display the point information on the screen. Press F3 store to record the point information into the memory. If you would like to measure the point and record it in one go, press F1 Meas.

11. To view the eastings, northings and height of the point either used the arrow pad on the keyboard or the scroll bar on the screen to scroll right down to the bottom.

Measure or calculate the distance and level difference between two points ('Tie distance')

To measure between two points, accurately level the total station at a point where you can sight to the two points you wish to measure between. Alternatively, to calculate the distance and level difference between two previously input or measured points, there is no need to level the instrument.

1. >>Main menu

2. >>Programs

3. >>Survey+ (tab at the top)

4. >>Tie Dist > "Compute distance, height difference and azimuth between two target points"

5. >> F4 Cont > Use left/ right arrows to select Radial

6. >>Radial > "Values are computed between one point and several others"

7. >>F4 Cont

You now have three options for entering the start point of the line:

d. Option 1 (To measure between two points) – Sight the point and press F1 Meas

e. Option 2 (To calculate between two previously measured or input points) – Press F2 List and select the point from the memory then press F4 Cont

f. Option 3 (To calculate between two points which you know the coordinates of) – Press F2 ENH and enter the coordinates of the point manually then press F4 Cont

8. Now enter the end point of the line using one of the three options.

9. The horizontal distance, slope distance and level change are displayed at the bottom of the screen.

10. If you have finished press F3 (End).

11. If you would like to measure to a new point from the start of the line you have just measured press F2 (NewPt 2).

12. If you would like to measure to a new point from the end of the line you have just measured press F2 (NewPt 1).

Stake out points of known coordinates ('Stakeout')

Set up the total station using 'Ori with coords' or 'resection' to determine the position and orientation of your instrument.

1. >>Main menu

2. >>Programs

3. >>Survey (tab at the top)

4. >>Stakeout> "Make marks in the field at predefined points. Predefined points are points to be staked"

5. >>F4 Cont

6. Use the left/ right arrows to select the point you wish to set out. Pt type/ID indicates whether it is a Fixpoint (Fixpt.) or a measured point (Meas). Press F1 View if you would like to view and check the point data before continuing.
Alternatively press F2 ENH if you would like to manually enter the coordinates of the point.

There are four tabs along the top – Polar, Local, Coord 1 and Coord 2

Tab 1- Polar.

1. Manually turn the horizontal tangent screw to set ΔHz to 0° 0' 0"

2. Looking through the telescope, direct your assistant into the correct line.   Press F2 Dist. The arrows on the screen will indicate how far the prism needs to be moved. 'Up' means away from you and 'down' means towards you.

3. Repeat until you are within the required tolerance. Once point is set out, sight to it and press F1 Meas to record the as-set-out position.

4. It will automatically jump to the next point in the list.

Tab 2 - Local

1. Manually turn the horizontal tangent screw to set ΔL to 0.000m

2. Looking through the telescope, direct your assistant into the correct line. Press F2 Dist. The arrow on the screen will indicate how far the prism needs to be moved. 'Up' means away from you and 'down' means towards you.

3. Alternatively, sight to the prism, press F2 Dist and direct your assistant according to the arrows. ΔL and ΔO will match the distances given on the arrows.

4. Repeat until you are within the required tolerance. Once point is set out, sight to it and press F1 Meas to record the as-set-out position.

Tab 3 – Coord 1

→ The process is very similar to the angle and distance method, but in this option, the relative eastings and northings of the target are displayed. This is useful if you have a gridlines system in place in which case it may be more convenient to give your assistant directions in relation to grid lines.

Tab 4 - Coord 2

→ This option is not really convenient for staking out a point, but is useful for checking the coordinates of a point you are setting out, or nearby points, without having to exit the Stakeout function and go into survey mode.

Stake out points in relation to a defined line ('Ref. line')

Set up the total station. If you are using coordinates, use 'Ori with coords' or 'resection' to determine the position and orientation of your instrument. If you are programming in a line by sighting directly to the start and end points you only need to level the instrument.

1. >>Main menu
2. >>Programs
3. >>Ref. Line "Define reference line. Stake out and check points relating to the line"
4. >>F4 Cont

You now have three options for defining the start point of the line:

a. Option 1 – Sight the point and press F1 Meas

b. Option 2 – Press F2 List and select the point from the memory then press F4 Cont

c. Option 3 – Press F2 ENH and enter the coordinates of the point manually then press F4 Cont

5. Now enter the end point of the line using one of the three options.
6. The length of the line will be displayed so you can check this against its expected value.
7. You can now stake out a single point (Stake), multiple points at regular increments (Grid) or measure the perpendicular distance from any point to the defined line (Meas Pt).
8. If you wish to change the defined line at any time, select F1 New BL.

Next you there are three different options for setting out in relation to the line you have defined. These are:

a. F1 Grid - To set out multiple points at regular chainages and offsets in relation to the start of the line)

b. F2 Meas Pt - To measure the perpendicular distance from any point to the defined line

c. F3 Setout – To set out a single point on the line on or offset from the line

Set out a grid of points

1. >>F1 Grid (To set out multiple points at regular chainages and offsets in relation to the start of the line)
2. Enter the start chainage of the grid.
3. Enter the chainage and offset increments.
4. >>F4 Cont
5. Input the PtID
6. You can now use the left/ right arrows to define the chainage (Chn) and offset (Offs) in relation to the defined line.

7. Sight to the prism and press F2 Dist. The arrows on the screen will display the location of the point relative to the total station.

8. Select the 'Polar' tab to display horizontal angle and distance or the 'Local' tab to display ΔL, ΔO ΔH.

9. Once you have set out the point, sight to it and press F1 Meas to record its as-set-out position.

10. Input the next PtID and select the chainage and offset.

OR Measure a perpendicular distance from a point to the defined line

1. >> F2 Meas Pt (To measure the perpendicular distance from any point to the defined line)

2. Sight to the point and press F2 Dist to display the position on the screen. The position is defined as a distance along the line (ΔL), an offset from it (ΔO) and a height difference (ΔH).

OR Set out a single point on the line

1. >>F3 Setout (To set out a single point on the line)

2. Enter the PtID (only if you need to keep a record of the point you are setting out)

3. If you are working with heights as well as positions, enter the correct reflector height, hr.

4. In 'Line' enter the chainage of the point in relation to the start point of the line.

5. In 'Offset' enter the perpendicular distance from the defined line. In the direction P1 to P2, a negative offset is to the left and a positive offset is to the right.

6. In 'Height' enter the level of the point (only if you are interested in the level)

7. >>F2 Dist. The arrows on the screen will display the location of the point relative to the total station. ΔL and ΔO will give the relative position in relation to the defined line.

8. Once you are within the required tolerance of the design point, mark the point. Sight to it and press F1 Meas to record its as-set-out position.

Measure an area

There is no need to set the position and orientation of the instrument to simply measure an area. You only need to level it. If you need to record the area perimeter in relation to your coordinate system, you need to set the position and orientation.

1. >>Main menu

2. >>Programs

3. >>Survey+ (tab at the top)

4. >>Area&Vol > "Compute online areas, split areas and calc. volume by creating a digital terrain model"

5. >>F4 Cont > Use left/right arrows to select 2D/3D Area

6. >>2D/3D Area> "Calculate 2D-/3D- Area"

7. >>F4 Cont

8. Select correct EDM settings and input the reflector height, hr.

9. Sight to points around the perimeter of the area. Each time press F1 Meas. The cumulative distance will show on the display as each new point it added. The first and last points will automatically be connected. Avoid crossing back over the line you have created as this will cause an error message 'General polygon definition error! Please remeasure point!"

Measure a volume

There is no need to set the position and orientation of the instrument to simply measure a volume. You only need to level it. If you need to record the boundary and surface points in relation to your coordinate system, you need to set the position and orientation.

1. >>Main menu

2. >>Programs

3. >>Survey+ (tab at the top)

4. >>Area&Vol > "Compute online areas, split areas and calc. volume by creating a digital terrain model"

5. >>F4 Cont > Use left/right arrows to select 'DTM Volume"

6. >>DTM Volume> "Calculate Volume by creating a digital terrain model and weight calculation"

7. >>F4 Cont

8. Select correct EDM settings and input the reflector height, hr.

9. Sight to points around the perimeter of the base of the stockpile. Each time press F1 Meas. The cumulative distance will show on the display as each new point it added. The first and last points will automatically be connected. Avoid crossing back over the line you have created as this will cause an error message 'General polygon definition error! Please remeasure point!"

10. >>F2 Calc

11. >>F2 New BL (new break line)

12. To create the new break line, sight to a series of points around the inside of the perimeter. Imagine you are tracing contour lines.

13. >>F2 Calc. The volume will be displayed, based on the measurements you have taken so far. Keep adding new break lines until you have enough detail to calculate the volume to the required accuracy.

Import data from an Excel file

1. Open your Excel file

2. Select the cells you wish to transfer to the TS and paste them into an empty Excel file so they are the only information in it.

3. Click 'File'

4. Click' Save As'. Give your file a name which you will recognise.

5. File type 'CSV, Comma delimited'

6. Save onto the memory stick, creating a new folder with a job name you will recognise.

7. You will get a message saying 'Some features may be lost'. Click the 'X' to close this message.

8. Close your excel file

9. Click 'Safely remove hardware' and safely eject your memory stick.

10. In your TS06, make sure you are in the main menu.

11. Insert memory stick

12. On the main menu screen select 4. Transfer

13. Select 2. Import

14. Use left and right arrows to select 'Single File'

15. F4. Cont

16. Select the job file you created in step 6. Press enter (the red button with a white arrow)

17. Select the CSV file you created from your Excel file. Press enter.

18. F1. Cont.

19. Type in a Job Name of select the default name which you created in step 4.

20. F4. Cont

Top of screen should say 'Define ASCII Export'

→ Delimiter: Comma

→ Unit: meter   Incl. Header: No

→ Data Fields: Pt ID, East, North, Height

21. F1. View to check it is the correct data

22. F4. Back

23. F4. Cont

24. Screen should say 'Transfer of Data Successful completed. Transfer more data?'

25. Select F1. No

26. On the main menu screen, select 3. Manage

27. Select 2. Fixpoints

28. On the top line, use left and right arrows to select the Job

29. On the next line down, use the left and right arrows to scroll through the points you have just imported to check them.

Export data from the instrument to an Excel file

1. On the main menu screen select 4. Transfer

2. Select 1. Export

3. In the top line use the left and right arrows to select To: 'USB Stick'

4. In the next line down, use the left and right arrows to choose the type of data you wish to export e.g. Data Type: Fixpoints

5. In the third line down, use the left and right arrows to select 'Single job'

6. In the bottom line, use the left and right arrows to select the job where your points are stored.

7. Select F4. Cont

8. Top of screen should say 'Select Dest. Folder! 1/1'

9. Select a folder or create a new one using F2. New.

10. F1. Cont

11. Top of screen should say 'Save all jobs as..'

12. In the top line, use left and right arrows to select Format: ASCII

13. F4. Cont

14. Top of screen should say 'Define ASCII Export'

15. Delimiter: Comma

16. Unit: meter   Incl. Header: No

17. Data Fields: Pt ID, East, North, Height, None, None

18. F4 Cont.

19. Screen should say 'Transfer of Points successfully completed! Transfer more data?'

20. Select F1. No

21. This should take you back to the main menu, but in case it doesn't always esc. Back to main menu before removing the memory stick.

22. Remove the memory stick from the TS06 and plug into computer

23. Open Excel

24. On row of tabs across the top of the screen, click the 'Data' tab.

25. In the top ribbon, click on 'From Text/CSV'

26. Choose 'From text'

27. Select the relevant file from the memory stick. Click 'Jobs' and then select the file within the 'Jobs' section.

28. Double click the CSV file you wish to transfer.

29. The data will be displayed.

30. At the bottom of the panel, click 'Load'

31. The data will appear in your excel spreadsheet.

32. 'Save As' whatever you want to call the file.

33. Remember to 'Safely Eject Hardware' before removing the USB stick.

34. Make sure you are on the main menu of your TS06 before reinserting the USB stick.

Appendix B

LEICA TS07

USEFUL INFORMATION

Toggling the options menu along the bottom of the screen

When you are in a program, pressing F4 will scroll through different menu options along the bottom of the screen. If you cannot see what you are looking for, it is probably hidden. Press F4 until you find it.

Changing the target type, prism constant and reflector height

'EDM settings' is the instrument's terminology for target type, prism constant and reflector height.

Within each function, one of the 'F' keys along the bottom will be 'EDM'. Remember to set the correct EDM settings for the target type you are sighting to. If you are working with heights the reflector height must be input manually each time you change from one EDM type to another. If you are not working with heights, you can ignore the reflector height.

SCREEN ARRANGEMENT

Icons across the top of the homescreen

Furthest right – Batter icon - TS battery level indicator. Tap on this icon to check how much battery and internal memory is available.

Second from the right – USB icon – Connectivity. Tap on this icon to check Bluetooth status and to adjust interface settings.

Third from the right – Instrument status – Will be one of the following:

- → 'I' or 'II' to indicate the face when the bubble is level
- → Bubble icon with yellow triangle if instrument is off level
- → Bubble icon with orange cross when tilt sensors are switched off
  Tap on this icon to check current Station details and to check if bubble is level.

Fourth from the right – Target type – Indicates what target type the instrument is currently set to. It will be one of the following:

- → Reflectorless mode (A house with a laser beam bouncing off it)
- → Round prism (Picture of a round prism)
- → Mini prism (Picture of a mini prism).

> → Tape (Picture of a retro target)
> Click on this icon to select Target type, EDM settings and to switch the laser pointer on or off.

MENUS

'Start'

This is the main menu where you will find all the measuring and setting out functions.

'Manage'

Job - Create, edit, view or delete jobs

Fixpoints - Find, edit, add or delete Fixpoints

Meas. Data – View measured data. Every measurement taken is assigned a block number

Data Transfer – Export and import data

Del. Data – Delete data

> - Takes you to the next page of the 'Manage' menu

< - Takes you to the previous page of the 'Manage' menu

USB Stick – Takes you to the USB stick

SD Card – Takes you to the SD card

Internal memory – Takes you to the internal memory

Codes – Create, edit, view or delete point codes

Formats – Manage formats

Screenshots – Takes you to screenshots you have stored

'Settings'

Work – Set trigger keys, set user keys (third and fourth round buttons), set the order of the fields on the screen, set map parameters (e.g. which types of points to show, labels etc), manage icons in the top icon bar.

Regional – Set units, horizontal circle direction, language, time, date etc

Data – order how data is stored and select where it is stored (internal memory or interface)

Screen – Set keyboard illumination and intensity of screen illumination, screensaver, touchscreen, switch off stakeout bleep

EDM – Set atmospheric pressure, prism type, switch laser pointer and guide light on/off.

Interface – Select the interface e.g. USB, Bluetooth etc

'Tools'

Adjust – Carry out instrument checks and adjust accordingly

Startup – Configure start up sequence

Info – Software, memory, system, maintenance information

Licence – Install licence keys for software

PIN – Set a PIN code for access to the instrument

Load FW – Load firmware

INSTRUMENT SET-UP

Set up over a known point ('Ori. With Coord')

1. >> 'Start' tab on home screen
2. >> Setup
3. Station Setup - "Set up a station and determine its coordinates and orientation"
4. >>F4 Cont.
5. Use the left/ right arrows to select 'Ori. With Coord.
6. >> Ori. With Coord. > "Station is known. Measure 1 or more known targets to compute orientation and height"
7. >>F4 Cont
8. The top of the screen says 'Enter Station Data".
9. >>F2 List and select the point you are set up over. Alternatively select F2 ENH and manually enter the coordinates.
10. >>F4 Cont.
11. If you are happy with the station name, press F3 Cont.
12. The top of the screen says 'Enter Target Data".
13. >>F1 List and select the point you are backsighting to.
14. The top of the screen says 'Sight target point!".
15. Line up on the target. Press F1 Meas.
16. You can now choose from the following:
    a. F1 Measure more points *(to measure more backsights)*
    b. F2 Measure in other face *(to improve the accuracy of your set up)*
    c. F3 Access tolerances
    d. F4 Compute *(once you are satisfied that you have taken enough backsights)*
17. >>F4 Compute

18. The top of the screen says 'Station setup result 1/2". The coordinates displayed here are the coordinates you entered earlier.

19. >>F4 Set

20. >>F4 New

21. The screen will say 'Station and orientation set' then it will return to the survey menu.

Set up at an unknown point ('Resection')

1.   >> 'Start' tab on home screen

2.   >> Setup

3.   Station Setup - "Set up a station and determine its coordinates and orientation"

4.   >>F4 Cont.

5.   Use the left/ right arrows to select 'Resection'.

6.   >> Resection. > "Station is unknown. Measure 2 or more known targets to compute station and orientation"

7.   >>F4 Cont

8.   The top of the screen says 'Enter Station Data". Give the station a name.

9.   >>F3 Cont

10. The top of the screen says 'Enter Target Point".

11. >>F1 List and select the point you are backsighting to.

12. >>F4 Cont

13. The top of the screen says 'Sight target point!".

14. Line up on the target. Press F1 Meas.

15. You can now choose from the following:

    a. F1 Measure more points *(to measure more backsights)*

    b. F2 Measure in other face *(to improve the accuracy of your set up)*

    c. F3 Access tolerances

16. >>F1 Measure more points

17. The top of the screen says 'Enter Target Point".

18. >>F1 List and select the point you are backsighting to.

19. The top of the screen says 'Sight target point!".

20. Line up on the target. Press F1 Meas.

21. The accuracy of your set up. Accur. Posit. is displayed. If this is within acceptable limits press F4 Compute. If you would like to improve the accuracy of your set up press F1 Measure more points.

22. Repeat until you have an acceptable error.

23. >>F4 Compute

24. The eastings, northings and height of your total station position will be displayed on the screen.

25. >>F4 Set

26. The screen will say 'Station and orientation set' then it will return to the survey menu.

INSTRUCTIONS

Measure points/ take a survey ('Survey')

1.   >> 'Start' tab on home screen

2.   >> Survey

3.   Enter the PtID

4.   Use the left/right arrows to select a Code from the code list. If there is no code list stored, and you would like to create codes as you go then tap in the empty green box then F1 New. However, it is much easier to add codes in Manage>Codes or by importing them in Transfer>Import.

5.   Sight to the point you would like to survey. Press F2 Dist to display the point information on the screen. Press F3 store to record the point information into the memory. If you would like to measure the point and record it in one go, press F1 Meas.

6.   To view the eastings, northings and height of the point either used the arrow pad on the keyboard or the scroll bar on the screen to scroll right down to the bottom.

Measure or calculate the distance and level difference between two points ('Tie distance')

To measure between two points, accurately level the total station at a point where you can sight to the two points you wish to measure between. Alternatively, to calculate the distance and level difference between two previously input or measured points, there is no need to level the instrument.

1.   >> 'Start' tab on the home screen

2.   >> Apps

3.   >>Survey+ (tab at the top)

4.   >>Tie Dist > "Compute distance, height difference and azimuth between two target points"

5.   >> F4 Cont > Use left/ right arrows to select Radial

6.   >>Radial > "Values are computed between one point and several others"

7.   >>F4 Cont

8.   You now have three options for entering the start point of the line:

     a.   Option 1 (To measure between two points) – Sight the point and press F1 Meas

     b.   Option 2 (To calculate between two previously measured or input points) – Press F2 List and select the point from the memory then press F4 Cont

    c. Option 3 (To calculate between two points which you know the coordinates of) – Press F2 ENH and enter the coordinates of the point manually then press F4 Cont

9. Now enter the end point of the line using one of the three options.

10. The horizontal distance, slope distance and level change are displayed at the bottom of the screen.

11. If you have finished press F3 (End).

12. If you would like to measure to a new point from the start of the line you have just measured press F2 (NewPt 2).

13. If you would like to measure to a new point from the end of the line you have just measured press F2 (NewPt 1).

Stake out point of known coordinates ('Stake out')

Set up the total station using 'Ori with coords' or 'resection' to determine the position and orientation of your instrument.

1. >> 'Start' tab in home screen

2. >> Setout

3. Stakeout>> "Place marks in the field at predefined points. Predefined points are points to be staked"

4. >>F4 Cont

5. Use the left/right arrows to select the point you wish to set out. Pt type/ID indicates whether it is a Fixpoint (Fixpt.) or a measured point (Meas). Press F1 View if you would like to view and check the point data before continuing.

6. Alternatively press F2 ENH if you would like to manually enter the coordinates of the point.

7. There are four tabs along the top – Polar, Local, Coord 1 and Coord 2

Tab 1- Polar.

1. Manually turn the horizontal tangent screw to set ΔHz to 0° 0' 0"

2. Looking through the telescope, direct your assistant into the correct line. Press F2 Dist. The arrows on the screen will indicate how far the prism needs to be moved. 'Up' means away from you and 'down' means towards you.

3. Repeat until you are within the required tolerance. Once point is set out, sight to it and press F1 Meas to record the as-set-out position.

4. It will automatically jump to the next point in the list.

Tab 2 - Local

1. Manually turn the horizontal tangent screw to set ΔL to 0.000m

2. Looking through the telescope, direct your assistant into the correct line. Press F2 Dist. The arrow on the screen will indicate how far the prism needs to be moved. 'Up' means away from you and 'down' means towards you.

3. Alternatively, sight to the prism, press F2 Dist and direct your assistant according to the arrows. ΔL and ΔO will match the distances given on the arrows.

4. Repeat until you are within the required tolerance. Once point is set out, sight to it and press F1 Meas to record the as-set-out position.

Tab 3 – Coord 1

The process is very similar to the angle and distance method, but in this option, the relative eastings and northings of the target are displayed. This is useful if you have a gridlines system in place in which case it may be more convenient to give your assistant directions in relation to grid lines.

Tab 4 - Coord 2

This option is not really convenient for staking out a point, but is useful for checking the coordinates of a point you are setting out, or nearby points, without having to exit the Stakeout function and go into survey mode.

Stake out point in relation to a defined line ('Ref. Line')

Set up the total station. If you are using coordinates, use 'Ori with coords' or 'resection' to determine the position and orientation of your instrument. If you are programming in a line by sighting directly to the start and end points you only need to level the instrument.

1. >>'Start' tab on home screen

2. >>Ref. El (tab at the top)

3. >>Ref. Line "Define reference line. Stake out and check points relating to the line"

4. >>F4 Cont

5. You now have three options for defining the start point of the line:

    a. Option 1 – Sight the point and press F1 Meas

    b. Option 2 – Press F2 List and select the point from the memory then press F4 Cont

    c. Option 3 – Press F2 ENH and enter the coordinates of the point manually then press F4 Cont

6. Now enter the end point of the line using one of the three options.

7. The length of the line will be displayed so you can check this against its expected value.

8. You can now stake out a single point (Stake), multiple points at regular increments (Grid) or measure the perpendicular distance from any point to the defined line (Meas Pt).

9. If you wish to change the defined line at any time, select F1 New BL.

10. Next you there are three different options for setting out in relation to the line you have defined. These are:

a. F1 Grid - To set out multiple points at regular chainages and offsets in relation to the start of the line)

b. F2 Meas Pt - To measure the perpendicular distance from any point to the defined line

c. F3 Setout – To set out a single point on the line on or offset from the line

Set out a grid of points

1. >>F1 Grid (To set out multiple points at regular chainages and offsets in relation to the start of the line)

2. Enter the start chainage of the grid.

3. Enter the chainage and offset increments.

4. >>F4 Cont

5. Input the PtID

6. You can now use the left/ right arrows to define the chainage (Chn) and offset (Offs) in relation to the defined line.

7. Sight to the prism and press F2 Dist. The arrows on the screen will display the location of the point relative to the total station.

8. Select the 'Polar' tab to display horizontal angle and distance or the 'Local' tab to display ΔL, ΔO ΔH.

9. Once you have set out the point, sight to it and press F1 Meas to record its as-set-out position.

10. Input the next PtID and select the chainage and offset.

Measure a perpendicular distance from a point to the defined line

1. >> F2 Meas Pt (To measure the perpendicular distance from any point to the defined line)

2. Sight to the point and press F2 Dist to display the position on the screen. The position is defined as a distance along the line (ΔL), an offset from it (ΔO) and a height difference (ΔH).

Set out a single point on the line

1. >>F3 Setout (To set out a single point on the line)

2. Enter the PtID (only if you need to keep a record of the point you are setting out)

3. If you are working with heights as well as positions, enter the correct reflector height, hr.

4. In 'Line' enter the chainage of the point in relation to the start point of the line.

5. In 'Offset' enter the perpendicular distance from the defined line. In the direction P1 to P2, a negative offset is to the left and a positive offset is to the right.

6. In 'Height' enter the level of the point (only if you are interested in the level)

7. >>F2 Dist. The arrows on the screen will display the location of the point relative to the total station. ΔL and ΔO will give the relative position in relation to the defined line.

8. Once you are within the required tolerance of the design point, mark the point. Sight to it and press F1 Meas to record its as-set-out position.

Measure an area

There is no need to set the position and orientation of the instrument,

1. >> 'Start' tab on home screen
2. >> Apps
3. >> Survey+ (tab at the top)
4. >> Area&Vol > "Compute online areas, split areas and calc. volume by creating a digital terrain model"
5. >> F4 Cont > Use left/right arrows to select 2D/3D Area
6. >> 2D/3D Area> "Calculate 2D-/3D- Area"
7. >> F4 Cont
8. Select correct EDM settings and input the reflector height, hr.
9. Sight to points around the perimeter of the area. Each time press F1 Meas. The cumulative distance will show on the display as each new point it added. The first and last points will automatically be connected. Avoid crossing back over the line you have created as this will cause an error message 'General polygon definition error! Please remeasure point!"

Measure a volume

There is no need to set the position and orientation of the instrument,

1. >> 'Start' tab in home screen
2. >> Apps
3. >> Survey+ (tab at the top)
4. >> Area&Vol > "Compute online areas, split areas and calc. volume by creating a digital terrain model"
5. >> F4 Cont > Use left/right arrows to select 'DTM Volume"
6. >> DTM Volume> "Calculate Volume by creating a digital terrain model and weight calculation"
7. >> F4 Cont
8. Select correct EDM settings and input the reflector height, hr.
9. Sight to points around the perimeter of the base of the stockpile. Each time press F1 Meas. The cumulative distance will show on the display as each new point it added. The first and last points will automatically be connected. Avoid crossing back over the line you have created as this will cause an error message 'General polygon definition error! Please remeasure point!"
10. >> F2 Calc

11. \>> F2 New BL (new break line)

12. To create the new break line, sight to a series of points around the inside of the perimeter. Imagine you are tracing contour lines.

13. \>> F2 Calc. The volume will be displayed, based on the measurements you have taken so far. Keep adding new break lines until you have enough detail to calculate the volume to the required accuracy.

Import data from an excel file

1. Open your Excel file

2. Select the cells you wish to transfer to the TS and paste them into an empty Excel file so they are the only information in it.

3. Click 'File'

4. Click' Save As'. Give your file a name which you will recognise.

5. File type 'CSV, Comma delimited'

6. Save onto the USB stick or SD card, creating a new folder with a job name you will recognise.

7. You will get a message saying 'Some features may be lost'. Click the 'X' to close this message.

8. Close your excel file

9. Click 'Safely remove hardware' and safely eject your USB stick or SD card.

10. In your TS07, make sure you are in the main menu.

11. Insert USB stick or SD card

12. \>> 'Manage' tab on the home screen

13. \>> Data Transfer

14. Select 2. Import

15. Use left and right arrows to select 'USB' or 'SD Card'

16. Use left and right arrows to select 'Single File'

17. F4. Cont

18. Select the job file you created in step 6. Press enter (the red button with a white arrow)

19. Select the CSV file you created from your Excel file. Press enter.

20. F1. Cont.

21. Type in a Job Name of select the default name which you created in step 4.

22. F4. Cont

23. Top of screen should say 'Define ASCII Export'

    a. Delimiter: Comma

    b. Unit: meter   Incl. Header: No

    c. Data Fields: Pt ID, East, North, Height

24. F1. View to check it is the correct data

25. F4. Back

26. F4. Cont

27. Screen should say 'Transfer of Data Successful completed. Transfer more data?'

28. Select F1. No

29. On the main menu screen, select 3. Manage

30. Select 2. Fixpoints

31. On the top line, use left and right arrows to select the Job

32. On the next line down, use the left and right arrows to scroll through the points you have just imported to check them.

Export data to an Excel file

1. >> 'Manage' tab on home screen

2. >> Data Transfer

3. >> Export

4. In the top line use the left and right arrows to select To: 'USB Stick' or 'SD card'

5. In the next line down, use the left and right arrows to choose the type of data you wish to export e.g. Data Type: Fixpoints

6. In the third line down, use the left and right arrows to select 'Single job'

7. In the bottom line, use the left and right arrows to select the job where your points are stored.

8. Select F4. Cont

9. Top of screen should say 'Select Dest. Folder! 1/1'

10. Select a folder or create a new one using F2. New.

11. F1. Cont

12. Top of screen should say 'Save all jobs as..'

13. In the top line, use left and right arrows to select Format: ASCII

14. F4. Cont

15. Top of screen should say 'Define ASCII Export'

    a. Delimiter: Comma

    b. Unit: meter   Incl. Header: No

    c. Data Fields: Pt ID, East, North, Height, None, None

16. F4 Cont.

17. Screen should say 'Transfer of Points successfully completed! Transfer more data?'

18. Select F1. No

19. This should take you back to the main menu, but in case it doesn't always esc. Back to main menu before removing the USB stick or SD card.

20. Remove the USB stick or SD Card from the TS07 and plug into computer

21. Open Excel

22. On row of tabs across the top of the screen, click the 'Data' tab.

23. In the top ribbon, click on 'From Text/CSV'

24. Choose 'From text'

25. Select the relevant file from the memory stick. Click 'Jobs' and then select the file within the 'Jobs' section.

26. Double click the CSV file you wish to transfer.

27. The data will be displayed.

28. At the bottom of the panel, click 'Load'

29. The data will appear in your excel spreadsheet.

30. 'Save As' whatever you want to call the file.

31. Remember to 'Safely Eject Hardware' before removing the USB stick or SD card

32. Make sure you are on the main menu of your TS07 before reinserting the USB stick or SD card.

Appendix C

LEICA TS09

USEFUL INFORMATION

Toggling the options menu along the bottom of the screen

When you are in a program, pressing F4 will scroll through different menu options along the bottom of the screen. If you cannot see what you are looking for, it is probably hidden. Press F4 until you find it.

Changing the target type, prism constant and reflector height

'EDM settings' is the instrument's terminology for target type, prism constant and reflector height.

Within each function, one of the 'F' keys along the bottom will be 'EDM'. Remember to set the correct EDM settings for the target type you are sighting to. If you are working with heights the reflector height must be input manually each time you change from one EDM type to another. If you are not working with heights, you can ignore the reflector height.

SCREEN ARRANGEMENT

Icons across the top of the screen

Top left – Indicates the EDM mode the instrument is currently set to. It will be one of the following:

> → Reflectorless mode (A house with a laser beam bouncing off it)

> → Prism (Picture of a prism). Remember to check that it is the correct prism constant!

> → Tape (Picture of a retro target)

Second from the left – Long bubble indicates if the instrument has gone off level

Third from the left – What looks like a capital I is actually a roman numeral 1. This indicates that you are on face one. If you change faces the I will change to II, which indicates you are on face two. If you click on this icon it will display the digital bubble. 'Tilt correct' and 'Hz correct' should be set to on. If you change to the 'plummet' tab at the top, you can change the intensity of the laser plummet.

Forth from the left – Displays the time

Fifth from the left – Screenshot button. If you click this button it will take a screenshot and save it into Manage>> Data 2 (tab)>> Img- Notes. F1 Prev/ F2 Next will navigate through the images. You can add notes to images using the drawing tools down the right-hand side.

Sixth from the left – Interface settings icon indicates what port the instrument is set to. It will be one of the following:

> → USB (standard USB symbol)
>
> → Automatic (two- way horizontal arrows surrounded by yellow stars)
>
> → Bluetooth (standard Bluetooth symbol)
>
> → Default port (VGA cable symbol)

Seventh from the left – TS battery level indicator

Small round buttons immediately to the right of the screen

Top button – Menu button (looks like a piece of paper with writing on it) – This button will switch between the different tabs found at the top of the screen.

Second button down – Favourites (star) – Contains quick links to handy functions.

Third button down – User defined button (Person 1) – This key can be set by the user. By default it is usually set to the Level & Plummet screen.

Fourth button down – User defined button (Person 2) – This key can be set by the user. Many people set this to the laser pointer.

Fifth button down – Escape key (u-turn arrow) – This button (or the equivalent button at the right hand of the blue bar at the top of the screen) will always take you back to the previous screen.

MENUS

'Manage'

Job - Create, edit, view or delete jobs

Fixpoints - Create, edit, view or delete Fixpoints

Meas. Data – View measured data. Every measurement taken is assigned a block number.

Codes – Create, edit, view or delete point codes

Formats – Manage formats

Del. Data – Delete data

'Settings'

Work – Set trigger keys, set user keys (third and fourth round buttons), set the order of the fields on the screen, set map parameters (e.g. which types of points to show, labels etc), manage icons in the top icon bar.

Regional – Set units, horizontal circle direction, language, time, date etc

Data – order how data is stored and select where it is stored (internal memory or interface)

Screen – Set keyboard illumination and intensity of screen illumination, screensaver, touchscreen, switch off stakeout bleep

EDM – Set atmospheric pressure, prism type, switch laser pointer and guide light on/off.

Interface – Select the interface e.g. USB, Bluetooth etc

'Tools'

Adjust – Carry out instrument checks and adjust accordingly

Startup – Configure start up sequence

Info – Software, memory, system, maintenance information

Licence – Install licence keys for software

PIN – Set a PIN code for access to the instrument

Load FW – Load firmware

INSTRUMENT SET-UP

Set up over a known point ('Ori. With Coord')

1. >>Main menu
2. >>Programs
3. >>Survey (tab at the top)
4. >>Stn. Setup > "Set up a station and determine its coordinates and orientation"
5. >>F4 Cont.
6. Use the left/ right arrows to select 'Ori. With Coord.
7. >> Ori. With Coord. > "Station is known. Measure 1 or more known targets to compute orientation and height"
8. >>F4 Cont
9. The top of the screen says 'Enter Station Data".
10. >>F2 List and select the point you are set up over. Alternatively select F2 ENH and manually enter the coordinates.
11. >>F4 Cont.
12. If you are happy with the station name, press F3 Cont.
13. The top of the screen says 'Enter Target Data".
14. >>F2 List and select the point you are backsighting to.
15. The top of the screen says 'Sight target point!".
16. Line up on the target. Press F1 Meas.
17. You can now choose from the following:
18. F1 Measure more points *(to measure more backsights)*
19. F2 Measure in other face *(to improve the accuracy of your set up)*

20. F3 Access tolerances

21. F4 Compute *(once you are satisfied that you have taken enough backsights)*

22. >>F4 Compute

23. The top of the screen says 'Station setup result 1/2". The coordinates displayed here are the coordinates you entered earlier.

24. >>F4 Set

25. >>F4 New

26. The screen will say 'Station and orientation set' then it will return to the survey menu.

Set up at an unknown point ('Resection')

1. >>Main menu

2. >>Programs

3. >>Survey (tab at the top)

4. >>Stn. Setup > "Set up a station and determine its coordinates and orientation"

5. >>F4 Cont.

6. Use the left/ right arrows to select 'Resection."

7. >> Resection. > "Station is unknown. Measure 2 or more known targets to compute station and orientation"

8. >>F4 Cont

9. The top of the screen says 'Enter Station Data". Give the station a name.

10. >>F3 Cont

11. The top of the screen says 'Enter Target Point".

12. >>F1 List and select the point you are backsighting to.

13. >>F4 Cont

14. The top of the screen says 'Sight target point!".

15. Line up on the target. Press F1 Meas.

16. You can now choose from the following:

    d. F1 Measure more points *(to measure more backsights)*

    e. F2 Measure in other face *(to improve the accuracy of your set up)*

    f. F3 Access tolerances

17. >>F1 Measure more points

18. The top of the screen says 'Enter Target Point".

19. >>F1 List and select the point you are backsighting to.

20. The top of the screen says 'Sight target point!".

21. Line up on the target. Press F1 Meas.

22. The accuracy of your set up. Accur. Posit. is displayed. If this is within acceptable limits press F4 Compute. If you would like to improve the accuracy of your set up press F1 Measure more points.

23. Repeat until you have an acceptable error.

24. >>F4 Compute

25. The eastings, northings and height of your total station position will be displayed on the screen.

26. >>F4 Set

27. The screen will say 'Station and orientation set' then it will return to the survey menu.

INSTRUCTIONS

Measure points/ take a survey ('Survey')

Set up the instrument

1. >>Main menu

2. >>Q-Survey will take you straight into the survey function and jump to Enter PtID below.

Alternatively:

1. >>Programs

2. >>Survey (tab at the top)

3. >>Survey > "Measure unlimited number of points. Pre-settings for job and stations can be done"

4. >>F4 Cont

5. Enter the PtID

6. Use the left/right arrows to select a Code from the code list. If there is no code list stored, and you would like to create codes as you go along press F2 Code then F1 New. However it is much easier to add codes in Manage>Codes or by importing them in Transfer>Import.

7. Sight to the point you would like to survey. Press F2 Dist to display the point information on the screen. Press F3 store to record the point information into the memory. If you would like to measure the point and record it in one go, press F1 Meas.

8. To view the eastings, northings and height of the point either used the arrow pad on the keyboard or the scroll bar on the screen to scroll right down to the bottom.

Measure or calculate the distance and level difference between two points ('Tie distance')

Set up and accurately level the total station at a point where you can sight to the two points you wish to measure between.

1. >>Main menu

2. >>Programs

3. >>Survey+ (tab at the top)

4. >>Tie Dist > "Compute distance, height difference and azimuth between two target points"

5. >> F4 Cont > Use left/ right arrows to select Radial

6. >>Radial > "Values are computed between one point and several others"

7. >>F4 Cont

8. You now have three options for entering the start point of the line:

    a. Option 1 – Sight the point and press F1 Meas

    b. Option 2 – Press F2 List and select the point from the memory then press F4 Cont

    c. Option 3 – Press F2 ENH and enter the coordinates of the point manually then press F4 Cont

9. Now enter the end point of the line using one of the three options.

10. The horizontal distance, slope distance and level change are displayed at the bottom of the screen.

11. If you have finished press F3 End

12. If you would like to measure to a new point from the start of the line you have just measured press F2 NewPt 2.

13. If you would like to measure to a new point from the end of the line you have just measured press F2 NewPt 1.

Stake out points of known coordinate ('Stakeout')

1. Set up the total station using 'Ori with coords' or 'resection' to determine the position and orientation of your instrument.

2. >>Main menu

3. >>Programs

4. >>Survey (tab at the top)

5. >>Stakeout "Make marks in the field at predefined points. Predefined points are points to be staked"

6. >>F4 Cont

7. Use the left/ right arrows to select the point you wish to set out. Pt type/ID indicates whether it is a Fixpoint (Fixpt.) or a measured point (Meas). Press F1 View if you would like to view and check the point data before continuing.

8. Alternatively press F2 ENH if you would like to manually enter the coordinates of the point.

9. There are four tabs along the top – Polar, Local, Coord 1 and Coord 2

Tab 1- Polar.

1. Manually turn the horizontal tangent screw to set ΔHz to 0° 0' 0"

2. Looking through the telescope, direct your assistant into the correct line.   Press F2 Dist. The arrows on the screen will indicate how far the prism needs to be moved. 'Up' means away from you and 'down' means towards you.

3. Repeat until you are within the required tolerance. Once point is set out, sight to it and press F1 Meas to record the as-set-out position.

4. It will automatically jump to the next point in the list.

Tab 2 - Local

1. Manually turn the horizontal tangent screw to set ΔL to 0.000m

2. Looking through the telescope, direct your assistant into the correct line. Press F2 Dist. The arrow on the screen will indicate how far the prism needs to be moved. 'Up' means away from you and 'down' means towards you.

3. Alternatively, sight to the prism, press F2 Dist and direct your assistant according to the arrows. ΔL and ΔO will match the distances given on the arrows.

4. Repeat until you are within the required tolerance. Once point is set out, sight to it and press F1 Meas to record the as-set-out position.

Tab 3 – Coord 1

1. The process is very similar to the angle and distance method, but in this option, the relative eastings and northings of the target are displayed. This is useful if you have a gridlines system in place in which case it may be more convenient to give your assistant directions in relation to grid lines.

Tab 4 - Coord 2

1. This option is not really convenient for staking out a point, but is useful for checking the coordinates of a point you are setting out, or nearby points, without having to exit the Stakeout function and go into survey mode.

Stake out point in relation to a defined line ('Ref. Line')

Set up the total station. If you are using coordinates, use 'Ori with coords' or 'resection' to determine the position and orientation of your instrument. If you are programming in a line by sighting directly to the start and end points you only need to level the instrument.

1. >>Main menu

2. >>Programs

3. >>Ref. El (tab at the top)

4. >>Ref. Line "Define reference line. Stake out and check points relating to the line"

5. >>F4 Cont

6. You now have three options for defining the start point of the line:

    a. Option 1 – Sight the point and press F1 Meas

    b. Option 2 – Press F2 List and select the point from the memory then press F4 Cont

    c. Option 3 – Press F2 ENH and enter the coordinates of the point manually then press F4 Cont

7. Now enter the end point of the line using one of the three options.

8. The length of the line will be displayed so you can check this against its expected value.

9. You can now stake out a single point (Stake), multiple points at regular increments (Grid) or measure the perpendicular distance from any point to the defined line (Meas Pt).

10. If you wish to change the defined line at any time, select F1 New BL.

11. Next you there are three different options for setting out in relation to the line you have defined. These are:

    a. F1 Grid - To set out multiple points at regular chainages and offsets in relation to the start of the line)

    a. F2 Meas Pt - To measure the perpendicular distance from any point to the defined line

    a. F3 Setout – To set out a single point on the line on or offset from the line

Set out a grid of points

1. >>F1 Grid (To set out multiple points at regular chainages and offsets in relation to the start of the line)

2. Enter the start chainage of the grid.

3. Enter the chainage and offset increments.

4. >>F4 Cont

5. Input the PtID

6. You can now use the left/ right arrows to define the chainage (Chn) and offset (Offs) in relation to the defined line.

7. Sight to the prism and press F2 Dist. The arrows on the screen will display the location of the point relative to the total station.

8. Select the 'Polar' tab to display horizontal angle and distance or the 'Local' tab to display ΔL, ΔO ΔH.

9. Once you have set out the point, sight to it and press F1 Meas to record its as-set-out position.

10. Input the next PtID and select the chainage and offset.

OR Measure a perpendicular distance from a point to the defined line

1. >> F2 Meas Pt (To measure the perpendicular distance from any point to the defined line)
2. Sight to the point and press F2 Dist to display the position on the screen. The position is defined as a distance along the line (ΔL), an offset from it (ΔO) and a height difference (ΔH).

OR Set out a single point on the line

1. >>F3 Setout (To set out a single point on the line)
2. Enter the PtID (only if you need to keep a record of the point you are setting out)
3. If you are working with heights as well as positions, enter the correct reflector height, hr.
4. In 'Line' enter the chainage of the point in relation to the start point of the line.
5. In 'Offset' enter the perpendicular distance from the defined line. In the direction P1 to P2, a negative offset is to the left and a positive offset is to the right.
6. In 'Height' enter the level of the point (only if you are interested in the level)
7. >>F2 Dist. The arrows on the screen will display the location of the point relative to the total station. ΔL and ΔO will give the relative position in relation to the defined line.
8. Once you are within the required tolerance of the design point, mark the point. Sight to it and press F1 Meas to record its as-set-out position.

Measure an area

1. There is no need to set the position and orientation of the instrument,
2. >>Main menu
3. >>Programs
4. >>Survey+ (tab at the top)
5. >>Area&Vol > "Compute online areas, split areas and calc. volume by creating a digital terrain model"
6. >>F4 Cont > Use left/right arrows to select 2D/3D Area
7. >>2D/3D Area> "Calculate 2D-/3D- Area"
8. >>F4 Cont
9. Select correct EDM settings and input the reflector height, hr.
10. Sight to points around the perimeter of the area. Each time press F1 Meas. The cumulative distance will show on the display as each new point it added. The first and last points will automatically be connected. Avoid crossing back over the line you have created as this will cause an error message 'General polygon definition error! Please remeasure point!"

Measure a volume

There is no need to set the position and orientation of the instrument,

1. >>Main menu
2. >>Programs
3. >>Survey+ (tab at the top)
4. >>Area&Vol > "Compute online areas, split areas and calc. volume by creating a digital terrain model"
5. >>F4 Cont > Use left/right arrows to select 'DTM Volume"
6. >>DTM Volume> "Calculate Volume by creating a digital terrain model and weight calculation "
7. >>F4 Cont
8. Select correct EDM settings and input the reflector height, hr.
9. Sight to points around the perimeter of the base of the stockpile. Each time press F1 Meas. The cumulative distance will show on the display as each new point it added. The first and last points will automatically be connected. Avoid crossing back over the line you have created as this will cause an error message 'General polygon definition error! Please remeasure point!"
10. >>F2 Calc
11. >>F2 New BL (new break line)
12. To create the new break line, sight to a series of points around the inside of the perimeter. Imagine you are tracing contour lines.
13. >>F2 Calc. The volume will be displayed, based on the measurements you have taken so far. Keep adding new break lines until you have enough detail to calculate the volume to the required accuracy.

Import data from Excel

1. Open your Excel file
2. Select the cells you wish to transfer to the TS and paste them into an empty Excel file so they are the only information in it.
3. Click 'File'
4. Click' Save As'. Give your file a name which you will recognise.
5. File type 'CSV, Comma delimited'
6. Save onto the memory stick, creating a new folder with a job name you will recognise.
7. You will get a message saying 'Some features may be lost'. Click the 'X' to close this message.
8. Close your excel file

9. Click 'Safely remove hardware' and safely eject your memory stick.

10. In your TS09, make sure you are in the main menu.

11. Insert memory stick

12. On the main menu screen select 4. Transfer

13. Select 2. Import

14. Use left and right arrows to select 'Single File'

15. F4. Cont

16. Select the job file you created in step 6. Press enter (the red button with a white arrow)

17. Select the CSV file you created from your Excel file. Press enter.

18. F1. Cont.

19. Type in a Job Name of select the default name which you created in step 4.

20. F4. Cont

21. Top of screen should say 'Define ASCII Export'
    → Delimiter: Comma
    → Unit: meter   Incl. Header: No
    → Data Fields: Pt ID, East, North, Height

22. F1. View to check it is the correct data

23. F4. Back

24. F4. Cont

25. Screen should say 'Transfer of Data Successful completed. Transfer more data?'

26. Select F1. No

27. On the main menu screen, select 3. Manage

28. Select 2. Fixpoints

29. On the top line, use left and right arrows to select the Job

30. On the next line down, use the left and right arrows to scroll through the points you have just imported to check them.

Export data to Excel

1. On the main menu screen select 4. Transfer

2. Select 1. Export

3. In the top line use the left and right arrows to select To: 'USB Stick'

4. In the next line down, use the left and right arrows to choose the type of data you wish to export e.g. Data Type: Fixpoints

5. In the third line down, use the left and right arrows to select 'Single job'

6. In the bottom line, use the left and right arrows to select the job where your points are stored.

7. Select F4. Cont

8. Top of screen should say 'Select Dest. Folder! 1/1'

9. Select a folder or create a new one using F2. New.

10. F1. Cont

11. Top of screen should say 'Save all jobs as..'

12. In the top line, use left and right arrows to select Format: ASCII

13. F4. Cont

14. Top of screen should say 'Define ASCII Export'

    → Delimiter: Comma

    → Unit: meter   Incl. Header: No

    → Data Fields: Pt ID, East, North, Height, None, None

15. F4 Cont.

16. Screen should say 'Transfer of Points successfully completed! Transfer more data?'

17. Select F1. No

18. This should take you back to the main menu, but in case it doesn't always esc. Back to main menu before removing the memory stick.

19. Remove the memory stick from the TS09 and plug into computer

20. Open Excel

21. On row of tabs across the top of the screen, click the 'Data' tab.

22. In the top ribbon, click on 'From Text/CSV'

23. Choose 'From text'

24. Select the relevant file from the memory stick. Click 'Jobs' and then select the file within the 'Jobs' section.

25. Double click the CSV file you wish to transfer.

26. The data will be displayed.

27. At the bottom of the panel, click 'Load'

28. The data will appear in your excel spreadsheet.

29. 'Save As' whatever you want to call the file.

30. Remember to 'Safely Eject Hardware' before removing the USB stick.

31. Make sure you are on the main menu of your TS06 before reinserting the USB stick.

Appendix D

LEICA TS15

USEFUL INFORMATION

Power Search

F12 triggers the total station to search for and lock on to the 360 degree prism.

Blue bar and 'back' button

The blue bar at the top of the screen displays the screen title.

The 'U-turn' arrow – takes you back to the previous screen you were on.

Function button 'FN'

Within any screen, menu or program, you can access the FN menu by pressing the 'FN' hard key. The FN menu which appears will relate to the screen you are on at the time. If you can't see what you are looking for in the options on the screen, it is most likely in the 'FN' menu.

Entering data

If you make a mistake when entering data and want to overwrite it, the 'left' joystick arrow will take you back to the first digit you want to overwrite. If you keep typing, it will continue to overwrite subsequent digits.

To delete digits, use the arrow which points to the right, immediately to the right of the '3' key.

At the very bottom left of the keypad, there is an arrow pointing upwards. This toggles between upper case and lower case.

The space bar is the same size as the other keys. It is second from the right on the very bottom row.

To enter a negative number, use the +/- key immediately to the left of the '1' key.

'Ref triplet' warning

Ref triplet – If you attempt to delete a resection station, you will get a message saying 'This REF triplet has one or more measurements belonging to it. I you delete this REF triplet you will also delete these measurements'. This means you will delete all the measurements taken from that resection set up point.

SCREEN ARRANGEMENT AND HARD KEYS

Icons across the top of the home screen

Top left – Prism lock type - Padlock, prism with a hand or prism in the cross hairs. The padlock displays when the target type is set to 360 degree prism. Closed padlock indicates that the total station is locked onto the prism. Open padlock indicates that it is not locked.

When the target type is set to 'any surface' a picture of the cross-hairs with a hand indicates that you need to sight through the telescope and manually adjust the horizontal and vertical circles to line up the cross-hairs on your target.

Tap on this icon to take you to the 'Leica TPS Favourites' menu.

Second from the left – Indicates the target type the instrument is currently set to. It could be one of the following:

→ Large 360 degree prism

→ Mini 360 degree prism

→ Standard round prism

→ Standard mini prism

→ Retro target

→ Reflectorless mode/ any surface (A house with a laser striking it. Reflectorless mode (A house with a laser beam bouncing off it)

Tap on this icon to go to the targets menu. In this menu, you can select the target type you are about to measure to.

Third from the left – Indicates measurement type mode. It could be one of the following:

→ Continuous (Grey circle with circular arrows with white oblong below it)

→ Continuous plus (Grey circle with circular arrows and a plus sign, with white oblong below it)

→ Single (Grey circle with a '1' with white oblong below it)

→ Single, fast (Running hare with white oblong below it)

→ Averaging (Algebraic x symbol with white oblong below it)

→ Long range (globe with an arrow round it with white oblong below it)

→ Long range average (globe with an algebraic x symbol and arrow round it and a white oblong below it)

The white oblong indicates when a measurement is taken. When it is white, the instrument is not measuring. When a red line moves back and forth along the oblong, a measurement is being taken.

Tap on this icon to go to the 'Measure and target settings' menu. In this menu you can select the measure and target type, measure mode, target aiming and visibility.

Fourth from left – Indicates which face you are in. Roman numeral '**I**' indicates your in face 1 or left face. Roman numeral '**II**' indicates you are in face 2 or right face.

Tap on this icon to go to the 'Level bubble and compensator menu'. In this menu, you can view the electronic bubble, adjust the intensity of the laser plumb (red dot indicating the position of the TS), and turn the tilt compensators or HZ correction on or off.

Fifth from left – Indicates which instrument mode you are in. It could be one of the following:

- → GPS only (Rover on a pole)
- → Total station only (total station)
- → Both GPS and total station (Total station and rover on a pole). Toggling this key will switch between giving preference to total station or GPS.

If you connect to GPS and want to go back to total station only, in the main menu screen press the 'Fn' hard key, then tap 'mode'. In this menu, you can select 'use total station only'.

Sixth from left – Camera (Camera icon)– Turns the camera on.

Second from right – Data (microchip icon) or diagram showing areas and lines – Takes you to the 'view and edit data' menu

Furthest right – Battery indicator. There are two batteries displayed:

- → CS – Controller battery
- → TS – Total station battery

Hard keys

F1 – Usually corresponds with 'OK'

House icon – Brings up the 'home screen' options across the top, with drop down menus so you can quickly navigate to what you are looking for

Star key – TPS favourites menu

U-turn arrow – takes you back to the previous screen you were on

Red ' Return arrow' selects whatever is selected on the screen (works the same as the 'OK' button at the centre of the joystick arrows)

Fn – Only works on some screens, but takes you to a new menu along the bottom of the screen.

Hold down the Fn key and use '4' and '7' to increase or decrease the screen brightness

Hold down the Fn key and use '6' and '9' to increase or decrease the keypad volume

Hold down the Fn key and press '.' to take a screenshot and store it to images

When in the map view, the '1' key will zoom extents

When in the map view, the '2' key will zoom in

When in the map view, the '3' key will zoom out

When in the map view, the '5' key will ???

MENUS AND TOOLBARS

Main menu/ Home page

In this menu you will find the following menus:

1. Go to Work! (Set up total station and stake points)
2. Jobs and Data (Point management, import and export)
3. Instrument (Status and Connection)
4. User (Software and settings, Screen and audio)

To minimise the SmartWorx software, press 'Fn' then 'minim'. To reinstate it, click the 'SmartWorx' tab on the bottom bar.

'Go to work!'

In this menu you can set up the instrument, carry out a survey, stake out points and access the COGO menu

'Jobs and data'

In this menu you can create new jobs (including code lists), view and edit data, create control data, view job properties, choose the working and control jobs, import and export data

'Create a new job'

Tabs:

General – Enter the job name, description, creator and storage device

Codelist – Select the Codelist. To add a new Codelist, tap in the codelist box and click 'New'. Enter the Codelist name, description and creator. To add codes, click 'Codes' at the bottom of the screen, then 'New'. Enter the Code letter, description, code group, code type (point, line or area). Select the line style and colour.

'View and edit data'

This menu can easily be accessed at any time by tapping the 'data' icon immediately to the left of the battery indicator icons at the top of the screen.

Tabs:

Points – Data about measured and input points. To view point data including date and time it was measured or entered, point code, eastings and northings and class, tap the 'More' button at the bottom of the screen. This will scroll through the table to the right.

To get more details about a point, select the line in the table then click 'edit'. This will open up a new screen with tabs for Coords, Obs (where it was observed from, the target height and the target type), Code, Annots (where you can add any notes you want) and Images (where you can view, mark up or edit photographs and sketches attributed to the point)

You can retrospectively change the target height if you realise you used the wrong target height at the time.

Lines – Press 'More' to scroll to the right to see more information. Click 'Edit' to view more information and edit.

Areas – Press 'More' to scroll to the right to see more information. Click 'Edit' to view more information and edit.

Images – Here you can view, mark up or edit photographs and sketches and attribute them to a point.

Map – Map view of all the points

To filter data:

1. >> 'Fn'
2. >> Filter

You can sort points, lines or areas (tabs at the top) by:

→ Ascending point ID

→ Descending point ID

→ Forward time

→ Backward time

The Zoom Toolbar

→ Top icon with arrows pointing outwards – Zoom extents

→ Second down – Zoom in

→ Third down – Zoom out

→ Forth down – Zoom window (drag red point to create a window around the points you want to zoom in on)

→ Fifth down

Create control data

Use this menu to enter control points, lines, polylines or arcs.

Map view configuration

This menu can be accessed and adjusted from the map view in any programme by pressing:

1. >> 'FN'
2. >> Config

Tabs:

General – Select which toolbars to display in the map view setting and whether to make the centre of the map the TS or the controller.

Points – Select what information to display about points, and how many points to display it for.

Lines & Areas – Select what information to display about lines and areas.

DTM – Select whether to display DTM in the map view

Alignments - Select whether to display DTM in the map view

CAD import – Select parameters for CAD import

Stakeout configuration menu

Whilst in the stakeout mode (but not whilst measuring), select the 'Stake' tab.

1. >> 'FN'
2. >> Config

Tabs:

General –Set up what happens once you have stored a point. It can automatically jump to the closest point rather than the next in the list, specify how points are stored, automatically stakeout on two faces, get directions from the point you are at to the next point or the get the total station to automatically turn to the next point.

Quality control – Here you can set your required tolerance. If you measure or store a point which is not within tolerance, the 'Stakeout limit exceeded' screen will appear. A yellow bar saying 'Stakeout limit exceeded' will appear at the bottom of the screen for a few seconds. The distances from the correct position will be displayed. If you would like the distances to be displayed for every point measured, set the tolerance to 0.000m.

Heights – Determine whether heights of points can be edited and whether to offset the heights of points being staked by a fixed amount.

Graphics - Depending on personal preference or the job you are doing, you can set whether the arrows are in relation to the total station, north, or a line. To set out the point in relation to the position of the total station (towards/ away from and left/right) select 'From instrument' or 'To instrument' depending on whether you are facing the instrument or have your back to it.

Report sheet – You can create a report of as-set-out (staked) positions versus their required positions.

'Setup configuration' menu

Whilst the in any of the setup programmes :

1. >> 'Fn'
2. >> Config

Tabs:

General – You can choose to be reminded of your last set up before measuring, select whether you want to use two face measurements in your setups or request that a message is shown when setup is complete

Known backsight – You can specify if you want backsight position and height to be checked

Advanced – You can select if you want to auto position setup targets, use on-the-fly setups, calculate scale from target observations, use Helmert method for resection and edit default station quality checks

Report sheet – You can create a report sheet for your setups

Stakeout reference line configuration menu

Whilst in the stakeout reference line mode (but not whilst measuring), select the 'Stake' tab.

1. >> 'FN'
2. >> Config

Tabs:

General – Here you can select:

→ Use chainages (instead of increments)

→ Allow stake definition in stake page (which means you can manually select any chainage or offset at any time)

→ View results page after staking a point (automatically)

→ Only update stakeout values when distance is measured

→ Turn to point automatically

Quality control – Here you can set your required chainage and offset tolerances. If you measure or store a point which is not within tolerance, you will get an alert saying 'Measured position exceeds check limits'. The distances from the correct position will be displayed. If you would like the distances to be displayed for every point measured, set the tolerance to 0.000m.

Design – You can select what other points you would like to stake in addition to the specified increments e.g. intersection points, mid curve, curve radius, low points, high points etc.

LEICA TS15

Graphics - Depending on personal preference or the job you are doing, you can set whether the arrows are in relation to the total station or the defined line. To set out the point in relation to the position of the total station (towards/ away from and left/right) select 'From instrument' or 'To instrument' depending on whether you are facing the instrument or have your back to it.

Info – Here you can set what information you would like to be shown on the 'Info' tab in the stakeout reference line mode.

Report sheet – You can create a report of as-set-out (staked) positions versus their required positions.

Leia TPS Favourites menu

Quick access to some useful functions:

1. Change target type to 'Measure any surface'
2. Toggle between 'Continuous measurement' on and off
3. Change faces
4. PowerSearch right
5. PowerSearch left
6. Toggle between red laser dot on and off
7. Toggle between auto aiming and manual aiming
8. Toggle between lock and unlock to target
9. Change to next page for useful functions (joy stick, turn to Hz/V, check point, compass, Bluetooth connection, sketch pad, active assist)

Measure and targets menu

You can set the default target type for setup and survey modes

Tabs:

Survey

> → 'Measure' – Select the default target type you require for survey mode
>
> → 'Target' – Select the default prism type you require for survey mode
>
> → 'Measure mode' – Select the default measure mode you require for survey mode
>
> → 'Target aiming ' – Select the default target aiming mode you require for survey mode
>
> → 'Visibility' – Select the default visibility you require for survey mode

You can set the visibility to:

> → Good
>
> → Rain and fog
>
> → Rain and fog always

→ Sun and reflections

→ Sun and reflections always

Setup

→ 'Measure' – Select the default target type you require for setup mode

→ 'Target' – Select the default prism type you require for setup mode

→ 'Measure mode' – Select the default measure mode you require for setup mode

→ 'Target aiming ' – Select the default target aiming mode you require for setup mode

Camera

Take a photograph and assign the image to a point

COGO

Various geometry calculations can be carried out here including:

→ Inverse (distance and height difference between two points)

→ Traverse

→ Shift, rotate and scale

→ Angles

→ Horizontal curves

→ Triangles

Regional settings

Tap on the time at the bottom right hand of the screen. Here you can set:

→ Units

→ Number of decimal places

→ Angle format and angle display parameters

→ Time and date format

→ Grid format e.g. whether eastings or northings is displayed first

INSTRUMENT SET-UP

Set up over a known point ('Known backsight')

First of all, accurately position and level the total station over point of known coordinates.

1. >> Main menu

2. >> Go to work

3. >> Setup

4. The top of the screen says 'Total station setup'

5. >> From the drop down menu select 'Known backsight'

6. >> OK

7. The top of the screen says 'Set station point'

8. Complete the boxes as follows:
    → 'Station point from' – If you have previously stored your control points, select 'Job'. If you wish to use a point which is not stored, select 'Enter new point'. Alternatively, you can select 'Last used station'
    → 'Job' - Select the job where your control data is stored. Points must be stored as control data or they will not appear in the list
    → 'Point ID' – Select the point which you are set up over
    → 'Instrument height' – If you need to measure levels, enter the instrument height. If you do not need to measure levels, you can leave this as 0.000m

9. >> OK

10. The top of the screen says 'Set Station Orientation'

11. Complete the boxes as follows:
    → 'Backsight ID' – Select the control point you are going to sight to
    → 'Target height' - If you need to measure levels, enter the target height. If you do not need to measure levels, you can leave this as 0.000m

12. Ensure that the target type is set correctly. The default target type for this setup mode can be set in the 'Measure and Targets Settings' menu.

13. You can select the tabs across the top of the page to check your backsight and station data.

14. Sight to the object if it is standard prism or reflective tape. If it is a 360 degree prism press 'F12' until the TS locks onto the target (padlock item at top left of screen)

15. Tap 'Meas' and then select the 'plot' tab. This will give display a visual indication of the points in relation to each other.

16. Press 'set'. This will return you to the home main menu, but will not give any other indication that the measurement was successful.

Set up at an unknown point ('Resection')

1. First of all, level the instrument.

2. >> Main menu

3. >> Go to work!

4. >> Setup

5. Top of screen says 'Total station set up'
    → 'Setup method' – Select 'Resection'

6. >> OK

7. Top of screen says 'Enter Station Information'

    → 'Station ID' – Give the station a name (it can be a temporary name)

    → 'Instrument height' – This can be left as 0.000m if you are doing a 3D resection as the elevation will be calculated

    → 'Point code' – You can select a point code here if you wish

If you tick 'Use control job for the target points' only control data will appear in the list. If you wish to use points which you have measured, untick this box.

8. >> OK

9. The top of the screen says 'Measure Target 1'

10. Set the correct target type (for retros select 'reflectorless')

11. Sight through the telescope and manually line up on target 1.

12. >> Meas

13. If you would like to take a reading on the opposite face, transit the telescope and rotate horizontally through 180 degrees

14. Sight through the telescope and manually line up on target 1.

15. >> Meas

16. The top of the screen says 'Measure Target 2'

17. Sight through the telescope and manually line up on target 2.

18. >> Meas

19. >> Calc

Tabs:

Results – Displays the eastings, northings and height of the instrument

Station – Displays Station ID and Instrument height

Quality – Displays the errors in easting, northing and height

Targets – Select whether you want each target to be used as 2D or 3D by clicking 'Use' to toggle

Plot – Shows a map view of the targets in relation to the total station

20. To measure to an additional target tap 'Trgt+'

21. If you are happy with the results, press 'Set'

INSTRUCTIONS

1. Measure points/ take a survey ('Survey')

2. First of all, set the position and orientation of your instrument.

3. Choose the working job in the 'Jobs and Data menu'

4. >> Main menu

5. >> Go to work

6. >> Survey

7. To take measurement, complete the boxes as follows:

    → 'Point ID' – Enter the name of the point you want to measure

    → 'Target height' – If you need to measure levels as well as position, enter the correct target height

8. >> To view the coordinates without storing, press 'Dist'. Then 'Store' if you are happy with them.

9. >> To record the coordinates without checking them first press 'Meas'

10. The point will be stored and the Point ID will automatically increase by 1.

Tabs across the top:

Survey – Measurement screen

Offset - If a point is not accessible, you can measure to perpendicular offsets.

Code – Attribute a code to a point. You can add new codes or select from a code list you have previously created.

Map – Shows a map view of all points

11. To add a new code, select the 'code tab'

    → 'Code' – Tap in the box

12. >> New

    → 'Code' – e.g. T

    → 'Description' e.g. Tree

    → 'Code group' – Default or 'New'

    → Code type – Point, line or area

13. 'Line' will form a continuous line of consecutive points of the same code. If you change codes then go back to the same code, a new line will be started.

14. 'Area' will form an area of consecutive points of the same code. If you change codes then go back to the same code, a new area will be stared.

Calculate the distance and level difference between two points ('Compute Inverse')

Method 1 (to measure distance)

1. In the survey program, measure the two points.

2. In the 'view and edit data' menu select the 'Map' tab

3. Tap the red pointer on both points

4. Hold the red pointer down on one of the points. A floating menu will appear. Select 'create line'

5. Make sure no points or lines or selected

6. Hold the red pointer down on the line. A floating menu will appear. Select 'Edit line'. The length of the line will be displayed.

Method 2 (to measure distance)

1. In survey mode, start a new line by attributing a code which has a code type 'line'

2. Measure to the two points

3. Select the 'Map' tab

4. Hold the red pointer down on the line. A floating menu will appear. Select 'Edit line'. The length of the line will be displayed.

Method 3 (to measure distance and level difference)

1. In the survey program, measure the two points.

2. Go back to the main menu

3. Press F6 'Map'

4. Tap on both points to highlight them.

5. Hold the stylus down on white space.

6. A floating menu will appear. Select 'Compute Inverse'. The length of the line and the level difference between the two points will be displayed.

Measure the distance and level difference between two points ('Compute Inverse')

1. >> Main menu

2. >> Go to work

3. >> COGO

4. >> Inverse

    → 'Inverse method' – Select 'Point to Point'

5. >> OK

6. The top of the screen says 'Inverse Point to Point'

    → 'From' – Tap in the box, select the start point from the list or map >> OK

    → 'To' – Tap in the box, select the end point from the list or map.

7. The distance, height difference, change in eastings and change in northings are displayed.

8. If you wish to store the line data, click 'Store'

Stake out points of known coordinate ('Stakeout')

First of all, set the position and orientation of your instrument.

1. >> Main menu

2. >> Go to work

3. >> Stakeout

   → 'Control job' – Select control job

4. >> OK

   → 'Point ID' – Select the point you want to set out

   → 'Target height' – Enter target height if you are also setting out the level

5. Select the target type and 'F12' to lock onto the target if you are using the 360-degree prism.

6. >> Dist

7. Arrows will appear on the screen, directing you towards the point. Adjust the position of the target until it is within your required tolerance. Once you have located and marked the required position, place the target on the point and press 'Meas' to record the as-set-out point in the memory.

8. To check your measured points, go to the 'View and edit data' menu. Staked points (as opposed to entered points) are listed with the prefix 'Stkd'. If you have taken several measurements to the same point ID, it will be listed once. To view the individual measurements, select the point then

9. >> Edit

   → 'Mean' tab. Here you can select which points are used and which are not. If you select 'Yes' in the 'Use' column the point will be included in the calculation of the mean. If you select 'No' the point will not be included. 'Auto' means that only points within a certain statistical distribution of the centre of the group of points will be included.

10. To set tolerance alerts, go to the Quality Control tab of the 'Configuration' menu whilst in the stakeout program.

11. You can go to straight to survey mode at any time by selecting 'Survy...' at the bottom of the screen.

12. In the 'config' menu, you can set the display to change to 'bullseye view' once you are within 0.5m of the target. This means that the large arrows disappear and instead displays the position of the target relative to the required position. The arrows are still shown, but to the side of the bullseye view.

13. In the 'config' menu, you can specify that the beep gets faster the closer it is to the required point and at what distance from the required point the beeping starts.

To set the direction of the arrows:

14. >> 'FN' (whilst in the stakeout program)

15. >> Config

16. The top of the screen will say 'Configuration'

17. Select the 'graphics' tab.

    → 'Navigate direction' – Select your preferred method

    → 'Navigate using' – Select 'In/Out'

Stake out points in relation to a defined line from their coordinates

First of all, set the position and orientation of your instrument.

1. >> Main menu

2. >> Go to work

3. >> Stakeout +

4. >> Stake to ref line

    → 'Control job' – Select control job

5. >> OK

6. Top of screen says 'Reference Line Task'

    → 'Stake' – For a straight line select 'Line'

7. >> OK

8. Top of screen says 'Define Line'

9. >> Create..

10. Top of screen says 'Choose segment to create'

    → 'Line/arc method' – Select 'Line – 2 points'

11. >> OK

    → 'Line ID' – Use the default name or create a new name

    → 'Start point' – If you have previously stored the point, select it from the list. If not tap 'New' and enter the name and co-ordinates of the start point.

    → 'End point' – If you have previously stored the point, select it from the list. If not tap 'New' and enter the name and co-ordinates of the start point.

    → 'Line length' – The length of the line will be displayed

    If your line is just one section:

    12. >> Finish

If you would like to create a polyline by adding another straight section or arc:

    12. >> Next

Create the next section of the line in the same way as above. The 'start point' of the next section is the 'end point' of the previous section. A polyline can consist of curved and straight lines end to end.

You can view the line on the 'Map' tab. Zoom extents if necessary.

Stake out points in relation to a defined line by sighting to the end points

Using this method, you can enter a defined line by sighting to the beginning and end point. There is no need to set the position and orientation of your instrument in relation to any site control. You can set an arbitrary position.

1. >> Main menu
2. >> Go to work
3. >> Stakeout +
4. >> Stake to ref line
   → 'Control job' – Select control job
5. >> OK
6. Top of screen says 'Reference Line Task'
   → 'Stake' – For a straight line select 'Line'
7. >> OK
8. Top of screen says 'Define Line'
9. >> Create..
10. Top of screen says 'Choose segment to create'
    → 'Line/arc method' – Select 'Line – 2 points'
11. >> OK
    → 'Line ID' – Use the default name or create a new name
    → 'Start point' – Press the 'down' arrow on joystick on the hard keys. An option to 'Meas' will appear. Sight to the point and press 'Meas'.
    → 'End point' – Press the 'down' arrow on joystick on the hard keys. An option to 'Meas' will appear. Sight to the point and press 'Meas'.
    → 'Line length' – The length of the line will be displayed

If your line is just one section:

12. >> Finish
13. >> OK

Top of screen says 'Define Stake'

   → 'Dist along line' – Enter the distance from the start point of the line of the first point you want to stake out (negative is backwards from the start, positive is forward)

→ 'Offset' – Enter the offset from the centre of the line (negative is to the left of the line, positive is to the right)

→ 'Height offset' – Enter the height offset (negative downwards, positive is upwards)

→ 'Use stake increments' – If you only want to set out the start point of the line, leave this box unchecked. If you would like to stake out at intervals along the line, check this box and the parameters will appear

→ 'Increment' – Enter the interval

→ 'After storing' – Select if you want it to automatically move forwards along the line by the increment, move backward along the line by the increment or 'do nothing' which means you can manually select each time whether you move forwards or backwards

→ 'Use different increment of curves' – Check this box if you would like to change the interval for curves under a specified radius

14. >>Stk- or Stk+ to move forwards or backwards to select the first point you want to set out.

15. >> OK

16. Sight or lock onto the prism

17. >> Dist

18. Follow the arrows on the screen to adjust position of the target until it is within your required tolerance.

19. You can click the 'Info' tab to get information about your position relative to the required position.

20. You can click the 'Map' tab to view the point to be staked out. As you click Stk- or Stk+ you will see the marker moving forwards or backwards along the line.

21. When the target is within the required tolerance, make a mark. Place the target on the mark.

22. >> Meas

23. The TS will automatically jump forward or backwards to the next point. If you wish to set out a different point, you can manually override this by clicking Stk- or Stk+

24. If you want to stake out a point a certain distance from the start of the line, click the 'back' arrow just underneath the batter indicator icons. This will take you back to the original display showing the 'Dist along line'.

Measure an area

Method 1

1. In the survey program, measure the points which form the boundary of the area

2. In the 'view and edit data' menu select the 'Map' tab

3. Tap to select and highlight the points which form the boundary

4. Hold the pointer down on one of the points and a floating menu will appear. Select 'Create area'

5. Hold the red pointer down on the boundary line and a floating menu will appear. Select 'Edit area'

6. The area bounded by the points and the perimeter will be displayed

Method 2

1. In survey mode, start a new are by attributing a code which has a code type 'area'

2. Measure to the boundary points.

3. Select the 'Map' tab

4. Hold the red pointer down on the perimeter line. A floating menu will appear. Select 'Edit area'. The area and the perimeter will be displayed.

Measure a volume

1. >> Main menu

2. >> Go to work

3. >> Survey +

4. >> Volume calculations

5. Select 'Create a new surface by measuring points'

6. >> OK

7. The top of the screen says 'New Surface'

    → 'Surface ID' – Give the surface a name

8. >> OK

9. Top of screen says 'Survey surface points'

10. Enter the target height

11. Lock to target and click 'Meas'

12. Take readings all over the surface of the stockpile, pressing 'store' at the required intervals

13. >> Boundary

14. Take readings all around the bottom boundary of the stockpile, pressing 'store' at the required intervals

15. When you have sufficient readings press 'Stop'

16. >> Done

17. Select 'Edit the boundary and triangulate the surface'

18. >> OK

19. The points will be listed. You can click the 'Map' tab to see them in map view

20. >> OK

21. Top of screen says 'Triangulation results'

22. Tabs:

23. Summary – Surface ID, Area, No. of triangles, No of surface points, No of boundary points
24. Details – Statistics about the results
25. Map – Map view showing boundary and triangulated surface points
26. >> OK
27. If you are happy with you measurements select ' Calculate the volume'
    → 'Calculate using' – Select 'Stockpile
28. >> OK
29. Top of screen says 'Volume Calculation Results'. Area and volume are displayed.
30. >> OK
31. Select 'Exit the volumes app'

Import data

1. >> Jobs & Data
2. >> Import data
3. Select the data type you want to import.
    → 'From' – SD card, USB or internal memory

Export data

1. >> Jobs & Data
2. >> Export data
3. Select the data type you want to export.
    → 'Export to' - SD card, USB or internal memory
4. >> Config
5. Select the fields you want to export and in what order. Select the delimiter as a comma.
6. >> OK
7. >> OK

Appendix E

LEICA TS16

USEFUL INFORMATION

Power Search

F12 triggers the total station to search for and lock on to the 360 degree prism.

Working within a job

In the top swiping menu, tap on a job. The panel will 'flip' to display the menu for that job. Here you can view and edit job properties, view and edit data, import, export and send data and delete the job. To use one of the apps in the bottom swiping menu, swipe the job you want to work in to the centre of the screen in the top swiping menu before tapping on the app.

'Back' button

The 'U-turn' arrow at the top left of the screen takes you back to the previous screen you were on.

Entering data

If you make a mistake when entering data and want to overwrite it, the 'left' joystick arrow will take you back to the first digit you want to overwrite. If you keep typing, it will continue to overwrite subsequent digits.

To delete digits, use the arrow which points to the right, immediately to the right of the F12 key.

To the left of the 'A' key on the keypad, there is an arrow pointing upwards. This toggles between upper case and lower case.

To enter a negative number, use the '-' key immediately to the left of the '0' key.

'Ref triplet' warning

Ref triplet – If you attempt to delete a resection station, you will get a message saying 'This REF triplet has one or more measurements belonging to it. I you delete this REF triplet you will also delete these measurements'. This means you will delete all the measurements taken from that resection set up point.

SCREEN ARRANGEMENT AND HARD KEYS

Icons across the top of the home screen

Top left – Back arrow – Tap to go back to the previous screen you were on.

Furthest right – Battery indicator. The controller batter is displayed. To check TS battery, tap this icon. In this menu, you can also change from TS to GS, access the home screen, camera or sketch pad.

Second from the right – '@' symbol – Internet connection and Bluetooth. An orange triangle with a '!' indicates that the TS is not connected to the internet. Tap on this icon for connections menu.

Third from the right – Horizontal and vertical angle readings - Tap to change this to slope distance

Fourth from the right – Total Station – The number '1' or '2' in a small circle indicate what face the instrument is in. An orange triangle with a '!' indicates the TS is not connected. Tap to access the 'instrument' menu. Here you can change face, check the level bubble, check the details of your current setup, turn the instrument to a specified horizontal or vertical angle or use the joystick arrows on the screen to adjust the TS.

Fifth from the right – Measure and target - Indicates the measurement type and target type the instrument is currently set to and whether the red laser dot is on or off. Target type could be one of the following:

> → Large 360 degree prism
>
> → Mini 360 degree prism
>
> → Standard round prism
>
> → Standard mini prism
>
> → Retro target
>
> → Reflectorless mode/ any surface (A house with a laser striking it. Reflectorless mode (A house with a laser beam bouncing off it)

Two arrows forming a circle indicate the TS continuous measurement mode is on. A number '1' indicates continuous measurement mode is off.

A static black line with a star at the end indicates the red laser dot is switched off. An orange line with a star, increasing in length indicates the red laser dot is on.

Tap on this icon to go to the 'Measure and targets' menu. In this menu, you can select the target type you are about to measure to, switch between continuous and single measurement and switch the red laser dot on or off.

Sixth from the right – Aiming and lock – May be one of the following:

> → Cross-hairs with a hand - manual aiming
>
> → Cross-hairs with prism - auto aiming
>
> → Two concentric circles with horizontal and vertical lines at the edge - Not locked onto prism)
>
> → Two concentric circles with horizontal and vertical lines at the edge and a prism in the centre -Locked onto prism)

> → Binoculars with a window – TS is searching a window for the prism
>
> → Binoculars with yellow beams – TS is power searching for the prism

Panels in top swiping menu

First on the left – Design data – Tap here to select the design data. As a general rule leave the 'Use points and lines data' box unchecked. If you check this box and select a job, when you are in stakeout mode in a different it will bring in the points from your selected job.

Second from the left – Create a new job – Tap here to create a new job and enter its properties. You can return to this menu at any time to edit the properties by tapping on the job in the swipe menu and selecting the top option 'View and edit job properties'

All other panels – Every other panel relates to a job. To select which jobs you would like to be visible in the menu, tap 'Jobs' next to 'Fn' at the bottom right of the screen and click on the eye icon to the left of the job name. A strike through the eye indicates that a job will not be visible in the swipe menu.

Hard keys

F1 – 'OK'

1. House icon – Takes you to the home screen with swipe menus
2. Camera button – Switches the camera on
3. Star key – 'My TS favourites' menu
4. U-turn arrow – takes you back to the previous screen you were on
5. 'Return arrow' selects whatever is selected on the screen (works the same as the 'OK' button at the centre of the joystick arrows)

Fn – Only works on some screens, but takes you to a new menu along the bottom of the screen.

1. Hold down the Fn key and use '1' and '4' to increase or decrease the screen brightness
2. Hold down the Fn key and use '3' and '6' to increase or decrease the keypad volume
3. Hold down the Fn key and press the camera button to take a screenshot and store it to images
4. Hold down the Fn key and press '7' to lock the key board
5. Hold down the Fn key and press '9' to lock the touch screen functionality
6. Hold down the Fn key and press '.' to switch on the in-built torch
7. Hold down the Fn key to use secondary keypad keys (green symbol above each letter)
8. When in the map view, the '1' key will zoom extents
9. When in the map view, the '2' key will zoom in
10. When in the map view, the '3' key will zoom out

11. When in the map view, the '5' key will centre the highlighted point

12. Arrow to the left (to the right of F12 key) deletes figures.

13. Arrow pointing up (above and to the left of the Z key) toggles between lower and upper case

14. If you make a mistake when entering data and want to overwrite it, the 'left' joystick arrow will take you back to the first digit you want to overwrite. If you keep typing, it will continue to overwrite subsequent digits.

15. To enter a negative number, use the '-' key to the left of the '0' key

MENUS AND TOOLBARS

'View and edit data'

This menu can easily be accessed at any time by tapping the job you wish to access in the swipe menu and selecting the second option in the list 'View and edit data'

Tabs:

Points – Data about measured and input points. To view point data including date and time it was measured or entered, point code, eastings and northings and class, tap the 'More' button at the bottom of the screen. This will scroll through the table to the right.

To get more details about a point, select the line in the table then click 'edit'. This will open up a new screen with tabs for Coords, Stake results (as-staked-out level and position Vs design level and position) Obs (where it was observed from, the target height and the target type), Code, Annots (where you can add any notes you want) and Images (where you can view, mark up or edit photographs and sketches attributed to the point)

You can retrospectively change the target height if you realise you used the wrong target height at the time.

Lines – Press 'More' to scroll to the right to see more information. Click 'Edit' to view more information and edit. To create a line click 'New'. Select the line style and colour and click 'Store'. Then select the line in the table and click 'Edit'. Select the 'Geometry' tab at the top. Click 'Add'. Select the type of line you wish to create. In 'Point' select the start point of the line. Click 'OK'. Select 'Add'. For a straight line select 'Continue line' in the Linework box. Select the end point of the line. Click 'OK'. To add more lines or arcs to create a polyline click 'Continue line'. If you have finished adding points click 'OK' then 'Store'. The length of the line will now display in the table.

Images – Here you can view, mark up or edit photographs and sketches and attribute them to a point.

Scans – You can import scan data

3D Viewer – Map view of all the points

You can sort points or lines (tabs at the top) by:

→ Ascending point ID

LEICA TS16

- → Descending point ID
- → Forward time
- → Backward time

Settings menu

Connections - Here you can find the TS connection wizard, Internet Wizard and 'All other connections' menu

TS Instrument Menu - Here_you can find the 'Measure and target', 'Target Search', 'Atmospheric corrections' 'Level & Compensator' and 'Lights' menus

Measure and targets menu

1. >> Settings (in bottom swipe menu)
2. >> TS Instrument
3. >> Measure and target

You can set the default target type for setup and survey modes

Tabs:

Survey

- → 'Target' – Select the default prism type you require for survey mode
- → 'Aim at target' – Select the default from **'with lock'**, 'automatically' or 'manually'
- → 'Enhance target recognition for misty rain conditions' – Tick this box if relevant
- → 'Measure distance' – Select the default measure mode you require for survey mode

Setup

- → 'Target' – Select the default prism type you require for setup mode
- → 'Aim at target' – Select the default from 'with lock', 'automatically' or
- → 'Enhance target recognition for misty rain conditions' – Tick this box if relevant
- → 'Measure distance' – Select the default measure mode you require for survey mode

Target search

Here you can set the default parameters for your target search

1. >> Settings (in bottom swipe menu)
2. >> TS Instrument

Level & compensator

Here you can view the level bubble, alter the laser plum intensity and turn the tilt compensator and Hz correction on or off. If you switch them off, its recommended that you select 'Off until restart'.

When the tilt compensators are switched on, the instrument will alert you if it goes off level.

1. >> Settings (in bottom swipe menu)
2. >> TS Instrument
3. >> Level & Compensator

Lights

In this menu you can select which lights/ lasers you wish to switch on including the red laser dot.

1. >> Settings (in bottom swipe menu)
2. >> TS Instrument
3. >> Lights

Point storage

In this menu you can decide what happens when you store duplicate points, select which information you want to be prompted to enter when taking a measurement and set default offsets when measuring to a point you cannot access directly.

1. >> Settings (in bottom swipe menu)
2. >> Points storage

Customisation

In this menu you can set your default working styles (recommended!) which takes you through all the different defaults, set the fields visible in user defined pages, set hot keys and favourites, set coding defaults and select which apps are visible

1. >> Settings (in bottom swipe menu)
2. >> Customisation

System

In this menu you can select which software to use when the controller starts up, set screen, audio and text input defaults, set units, angle display, time and data, restrict access to certain apps and calibrate internal sensors

1. >> Settings (in bottom swipe menu)
2. >> System

Tools

In this menu you can transfer user objects, update the software, load license keys, format the memory and use a basic scientific calculator

1. >> Settings (in bottom swipe menu)
2. >> Tools

3D viewer

There are different ways to access the 3D viewer. Select the job you want so it is at the centre of the top swipe menu, and then enter 3D viewer via the bottom swipe menu, there is greater functionality than if you enter it via the 'View and edit data' menu or within the individual programs.

In particular, COGO is only an option when entering via the bottom swipe menu. You can use this function to calculate the distance and height difference between any two points.

'View' toolbar (Picture of an eye at the top right of the screen)

Options change depending on the program you are in.

Options include:

> → 2D plan
>
> → 3D
>
> → Walk through view
>
> → Arrows only (directing you towards the point in stakeout mode)

'Zoom' toolbar (Magnifying glass to the right of the screen)

Zoom extents - Magnifying glass with arrows pointing outwards – Displays all of your points

Zoom in and out – Magnifying glass with up and down arrows – Swipe up to zoom in and swipe down to zoom out

Zoom window – Dotted rectangle with a magnifying glass at the bottom right – drag a rectangle around the area you want to zoom in on

Centre – Place a selected point at the centre of the screen

Pan – You can pan at any time by holding down and dragging (indicated by a hand)

3D viewer 'Display settings' menu

Here you can access layer management, the 'Object display' menu or select all points within a rectangle (dotted rectangle with an arrow at the bottom right)

Object display

When in 3D viewer:

1. >> Fn
2. >> Display

Here you can select what point and line information to display.

COGO

Various geometry calculations can be carried out here including:

- → Inverse (distance and height difference between two points)
- → Traverse
- → Shift, rotate and scale
- → Angles
- → Horizontal curves
- → Triangles

Regional settings

1. >> Settings
2. >> System
3. >> Regional

Here you can set:

- → Units
- → Number of decimal places
- → Angle format and angle display parameters
- → Time and date format
- → Grid format e.g. whether eastings or northings is displayed first

INSTRUMENT SET-UP

Set up over a known point ('Known backsight')

First of all, accurately position and level the total station over point of known coordinates.

1. >> Setup (in bottom swipe menu)
2. The top of the screen says 'Total station setup'
3. >> From the drop down menu select 'Known backsight'
4. >> OK
5. The top of the screen says 'Choose Setup Point'
6. Complete the boxes as follows:
   - → 'Setup point from' – If you have previously stored your control points, select 'Job'. Select the job you wish to use points from. To use a point which is not stored, select 'Enter new point'. Alternatively, you can select 'Last used station'
   - → 'Job' - Select the job where your control data is stored. Points must be stored as control data or they will not appear in the list

> → 'Point ID' – Select the point which you are set up over

> → 'Instrument height' – If you need to measure levels, enter the instrument height. If you do not need to measure levels, you can leave this as 0.000m

7. >> OK

The top of the screen says 'Set Station Orientation'

Complete the boxes as follows:

> → 'Backsight ID' – Select the control point you are going to sight to

> → 'Target height' - If you need to measure levels, enter the target height. If you do not need to measure levels, you can leave this as 0.000m

Ensure that the target type and height is set correctly. The default target type for this setup mode can be set in the 'Measure and Targets Settings' menu.

You can select the tabs across the top of the page to check your backsight and setup data.

Sight to the object if it is standard prism or reflective tape. If it is a 360 degree prism press 'F12' until the TS locks onto the target (padlock item at top left of screen)

Tap 'Meas' and then select the 'plot' tab. This will give display a visual indication of the points in relation to each other.

Press 'Set' at the bottom of the screen. A dialogue box will appear saying 'The station & orientation of the total station have been set'. Remember that if you change jobs, you may may ne to do the re-setup if the coordinate systems are different. A warning message will display if you try to work in a difrerent job to the one you have set up in.

Set up at an unknown point ('Resection')

First of all, level the instrument.

1. >> Setup (in bottom swipe menu)

2. The top of the screen says 'Total station setup'

3. >> From the drop down menu select 'Resection'

4. >> OK

5. Top of screen says 'Setup details'

> → 'Point ID' – Give the unknown point you are set up at a name

> → 'Instrument height' – This can be left as 0.000m if you are doing a 3D resection as the elevation will be calculated

> → 'Point code' – You can select a point code here if you wish

6. If you want to choose points from within the job you are working in, leave 'Choose target points from a different job' unticked

7. >> OK

8. The top of the screen says 'Measure Target 1'

9. Set the correct target type (for retros select 'reflectorless')

10. Sight through the telescope and manually line up on target 1.

11. >> Meas

12. If you would like to take a reading on the opposite face, transit the telescope and rotate horizontally through 180 degrees

13. Sight through the telescope and manually line up on target 1.

14. >> Meas

15. The top of the screen says 'Measure Target 2'

16. Sight through the telescope and manually line up on target 2.

17. >> Meas

18. >> Calc

Tabs:

Results – Displays the eastings, northings and height of the instrument

Setup – Displays setup point ID and Instrument height

Quality – Displays the errors in easting, northing and height

Targets – Select whether you want each target to be used as 2D or 3D by clicking 'Use' to toggle

Plot – Shows a map view of the targets in relation to the total station

19. To measure to an additional target tap 'Trgt+' (A minimum of 3 targets in total is recommended)

20. If you are happy with the results, press 'Set'

INSTRUCTIONS

1. Measure points/ take a survey ('Measure')

2. First of all, set the position and orientation of your instrument.

3. >> Measure (in bottom swipe menu)

4. To take measurement, complete the boxes as follows:

    → 'Point ID' – Enter the name of the point you want to measure

    → 'Target height' – If you need to measure levels as well as position, enter the correct target height

5. To view the coordinates without storing, press 'Dist'. Then 'Store' if you are happy with them.

6. To record the coordinates without checking them first press 'Meas'

7. The point will be stored and the Point ID will automatically increase by 1.

To assign codes to points:

To assign a code to a point you are about to measure, tap on the icon of a tree and a fence at the top left of the screen.

1. Tap in the 'Find and add codes' box twice.
2. From the drop-down menu select the code.
   Each time you select a code it will be added to the boxes down the left-hand side of the page.
3. If you want to add a new code click 'Define' then 'Code' at the bottom of the screen.
4. Click 'New' and enter the code details.
5. Click 'Store'.
6. Continue adding new codes or click 'OK' when you have finished adding codes.
7. To assign a code to a point, tap on the box and click 'Measure'.
8. If you want to clear any of the boxes from the list, press 'Fn' then 'Clear one' or 'Clear all'.

Create lines (strings) within your survey:

To create a line, select a code which has a line attribute e.g. CL (Centreline).

1. Measure points consecutively. The number in the top right of the code box indicate the CL line number (e.g.CL1). Even if you change to a different code and measure other points, then measure more points with the same code and as your line, the line will continue from the previous point on the line.
2. If you want to start a new CL, you can click on the + sign. If you want to measure several CLs within the same survey, you can create a code box for both. Then you can easily change between them.
3. If want to survey a road in a zig-zag you can set up code boxes KL1 (kerb line), CL, KL2. Remember to select the correct code before each measurement.

Measure the distance and height difference between 2 points already stored

1. >> Settings (in bottom swipe menu)
2. >> 3D viewer
3. Select 2 points by tapping on them with the red pointer. Hold the red pointer down on any white space. A floating menu will appear.
4. >> Calculate inverse
5. Data will be displayed including horizontal distance, height difference and slope distance.

Measure the distance and height difference between 2 points by sighting to them ('Inverse')

1. >> COGO (in bottom swipe menu)

2. >> Inverse

    → 'Method' – Point to point

3. Click 'Meas app' at bottom of screen, then sight to the point and click 'Measure'.

4. Press the down arrow on the joystick so the 'To' point is highlighted.

5. Click 'Meas app' then 'Measure'.

6. Data will be displayed including horizontal distance, height difference and slope distance. If you wish to store this data click 'Store'

Stake out points of known coordinates ('Stake points')

1. First of all, set the position and orientation of your instrument.

2. >> Stake points (in bottom swipe menu)

    → 'Point ID' – Select the point you want to set out

    → 'Target height' – Enter target height if you are also setting out the level

3. Select the target type and 'F12' to lock onto the target if you are using the 360-degree prism.

4. >> Dist

5. Arrows will appear on the screen, directing you towards the point. You can change the view by clicking on the eye icon at the top right of the screen and the arrow will always display down the bottom. Adjust the position of the target until it is within your required tolerance. Once you have located and marked the required position, place the target on the point and press 'Meas'.

6. To reverse the direction of the arrows in the graphics tab of the stake out 'Settings' menu.

7. To set tolerance alerts go to??

8. You can go to straight to measure app at any time by selecting 'Meas app' at the bottom of the screen.

9. To check your measured points, go to the 'View and edit data' menu. Staked points (as opposed to entered points) are by default listed with the prefix 'Stkd' and have the smbol of a rotated square with a dot. If you have taken several measurements to the same point ID, it will be listed once. To view the individual measurements, select the point then

10. >> Edit

    → 'Mean' tab. Here you can select which points are used and which are not. If you select 'Yes' in the 'Use' column the point will be included in the calculation of the mean. If you select 'No' the point will not be included. 'Auto' means it will calculate the mean position of all the points, and discard any that not within the specific tolerance of the centre of the group of points.

Stake out points in relation to a defined line

1. First of all, set the position and orientation of your instrument.

2. >> Stake to line (bottom swipe menu)

3. Top of screen says 'Task'

    → 'Stake' – For a straight line select 'Line'

4. >> OK

5. Top of screen says 'Define Line'

    → 'Line' – Select your line

6. If you want to create your line by sighting to the beginning and end points, navigate down to the start point using the joystick arrows and click 'Meas app'. Sight to the start point and click 'Measure'. Use the down arrow to move down to the end point and click 'Meas app' again. Sight to the end point and click 'Measure'.

7. >> Finish

8. >> If you are happy with the line displayed click 'OK'

9. Top of screen says 'Define Stake'

    → 'Dist along line' – Enter the distance from the start point of the line of the first point you want to stake out (negative is backwards from the start, positive is forward)

    → 'Offset' – Enter the offset from the centre of the line (negative is to the left of the line, positive is to the right)

    → 'Height offset' – Enter the height offset (negative downwards, positive is upwards)

    → 'Use stake increments' – If you only want to set out the start point of the line, leave this box unchecked. If you would like to stake out at intervals along the line, check this box and the parameters will appear

    → 'Increment' – Enter the interval

    → 'Increment after storing' – Select if you want it to automatically move forwards along the line by the increment, move backward along the line by the increment or 'No' which means you can manually select each time whether you move forwards or backwards

    → 'Use different increment of curves' – Check this box if you would like to change the interval for curves under a specified radius

10. >>Stk- or Stk+ to move forwards or backwards to select the first point you want to set out.

11. >> OK

Sight or lock onto the prism

1. >> Dist

2. Follow the arrows on the screen to adjust position of the target until it is within your required tolerance.

3. You can click the 'Info' tab to get information about your position relative to the required position.

4. You can click the 'Map' tab to view the point to be staked out. As you click Stk- or Stk+ you will see the marker moving forwards or backwards along the line.

5. When the target is within the required tolerance, make a mark. Place the target on the mark.

6. >> Meas

7. The TS will automatically jump forward or backwards to the next point depending on how you have set it. If you wish to set out a different point, you can manually override this by clicking Stk- or Stk+

Measure an area

This won't work in other programs so you must enter the 3D viewer via the bottom swipe menu ensuring that the correct job is at the centre of the screen in the top swipe menu. Tap to highlight the boundary points of the area you want to measure. Hold the red pointer down on the screen on whitespace. A floating menu will appear. Click 'Create closed line'. Click the 'General' tab. The area and perimeter of the line will be displayed. If you want to store the data, click 'Store'.

If you are measuring the points with the intention of calculating the area, remember to exit the 3D viewer in the measure function and enter it via the bottom swipe menu in order to do the calculation.

Measure a volume

1. >> Volume calc (bottom swipe menu)

2. Select 'Create a new surface by measuring individual points'

3. >> OK

4. The top of the screen says 'New Surface'

   → 'Surface name' – Give the surface a name

5. >> OK

6. Top of screen says 'Measure surface points'

7. Enter the target height

8. Lock to target and click 'Meas'

9. Take readings all over the surface of the stockpile, pressing 'store' at the required intervals

10. >> Boundary

11. Take readings all around the bottom boundary of the stockpile, pressing 'store' at the required intervals

12. When you have sufficient readings:

13. >> Done

14. Select 'Edit the boundary and triangulate surface'

15. Top of screen says 'Triangulation results'

16. >> OK

17. The points will be listed. You can click the '3D viewer' tab to see them in map view

18. >> OK

Tabs:

Summary – Surface ID, Area, No. of triangles, No of surface points, No of boundary points

Details – Statistics about the results

3D Viewer – Map view showing boundary and triangulated surface points

1. >> OK

2. If you are happy with you measurements select ' Calculate the volume'

    → Volume calculation method

3. >> OK

4. Top of screen says 'Volume Calculation Results'. Area and volume are displayed.

5. >> OK

6. Select 'Exit the app'

7. >> OK

Import data

1. Tap on the job in the swipe menu.

2. >> Import data

    → 'Data type to import' – ASCII data or GSI data

    → 'From' – SD card, USB or internal memory

    → 'From file' – select file name

    → 'To job' – select the job you want to import it into

3. Clicking 'display' in the bottom bar will let you see the original data in .txt form

4. Clicking 'Fn' then 'Settings' will let you select the order that the fields appear in

Export data

1. Tap on the job in the swipe menu.

2. >> export data

    → 'To device' - SD card, USB or internal memory

    → 'To folder'- Select the destination folder

    → 'Output file' – Give the file a name. It will default to the job name with a .txt extension. If you want to export it as a CSV (to open in Excel) change the extension to .csv

→ 'Example' – the order that the fields will be displayed in

3. Clicking 'Fn' then 'Settings' will let you select the order that the fields appear in

'Transfer Job'

4. Tap on the job in the swipe menu

5. >> Transfer job
   Here you can save an entire job onto the SD or a USB so that you can use it in another instrument.

Appendix F

LEICA ICON

INSTRUMENTS COVERED

The instructions are the same for both 'Build' and 'Site' softwares, however there is different terminology used in some of the apps and some icons may appear in different positions on the screen.

GETTING STARTED

Turn on the total station and level it up. Alternatively, you will be prompted to level the instrument when you open any app.

If you are using a robotic rather than a two- man, on the controller home screen, in the settings panel (bottom right of screen) tap 'Devices' and tap on the name of the total station you want to connect to.

USEFUL INFORMATION

Terminology

In terms of connection and battery, the total station is referred to as the sensor.

Resection is referred to as set up 'anywhere'

To select the display information

Within each app, select the display information you require for that app. At the bottom right hand of the screen there will be one, two or three grey boxes above the zoom toolbar. You can change the information which is displayed in these boxes. To do this, hold your finger down on one of the boxes. Three columns will appear. Select the information you require in each box.

To search for the prism

1. >> Controller/ Target menu (Prism and target height indicator in the middle of the blue bar at the top of the screen)
2. >> Move and search (Binoculars with beam)
3. Select search type from:
   - → Power search – Binoculars with beam and yellow star – Total station rotates horizontal and vertical circles fully to find the prism
   - → Search left - Binoculars with beam green arrow to the left – Total station rotates clockwise to find the prism

> → Search right - Binoculars with beam green arrow to the right – Total station rotates anti-clockwise to find the prism
>
> → Window search – Binoculars with window – Total station will search with a defined window (which can be defined in search settings)
>
> → Manual search – Joystick icon – Use the arrows pad to manually point the total station towards the prism

SCREEN ARRANGEMENT

Icons across the top of the screen

When in the home screen, these icons are small and you cannot open them. When an app is open, the icons will be slightly larger, and you can tap on them to display information or open another menu.

Top left – Previous app indicator – This icon will change depending on the app you used last. A dark grey box directly below will display the name of the app. From the home screen, tap on this box to open that app.

Second from the left – Total station/ measure mode/ bubble – Tap on this icon to change the measure mode, turn the red laser pointer on/ off, turn the guidelight on/off, check the level of the bubble, turn the laser plummet on/off, turn the tilt compensators on/ off or set the Horizontal Angle to zero. Icons are as follows:

> → Total station – Laser pointer and guidelight are off
>
> → Vertical red line with dot at the end coming out of total station – Red laser pointer is on
>
> → Red and yellow beams coming out of the total station – Guide lights are on
>
> → Grey circle with white circular arrows – Measurement type is set to 'Continuous'
>
> → Hand – Measurement type is set to 'Manual'
>
> → Cross hairs with round prism at the centre – Measurement type is set to 'Single Auton
>
> → Open padlock – Total station is not locked to target
>
> → Closed padlock – Total station is locked to target
>
> → Green level bubble indicator – Bubble is level (within set limits)
>
> → Green level bubble indicator – Bubble is off level (not within set limits)
>
> → White level bubble indicator with red cross – Tilt compensators are off

Third from the left – Target type/ target height indicator – Tap on this icon to open the Controller/ Target menu. From here you can select prism type, set prism height, turn the total station towards a point, check the total station and controller batteries, search for the target, change the search settings. Icons are as follows:

> → Prism icon – Target type is set to 'Prism'.
>
> → Wall with red line striking it – Target type is set to reflectorless

→ Target height – Number displayed indicates target height. Height will default to 0.000 when target type is set to 'reflectorless' and the previous height entered for the prism when target type is set to 'prism'.

Icons across the bottom of the screen

Furthest right – Layers with eye – Tap this icon to filter the jobs you want to display on the map and in the point list.

Second from right – Zoom in – Tap to zoom or hold down until button turns green then use your finger to tap on the screen to zoom.

Third from right – Zoom out – tap to zoom out

Forth from the right – Information/ view – Menu varies depending on which app is open.

Fifth from the right – Points table – Tap to view points in a table format. Select which control files and jobs you want to show points for and select the green tick.

Furthest left – Home – House icon. Will take you back to the home screen.

Second from the left – Tool box – Menu varies depending on which app is open.

Third from the left – Yellow star – Favourites. Commonly used functions. By default this menu contains instrument setup, calculator and camera.

Icons down the left of the screen

Top left – North arrow – To return to normal 2D view with north pointing vertically up the screen, tap this icon at any time.

Second down – White cross with arrows at the ends on a grey square – Tap on this to select 'pan' mode to use the touch screen to pan around.

Third down – 3D viewer mode – Tap on this to select 3D viewer mode. In this mode you can use the touch screen to rotate and view the points in 3D from any direction. Tap the red north arrow at the top left to return to normal 3D view at any point.

JOBS AND DATA

To create a new job

Data panel (bottom left of home screen)

4. >> Jobs
5. >> White '+' in green circle
6. Enter job name
7. >> Green tick
8. Enter description if required

9. >> Blue dot (top right of screen). Select reference data, control and code list.

10. >> Green tick

To input new points

Within any app, press the 'table' icon, fifth from the left at the bottom right of the screen.

1. >> Toolbox
2. >> New point
3. Enter point information

Or:

1. >> Sketching
2. >> Toolbox
3. >> New Point
4. Enter point information

To edit a point

1. >> Sketching
2. >> Edit point
3. Edit the point information at bottom right hand of screen then press the green tick.

Or:

Within any app, enter the 'Points Manager' (a table icon), fifth from the left at the bottom right of the screen. Select the jobs for the points you want to display then press green tick.

1. Highlight the point you want to edit by tapping on it.
2. >> Toolbox
3. >> Edit
4. Edit the point information then press the green tick. To exit the points manager, select 'File list' at the top left of the screen then the red cross.

To delete a point or line

1. Open any app.
2. >> Toolbox
3. >> Delete
4. Select all the elements you want to delete by tapping on them. Points and lines can be selected at the same time.
5. >> Green tick

To create a line

1. >> Sketching
2. >> Sketching functions menu (to the right of the toolbox button)
3. >> Points and lines
4. >> Toolbox
5. >> Connect points
6. Select the start point of the line by tapping on it. If several points are clustered together a list view will open so you can select the required point.
7. Select the end point of the line by tapping in it.
8. At the bottom right of the screen you can select:
   → Polyline (two straight lines connected by nodes) – Add extra sections onto the line
   → Close figure (square connected by nodes) – Joins the current end point to the start point to create an area
   → Eraser - Delete the line and stay in the function
   → Red cross – Delete the line and exit the function
   → Green tick – Accept the line

INSTRUMENT SET-UP

Set up over a known point ('Over a known point')

First of all, select the correct target type.

1. >> Setup
2. >> Over a known point (within the 'Coordinates' panel)
3. >> Check the level of the bubble. If you haven't already done so, set up the instrument over the known point using the laser plummet. Press the blue circle to continue.
4. If you need to work with levels, enter the instrument height and reflector height. Otherwise untick the box 'With height' or your station will be not within tolerance. Press the blue circle to continue. I
5. The top right of the screen will say 'Select station point'. Select the point by tapping.
6. The top right of the screen will say 'Select target point and measure'. Select the backsight point by tapping it, sight to it and press 'Start' then 'Store'. Press green tick to save the setup.
7. >> OK

To set up at an unknown point ('Anywhere')

1. First of all, select the correct target type.
2. >> Setup

3. >> Anywhere (within the 'Coordinates' panel)

4. >> Check the level of the bubble. Press the blue circle to continue.

5. Top of the screen will say 'Setup instrument'. Enter the station name. If you know the levels of your control points, or do not need to work with levels, leave the instrument height as 0.000m. If you are measuring to retro targets, ensure the reflector height to 0.000m. Unless you have accurate levels for your backsights, untick the 'With Height' box or your resection will not be within tolerance. Press the blue circle to continue.

6. Select your first backsight point on the map. Sight to the point and press 'Start' then 'Store'. Select your next backsight point on the map. Sight to it and press store. The top right of the screen will give a message:

7. 'Calculated station out of tolerance! Check residuals of measured points'

Or:

7. 'Calculated station within tolerance. Select additional point or accept/ set.' Press the green rosette icon to display residuals then green tick to accept.

8. >> OK

INSTRUCTIONS

Measure points/ take a survey ('Survey')

1. >> Measure

2. Sight to the point. To specify a point ID click on 'Point ID' in the information boxes at the bottom right (if Point ID is not one of the options hold down on one of the buttons to change the display).

3. >> Start

4. Eastings and northings will display at the top right of the screen.

5. >> Store

Measure or calculate the distance and level difference between two previously measured points

1. >> Checks

2. Select the two points by tapping on them. The line will be displayed on the screen with the plan distance marked next to the line. The grey information panel at the top right of the screen will display horizontal distance (triangle with an orange line along the bottom), slope distance (triangle with an orange line on the hypotenuse) and level difference (triangle with an orange line down the right hand side). To change the display, click arrow to the right of the panel.

Measure or calculate the distance and level difference between two points by sighting to them

1. >> Check

2. Sight to the first point and press start. Press 'Store' to record the point. If you have the total station locked to target and are in continuous measurement mode, as you move the prism, it's distance from the original point will be displayed next to the line. The level difference is displayed in information panel at the top right of the screen. Press store if you want to record the end point of the line. If you want to then measure a second point in relation to the original start point, press 'Clear'. Tap on the original start point and repeat. You can measure a distance from any point on your map to the prism in this way.

To measure point in relation to a defined line ('Point to line')

1. >> Checks

2. >> Tools

3. >> Point to line

4. Tap on the start point and then the end point of your reference line. Press the green tick to accept the line. Press start. The prism will be displayed in relation to the reference line. The number next to the perpendicular dotted orange line indicates the perpendicular distance from the reference line. The number at the end of the orange dotted line indicates the longitudinal distance from the start of the reference line. The longitudinal distance, offset and vertical offset (level difference) are displayed at the top right of the screen. Press 'Store' to store the point. If you want to clear the line and start a new task click the tool box, then delete.

Stake out points of known coordinates ('Stakeout')

1. >> Stakeout

2. You can select the point you want to stake out by tapping the point. Press start. The position of the prism will display in relation to the design point (yellow and black stake). Once the prism is within 2m of the deign point, the view will change from plan view to 3D. Arrows at the right of the screen indicate in/out and left/right in relation to the total station. The inner circle within the bullseye represents a radius of 0.5m from the design point. Once within 0.5m, the scale resets. Adjust the prism position until within required tolerances. A green circle will appear around the prism when it is within the set tolerances.

3. >> Store

4. A tick indicates the stored point is within tolerance. An exclamation mark means it is not within tolerance. Green indicates position, red indicates height.

To select the point you want to stake out by searching for it:

1. >> Points table icon

2. >> ID Search ('ID' in magnifying glass)

3. Use the search feature to find the point you require.

To select a list of points to stakeout:

1. >> Points table icon
2. Highlight the jobs for the points you want to show and press the green tick.
3. Highlight the points you want to add to the list by tapping on them.
4. >> Toolbox
5. >> Stakeout List
6. Give the list a name and press the green tick. You will get a warning message that the points will be added to the new Stakeout List. Press OK.
7. >> File List
8. >> Red cross (will take you back to the Stakeout app)

To set stakeout tolerances:

1. >> Home screen
2. >> Units (within settings panel)
3. >> Tolerances
4. >> User defined

Stake out points along a defined line by setting up the instrument

1. >> Stakeout
2. >> Toolbox
3. >> Reference
4. Select start and end point of line by tapping on them. At the bottom right of the screen enter how far from the start of your reference line you want the line to start (leave as 0.000m if you simply want to offset to the right or left of the line) and the required offset (negative for left and positive for right). Press the green tick.

Stake out points along a defined line by sighting to the beginning and end point (no instrument set-up needed)

1. >> Setup
2. >> Anywhere (within Control Line panel)
3. Check the level of the instrument. If OK, press the blue circle.
4. 'Possible data mismatch' warning. Press 'OK'
5. >> Start
6. Sight to start point and press 'Store'. Sight to end point point and press 'Store'
7. >> Green tick
8. >> Ok

9. >> Stakeout

10. >> Toolbox

11. >> Reference

12. Select start and end point of line by tapping on them. These are the two points you created by sighting to in the setup stage. At the bottom right of the screen enter how far from the start of your reference line you want the line to start (leave as 0.000m if you simply want to offset to the right or left of the line) and the required offset (negative for left and positive for right). Press the green tick.

13. Enter your required setup and stakeout tolerances.

Measure and area bounded by previously input or measured points

1. >> Checks

2. Select the boundary points by tapping on them. The bounded area will be highlighted light orange. The calculated area will be displayed at roughly the centre of the area. It will change as you add new points to the boundary. To 'release' a point, tap on it again.

Measure a volume

3. >> Volumes

4. >> New surface

5. >> Start

6. Measure to points all over the surface clicking store each time

7. >> Start boundary

8. Measure to points around the boundary by clicking store each time. The boundary must start and end at the same point.

9. >> Toolbox

10. >> End surface

11. Give the surface a folder name and file name.

12. >> Green tick

13. You can now edit the surface or calculate the volume.

14. >> Toolbox

15. >> Calculate volume

16. Select your calculation method. Select 'Stockpile' if you want to measure the volume of a stockpile in relation to where the boundary intersects with the existing ground. The stockpile volume, surface area and perimeter will be displayed.

17. >> Green tick

18. Select a name to save the surface as. Green tick.

Import data

1. Insert USB stick

2. >> Import & Delete (in data panel at bottom right)

3. >> 'Add' (White cross in green circle)

4. Select the type of data you want to import. Select the source as 'Removeable Disk (E:). When you insert the USB stick into your computer, the data you have copied will be in the file called 'Projects'

Appendix G

TOPCON - MAGNET

INSTRUMENTS COVERED

The instructions given in this appendix are for any Topcon total station running Magnet Field software. For many of the operations, there are several ways to achieve the same outcome, so the instructions given here may be one of a range of different processes which would work. These instructions have been written using a FC-500 controller, so instructions may vary slightly for other controllers.

GETTING STARTED

Turn the total station on and level it up using the bubble on the screen.

If you are using a robotic total station, turn the controller on. Click the 'Magnet Field' icon. In the connections screen select 'Optical' and 'My robotic'. Click 'Connect'. This will take you to the home screen.

USEFUL INFORMATION

General information

The instructions given in this book are given for the robotic TS controller screen in portrait. If the controller display is set to landscape or you are using a two-man total station, the positions of the elements as described here may be slightly different.

In the controller, to type into a field, tap in it and the key pad will appear. From this keypad menu you can tap on the '123' at the top to change to a number key pad.

On the two-man, to type into a field, tap the keypad then use the hard keys to enter your data. To toggle between numbers and letters press the 'a. hard key. BS means back space and will delete one character at a time. Sp means space.

The instructions given in this book are given for the controller screen in portrait. For landscape or for the display on a two-man total station the positions of elements as described here may be slightly different.

Set the target type

1. >> 'EDM' tab at the top of the screen.
2. In the second section of this menu you can select from prism, non-prism or sheet (for retro targets). At the very bottom left of the screen, an icon will show which mode you are currently in:
    → Non- prism (to measure to any surface) - A yellow brick wall with red line striking it.

→ Prism -A picture of the 360 degree prism.

→ Sheet – A picture of a retro target

3. To the right of the target type icon is the prism constant. You can tap in this box and edit constant at any time.

Setting 'Quick Codes'

1. >> Homepage
2. >> Configure
3. >> Codes
4. >> Add

Give the code group a name.

1. To add a code to the group, tick the top left of any box. Click the arrow to the right of the box and select the code from the drop-down list. For lines, click on the empty box to the right of the line icon and enter a string name or number. This allows you to easily switch between strings of the same code whilst in survey mode, whilst creating both lines separately.

2. Repeat until you have finished adding quick codes to the group then click the green tick. Then click the green tick with the home page icon. Now when you go into 'survey' the quick code boxes will be displayed on the screen.

3. To switch between code groups whilst in the survey program, hide the Quick Code boxes or to go to the Quick Codes menu:

4. >> M
5. >> Quick Codes
6. Select the code group or 'Edit Quick Codes' or 'Hide Quick Codes'

To create a 'Quick Code'

1. >> Edit
2. >> Codes
3. >> Add
4. 'Name' – Give the code a name
5. 'Description' – Give the code a description
6. 'Type' – Give the code a type (either point, line or area)
7. 'Layer' – Choose which layer you want to add the code to. You can create a new layer by clicking the three dots to the right of the layer list.
8. Style' – Adjust the default style if you wish
9. Green tick

To select the data you want to display on the screen

10. You can select the information fields you would like to display on the screen when you take a reading. At the bottom of the screen there are 6 fields. If no information has been selected yet, there will be a blank box. To select the information to be displayed in that field, tap on the box and select the required data type from the list. If there is already data displayed, tap in it to select a different data type. Most useful data fields include point name, eastings, northings and elevation.

SCREEN ARRANGEMENT

Icons are as follows:

→ **Green tick** - 'Accept data entry'

→ **Red cross** – 'Cancel data entry'

→ **Yellow back arrow** – 'Back to the previous menu'

→ **Red hexagon with a blue cross** – 'Cancel'.

→ **Red disk** – 'Store'

→ **Total station with a beam going to a prism on a pole** – 'Precise measurement' (takes the measurement and displays the results but does not store. To cancel the measurement without storing click 'cancel'

→ **Total station with a beam going to a prism on a pole with a red disk** – 'Quick measurement'

→ **Blue house** – 'Home page'

→ **M** – Magnet field menu (changes depending on the page you are on). The top option is always 'help'. This will display the information from the user manual, relating to the page you are on when you select it. To leave the help screen click the grey circle with four white dots at the bottom right hand of the screen then the circle with a cross.

Icons across the top of the screen in the home screen

Top left – M – Magnet Field menu. You can access the user manual here by clicking 'Help'

Second from the left – Job name- This is the job you are working in.

Furthest to the right – In the home screen, there is a yellow cross. Tap this cross to exit Magnet Field. On all other screens there is a picture of a house. Tapping this will bring you back to the home screen.

Second from the right – Total station – When the controller is searching for the TS there are yellow and grey arrows going around in a circle. When connected there is an icon of the total station.

Third from the right – Internet connection status. 'M' in a cloud with red cross if not connected and green tick if connected.

Fourth from the right – Battery indicator

Target connection toolbar (underneath the very top toolbar)

This toolbar displays when you are within a survey or stake program.

Top left – Yellow arrow – Toggles between different toolbars. Click until you see a red dot with four red lines around it, a green circular arrow, the joystick icon, cross hairs icon, padlock icon (or padlock with prism behind it) and red hexagon with a blue cross.

Second from the left – Red dot with four lines around it – turns red laser dot on or off.

Third from left – Circular green arrow – Rotate the total station to a chosen angle or point.

Fourth from the left – Joystick – Tap on this icon to enable the joystick arrows. If the arrows are green, you can control the total station from the arrow keys on the controller. If the arrows are grey the controller arrows are disabled.

Fifth from the left – Telescope icon – Tap this to get the total station to search and lock on to the prism. You must first use the joystick arrows to point the total station approximately at the prism. When the target type is set to 'sheet' the total station can lock onto the centre of a retro target.

Sixth from the left – Padlock icon – When the total station is locked onto the prism, there is a picture of a prism behind the padlock.

Seventh from left – Red hexagon with a blue cross – Cancel measurement or target search

MENUS AND TOOLBARS

'Configure' menu

Here you can change GPS and total station configuration, select coordinate system, change units, change instrument settings (temperature and pressure and turn tilt sensors on and off), change display settings (coord type, plane coord order etc), configure alarms, view and edit codes, create 'Stake Reports' and change enterprise settings (login, upload and download defaults)

JOBS AND DATA

To start a new job

1. >> Job
2. The top of the screen will say 'New job'. Enter job name and other relevant information. Click 'Browse' to select the where to store your new job or add a new folder to store it in. Green tick to save.
3. >> Next
4. The top of the screen will say 'Job configuration'. Here you can edit GPS and total station configurations.
5. >> Next
6. The top of the screen will say 'Coord System'. For normal site work leave projection as 'None' and Geoid as 'None'.

7. \>> Next

8. The top of the screen will say 'Units'. Select your units. Switch between distance, angle, coordinates and 'other' by clicking the tabs at the top of the page.

9. \>> Next

10. The top of the screen will say 'Display'.
    - → 'Coord type' – Select ground for normal site work using the total station
    - → 'Plane Coord Order' – Select 'Easting, Northing, Height'
    - → 'Geod. Az Origin' – Select 'North'
    - → 'Direction' – Select 'Azimuth'
    - → 'Display slope as' – Choose between percentage, run or angle
    - → 'Station prefix' – Select your station prefix.
    - → 'CL Position' – Select required the format of your centreline position

11. \>> Next

12. The top of the screen will say 'Alarms'. Select your required alarms. Switch between Main, Controller, GPS and Optical by clicking the tabs at the top of the page. Select the green tick.

13. You will get a message saying 'Import successfully finished'.

14. \>> Close

15. \>> Connect

Input new points

1. \>> Edit

2. \>> Points

3. \>> Add
    - → 'Point' – Give your point a name
    - → 'Code' – Give your point a code, if relevant
    - → 'Note'- Add any notes

4. Enter eastings, northings and elevation of the point.

INSTRUMENT SET-UP

Set up the instrument at a known point ('Occupation point')

1. \>> Setup

2. \>> Backsight

3. Occupation point:
    - → 'Point' – Enter the name of the point you are set up over
    - → 'HI' – Enter the instrument height if you will be measuring/ setting out levels

4. Backsight point:

   → 'Point' – Enter the name of the known backsight you will sight to

   → 'Prism icon' – Enter the target height if you will be measuring/ setting out levels

5. >> Next

6. Sight to the known backsight and press 'Set'

7. >> Home

Set up the instrument at an unknown point ('Resection')

1. >> Setup

2. >> Resection

   → 'Occupy' – Name the instrument position. This is a temporary name used only for this setup.

   → 'HI' – If you know the elevations of your control points, you can leave this as 0.000m and the instrument height (HI) will be calculated in the resection.

3. >> Next

4. 'Specify a control point'

   → 'Point' – Select the name of the first known point you will sight to. Click 'Quick measure'

5. 'Specify another control point'

   → 'Point' – Select the name of the first known point you will sight to. Click 'Quick measure'

6. Residual errors will be displayed. Press 'Add' if you want to measure additional points. Press 'Accept' if you have finished measuring to control points. This will return you to the home screen.

To measure points/ take a survey (Survey)

1. >> Survey

2. >> Topo

3. Sight to the point. Press 'Precise measurement' or 'Quick measurement'

4. To view the measured points in map view, click 'Normal' at the very top of the screen and change it to 'Map'.

To assign information to a survey point:

To attribute information to the next point you measure, tap the 'point information' button which is an icon of a page with an 'i' in a blue circle, directly above the target type icon. Here you can edit the point name, assign codes and relevant information, select the layer, attach a photo or add notes.

To assign a code to a survey point:

To the right of the 'point information' button, there is a box which may be empty or may display a code. To assign a code or edit the code, tap on the down arrow to the right of the box and select the code from the list. Now measure to your point.

Alternatively, for codes you use regularly you can select up to 9 'quick codes'. These are boxes which display at the top of the screen when in survey mode. Once they are set up, you can tap on the code box to measure and store a point and jump to the next point.

Measure the distance and level difference between two points by sighting to them

This function can be used without having to set the position and orientation of the instrument.

1. \>> Survey
2. \>> Missing line
    - → 'Method' – Select 'A-B, A-C'
    - → 'Start point' – Enter the name of the start point. Sight to the point. Click 'Quick measure'
    - → 'End point' – Enter the name of the end point. Click 'Quick measure'
3. \>> Calc
4. The horizontal, vertical distance and slope distance will be displayed.

Or:

1. \>> M
2. \>> Survey
3. \>> Missing line
    - → 'Method' – Select 'A-B, A-C'
    - → 'Start point' – Enter the name of the start point or select it from the map view or list.
    - → 'End point' – Enter the name of the end point or select it from the map view or list.
4. \>> Calc
5. The horizontal, vertical distance and slope distance will be displayed.

Alternatively, you can use map view:

1. \>> Home screen
2. \>> Map
3. Select and highlight the beginning and end point of the line by tapping on them. Hold down the stylus on the screen and a floating menu will appear. Select 'Calc Inverse'. The horizontal, vertical distance and slope distance will be displayed.

Measure the distance and level difference between two points by sighting to them

1. \>> Home screen
2. \>> Map
3. Select the boundary points of the area you wish to measure by tapping on them to highlight them. Hold down the stylus on the screen and a floating menu will appear. Select 'Calc Area'. The area and perimeter will be displayed.

4. Alternatively, select the area tool (yellow polygon) in the tool bar at the top of the screen. Select the boundary points on the map. Click the area tool again once you have finished selecting points. Tap on the perimeter line to select. Hold down the stylus on white space. Select 'Calc Area'. Area and perimeter will be displayed.

Stake out points of known coordinates (Stake)

1. >> Stake

2. >> Points

   → 'Design pt/ Nearest pt' – Toggle between 'design' and 'nearest' point by tapping on this button. Select 'Design Pt'. Enter the point name or select from map or list or type in a new point name. For a new point name, when you click 'Stake' you will be prompted to enter the coordinates of the point.

3. >> Stake

4. There are six fields to the bottom left of the screen. To select the information you want to display tp on the field and select from the drop down list. 'In/Out' and 'Left/ Right' are recommended. 'Delta East' and 'Delta North' can also be helpful.

5. Sight to the prism and click the total station without the disk. Directions are given:

   → 'To TS *(dist)*' or 'Away TS *(dist)*'

   → 'Left *(dist)*' or 'Right *(dist)*'

6. To change the view, click on the tab immediately to the right of the 'M' icon. You can select from:

   → Map view – Shows the prism in relation to the design point

   → Overhead – Gives a 3d overhead view

   → Normal – Displays arrows indicating which way the prism needs to move

7. Once you are within the required tolerance, click the total station with the disk. A warning will display if your tolerances are exceeded.

8. The point data will be displayed. Click the green tick. The as-staked position will be recorded as *point name_* stk

9. The point will automatically jump to the next one in the list.

Stake out points in relation to a defined line by selecting the beginning and end points from the list

1. Carry out instrument setup (Backsight or resection)

2. >> Home

3. >> Stake

4. >> Grid

   → 'Origin point' – Select the start point of your reference line from the map or list.

- → 'Azimuth'/ 'Az to pt' – Tap to select 'Az to pt'. Select the end point of your reference line from the map or list.
- → 'Along azimuth' – Select the spacing you require in the direction of your reference line.
- → 'Perpendicular' – Select the spacing you require perpendicular to your reference line.

5. >> Next

6. Define the grid boundary.

7. Click on 'points' and replace with 'Extents'

8. Define your grid area in relation to the reference line.
   - → 'Along azimuth' – Define the start of the grid. Negative is going backwards from the start point of your reference line, positive is forwards along the line.
   - → 'To' - Define the end of the grid. Negative is going backwards from the start point, positive is forwards along the line.
   - → 'Perpendicular' – Define the start of the grid. Negative is to the left of the reference line. Positive is to the right.
   - → 'To' - Define the end of the grid. Negative is to the left of the reference line. Positive is to the right.

9. >> Stake

10. Zoom in on your grid. Select the point you want to stake out next by moving the prism towards it. The closest point on the grid will automatically be selected. To record the as-staked position of the point, click 'Quick measure'. The point name will increase to the next one in the sequence.

Stake out points in relation to a reference line, without the need to set up the instrument

This method involves giving the first point on your line an arbitrary co-ordinate e.g. (0.000mE, 0.000mN). If you prefer to avoid negative numbers, use a larger number e.g. (100.00mE, 300.00mN). Before going into this function you must first create the point.

1. >> Setup

2. >> Reference direction
   - → 'Occupy' – Name the instrument position. This is a temporary name used only for this setup.
   - → 'HI' – Enter instrument height if you need to work with levels and you know the Reduced Level of the point you are set up over. If not, leave as 0.000m

3. >> Next

4. Select the start point of your line:
   - → 'Point' – Select the start point from the map or list.

5. >> Next

6. 'Observe the point on the reference line'. Sight to the start point and click 'Quick measurement'

7. 'Enter the name for point on the reference line'.

    → 'Point' -   Enter the name of another point on the line. You do not need to enter co-ordinates for this point.

    → 'Line Azimuth' – Enter the line azimuth. Vertical upwards is 0 0 00, left to right is **90 0 0,** vertical downwards is 180 00 00, right to left is 270 00 00.

8. >> Next

9. 'Observe the point on the reference line'. Sight to the end point and click 'Quick measurement'

10. >> Accept

11. >> Home

12. >> Stake

13. >> Grid

14. 'Origin point' – Select the start point of your line as defined in the 'Reference direction' setup above.

    → 'Azimuth' – Select the clockwise angle of your required reference line in relation to north. Set 90 0 0 if you want the line to be horizontal left to right.

    → 'Along azimuth' – Select the spacing you require in the direction of your reference line

    → 'Perpendicular' – Select the spacing you require perpendicular to your reference line

15. >> Next

16. Define the grid boundary.

17. Click on 'points' and replace with 'Extents'

18. Define your grid area in relation to the reference line.

    → 'Along azimuth' – Define the start of the grid. Negative is going backwards from the start point of your reference line, positive is forwards along the line.

    → 'To' - Define the end of the grid. Negative is going backwards from the start point, positive is forwards along the line.

    → 'Perpendicular' – Define the start of the grid. Negative is to the left of the reference line. Positive is to the right.

    → 'To' - Define the end of the grid. Negative is to the left of the reference line. Positive is to the right.

19. >> Stake

20. Zoom in on your grid. Select the point you want to stake out next by moving the prism towards it. The closest point on the grid will automatically be selected. To record the as-

staked position of the point, click 'Quick measure'. The point name will increase to the next one in the sequence.

Measure an area

In the map view, select the boundary points of the area you want to calculate. Hold the stylus on white space anywhere on the screen. A floating menu will appear. Select 'calculate area'.

Measure a volume

1. >> Home screen
2. >> Survey
3. >> Surface

There are several options for measuring a volume:

→ Boundary

→ Min elevation

→ Max elevation

→ Fixed elevation

→ Plane

→ Design surface

To measure a stock pile to the point where it intersects with the existing ground at the boundary points, select 'Boundary'.

Measure to points all over the stockpile. If the whole stockpile is not visible from your setup position, create a boundary across spine of the pile. Calculate the volume of the half of the stockpile you can see from that position. Move the total station to the other side of the stockpile and repeat. Add the two volumes together to calculate the full volume.

To export data to a file

1. >> Exchange
2. >> To file

   → 'Data' - Select data type you want to export e.g. points, lines, raw data, stake reports.

   → 'Format' – Select the format e.g. dwg, dxf. To read in Excel, select .csv

3. >> Next
4. Select the destination folder (/Hard Disk for USB stick) and select the job name. Click the green tick.
5. >> Next
6. The top of the page will say 'Coord System'. Click 'Next'
7. You will get a message saying 'Export successfully finished'.
8. >> Close

To import a data file

1. >> Exchange
2. >> From file
   - → 'Data' - Select data type you want to export e.g. points, lines, raw data, stake reports.
   - → 'Format' – Select the format e.g. dwg, dxf. To export from Excel, select .csv file extension.
3. Select file units or layers if relevant.
4. Insert the USB stick into the bottom of the controller.
5. >> Next
6. Click on the favourites folder (Folder with a star).
7. Select 'Hard Disk'. Select the .csv file you want to import. Click the green tick.
8. The top of the screen will say 'Coord System'. Click the green tick. You will get a message saying 'Import successfully finished'.
9. >> Close
10. >> Home screen icon
11. >> Edit
12. You should now be able to find the imported data.

Appendix H

TRIMBLE – ACCESS

GETTING STARTED

1. Switch on the total station – The Trimble icon appears (if not check battery)

2. Switch on the controller

3. Select software – Trimble icon at top left

4. The cross at the top right takes you to the main screen of Trimble (or it may be on that already). Press windows icon (bottom left). Select 'Access'.

5. >>General survey (contains all the functions you will need)

6. The first time you enter the 'Measure' or 'Stakeout' program you will be prompted to carry out an instrument setup.

USEFUL INFORMATION

General useful information

1. Changing page – A square in the bottom right of the screen indicates whether there are other pages within the menu e.g. 1/3 Click on the icon to move to the next page.

2. Once you have entered the right data press 'accept'. If you don't want to save the changes, press 'Esc'

3. If you exit 'General survey' you will lose your setup and connection with the total station and will have to start again. To minimise it rather than shut it down, use the minimise button ('-' next to the 'X' at the top right of the screen)

Select target type

1. Tap on the target icon (third down in the panel to the right of the screen). A box will display showing user selected prism and DR.

2. To change the prism type, click on the target height (which will be '?.???m' if you haven't entered a height yet). This will take you to the prism settings. Here you can enter the target height, prism type, display name and radio channel (for active prism). You can use this method to select the new prism type each time. Alternatively, if you have certain prisms you use regularly you can include them in the display box. To do this, once you are in the prism details page click 'Add' at the bottom of the screen. Here you can add a new prism to the list and give it a display name. This means to switch between prism types you can do it within the display box.

3. To adjust DR (reflectorless mode) settings, click on the target height (which will be '?.???m' if you haven't entered a height yet). Here you can enter the target height, display name and prism constant. For measuring to surfaces the prism constant should be 0.0mm. You have the option to change this if you need to e.g. if you are using this mode to measure to a prism.

SCREEN ARRANGEMENT

Icons down the righthand side of the screen

Top right – Battery indicator – Top battery is the controller battery. Bottom battery is the total station battery.

Second from the top – Total station/ Instrument functions – Picture of total station indicates controller is configured to total station. To the top right there will either be a 'T' for tracking mode or 'S' for single measurement mode. Below that is the height of the total station. If no height has been entered yet there will be a question mark.

Measurement type and connection status (space between total station icon and target type/ target instrument height indicator):

→ No line or icon between total station icon and target type/ instrument height indicator – Instrument is in single measurement mode.

→ A red rectangle going back and forth - The instrument is in tracking mode in DR mode.

→ Curved line going back and forth (no horizontal line) – Target type is set to active prism mode. Total station is not locked to target.

→ Horizontal line – Instrument is set to single measurement mode and locked to target

→ Horizontal line with a 'star' at the end and curved line going back and forth – Target type is set to active prism mode. Total station is locked to target.

Third from top – Target type – Displays target icon, prism constant and target height. Vertical red line with a flash at the top indicates DR (reflectorless mode). Icon shows prism type. The number indicates the user selected prism.

The top lines refers to the prisms. The bottom line refers to reflectorless mode (DR). The information is given in the order 'Display name', 'Target height', 'Prism constant' 'Radio channel'.

Icons across the top of the screen

Top left – Trimble icon- Menu changes depending on the screen you are on. Offers quick access to certain menus.

Furthest right – Cross – Shuts down general survey (including instrument setups and connections) and takes you back to Trimble Access home screen

Second from right – '-' Takes you back to the Trimble Access home screen without shutting down general survey

Third from right –'?' – Takes you to the 'help' pages for the whole of Trimble Access

Forth from the right – White circle with diagonal line – Alerts

Fifth from the right – White circle – Internet set up status

JOBS AND DATA

Jobs menu

Here you can add a new job, open a job, edit the properties of a job, view points in a list format (Point manager), copy information between jobs and import/ export information.

To start a new job

1. >> Jobs
2. >> Open job
3. Click on the folder in which you want to create your new folder.
4. >> New folder icon (fifth from the left). Enter folder name. Highlight the folder. Click 'Ok' at top right of the screen.
5. Click on the folder in which you want to add your new job. Click the new job icon which is a sheet of paper with the top right corner folded over.
   → 'Job name' – Enter job name then click the folder icon to the right of the field and select the folder you want to put that job into
   → 'Template' – Select the coordinate system you require (If you are working to a local grid, it doesn't matter what you select here. If you are working to a global coordinate system select the system that is there and then change to your required system in the properties section below. Here you can also change units, add linked files, select an active map, bring in a feature code library and change Cogo settings.
6. >>Accept

To enter point data

1. >> Key in
2. >> Points
3. Enter your point data
4. >> Enter
5. >> Store
6. Repeat until you have entered all your points.
7. >> Esc

To enter line data

1. >> Key in

2. >> Lines

   → 'Line name' – Give your line a name

   → 'Method' – Select 'Two points'

   → 'Start point' – Tap the arrow to the right of the box. Select 'List' if you have previously stored the point, 'Key in' if you want to enter the data now or 'Measure' if you would like to create the start point by measuring to it.

   → 'End point- Enter the end point as above.

   → 'Start station' – Enter the start station of the line. Leave this as zero unless you wish to start at a different point.

   → 'Station interval' – If you wish to mark points along the line at specified intervals, enter the interval here. If not, leave the question mark.

3. >> Enter

4. Line data will display.

5. >> Store

6. Repeat until you have finished entering lines.

7. >> Esc

View point data

1. >> Jobs

2. >>Point manager

3. You can choose what order to present the points in by tapping on the column heading. Tapping again on the same column heading will reverse the order.

4. To view the details of a point, tap 'Details'.

Create a line in map view

1. >> Map (fourth button from the bottom on the right hand panel)

2. Tap to highlight the start point of the line. Tap to highlight the end point of the line. Hole the stylus down anywhere on the screen. A floating menu will appear. Click 'Key in line'.

   → 'Line name' – Give your line a name

   → 'Method' – Select 'Two points'

   → 'Start point' – Will automatically be the first point you highlighted.

   → 'End point- Will automatically be the second point you highlighted.

   → 'Start station' – Enter the start station of the line. Leave this as zero unless you wish to start at a different point.

→ 'Station interval' – If you wish to mark points along the line at specified intervals, enter the interval here. If not, leave the question mark.

INSTRUMENT SET-UP

Set up the instrument at an unknown point ('Resection')

3. Select 'Measure' or 'Stakeout'

4. >> VX & S Series

5. >> Resection

6. >> Accept (corrections)

7. Top of screen will say 'Resection'

    → 'Instrument point name' – Give the instrument setup position a name. This is a temporary name and will not be used again.

    → 'Code'- If required, select the feature code list and select a code from the list. Or leave as '?'

    → 'Instrument height' – Leave as '?' and untick 'Compute station elevation'

8. >> Enter

9. >> Accept

10. Top of screen will say ' Resection – Face 1'

    → 'Point name' – Enter the name of the first known point you will sight to.

    → 'Code' - If required, select the feature code list and select a code from the list. Or leave as '?'

    → 'Method' – Select 'Angles and distance'

11. >> Measure

12. Repeat the above for the next known point.

13. Horizontal angles will display.

14. >> Results

15. The coordinates of the instrument position and standard errors will display. Change the page to view all errors.

16. >> Store

17. >> Enter. Station setup is completed and you can select what you want to do next.

Set up the instrument over a known point ('Station setup')

1. First of all, set up and level the instrument over a point of known coordinates.

2. Select 'Measure' or 'Stakeout'

3. >> VX & S Series

4. >> Station setup

5. >> Next

6. >> Accept (corrections)

7. Top of screen will say 'Station setup'

   → 'Instrument point name' – Select from the list, or key in, the name of the station you are set up over

   → 'Code'- The code of the point selected from the list will be displayed

   → 'Instrument height' – Leave as '?'

8. >> Accept

   → 'Backsight point name' – Select from the list, or key in, the name of the station you will backsight to.

   → 'Code'- The code of the point selected from the list will be displayed

   → 'Backsight height' – Leave as '?'

   → 'Method' – Angles and distance

9. >> Measure

10. Azimuth and H. Dist will display.

11. >> Store

INSTRUCTIONS

Measure points/ take a survey (Survey)

1. >> Survey

2. >> Measure

3. >> Measure topo

   → 'Point name' – Enter point name

   → 'Code' – Click on the arrow to the right of the box and you can select the code from the drop down list

   → 'Method' – Angles and distance

   → 'Target height' – Enter target height

4. >> Enter

5. >> Measure

6. >> Store

7. Point name will automatically jump to the next in the sequence. Repeat the above steps. When you have finished taking readings click 'Esc'

Measure a distance and level difference between two point by measuring to them

Note! You must have completed a station setup for this function to work. If you have not yet set up the station, you will be prompted to do so when you measure the first point. You must enter the instrument height (even if it an arbitrary number) if you would like to calculate level difference as well as distance.

1. >> Cogo
2. >> Compute inverse
   - → 'From point' – Click on arrow to the right of the box and click 'Measure'. The point will automatically be given a name which you can change if you want. Leave 'Method' as 'Angles and distance'. Click 'Measure'. Angles and slope distance will be displayed. Click 'Store'
   - → 'To point' – Repeat the same process again for the end point of your line.
3. Horizontal distance (H.Dist) and Vertical distance (V.Dist) will be displayed. To change between displays, click the arrow to the left of the information display panel.
4. >> Store
5. >> Esc

Stake out point of known coordinates (Stakeout)

1. >> Stakeout
2. >> VX & S Series
3. Station setup will be prompted if one has not been completed since you last opened 'General survey'. Once the station setup is completed, it will return you to the menu.
4. >> Stakeout
5. >> Points
6. Click 'Remove' to remove any points you don't want to stake out. Click 'Add' to add points to the list. Alternatively click on 'map' and select the points you want to stakeout and click 'Stakeout' (bottom right of the screen). Select the point you want to stake out from the list.
7. >> Stakeout
8. Follow the on screen instructions until you are within the required tolerance. Click 'Accept' Give the point an 'as-staked name' and as-staked code'. Delta East and Delta North display the distance between the design position and the as-staked position. Click 'Store'.
9. The point will disappear from the list and jump to the next one. If this is not the point you want to stake out, select a different and 'Stakeout'. When you have finished staking out points, click 'Esc'

Stake out points in relation to a defined line

1. >> Stakeout

2. Select either 'Line name' or 'Two points' by clicking the arrow to the right of the top information box. This example gives instruction for staking out the line by 'Two points' i.e. entering the start and end point of the line.

   → 'Start point' – Click 'List' to select from all the point data or 'Map selections' to display only the points you have currently highlighted in the map view.

   → 'End point' - Click 'List' to select from all the point data or 'Map selections' to display only the points you have currently highlighted in the map view.

   → 'Stake' - Select whether you want to stake 'To the line', or a 'Station/offset from the line'. If you select 'Station/offset from the line' further information boxes will appear.

   → 'Station' – Enter how far along the line you want your first point to be. Negative is back from the start of the line. Positive is forward along the line.

   → 'Station interval' – Enter the required interval of points you want to stake out. Leave as a '?' if you do not want to stake our intervals.

   → 'H. offset' – Enter the required horizontal offset. Negative is to the left of the line. Positive is to the right of the line.

   → 'V. offset' – Enter the required vertical offset. Negative is below the line. Positive is above it.

3. >> Start

4. Follow the on screen instructions until you are within the required tolerance. Click 'Accept' Give the point an 'as-staked name' and as-staked code'. Delta East and Delta North display the distance between the design position and the as-staked position. Click 'Store'.

5. >> Store

6. If you entered intervals, once you have stored a point, you can move forwards or backwards along the line, by the specified interval by pressing 'Sta-' or 'Sta+'. Keep clicking the button for the number of intervals you want to jump to. When you have finished staking out the line click 'Esc'.

Stake out points in relation to a defined line (without the need for coordinated control points)

1. >> General survey

2. >> Stakeout

3. >> VX & S Series

4. >> Refline

5. >> Accept (Corrections page)

6. Top of the screen will say Refline

   → 'Instrument point name' – Give your temporary setup position a name

   → 'Instrument height' – You can leave as '?' if you are only interested in positions and not levels

7. >> Enter

8. >> Accept

→ 'Point 1 name' – Give the first point of your reference line a name

→ 'Target height' - You can leave as '?' if you are only interested in positions and not levels

→ 'Method' – Angles and distances

9. >> Meas 1

→ 'Point 2 name' – Give the end point of your reference line a name

→ 'Target height' - You can leave as '?' if you are only interested in positions and not levels

→ 'Method' – Angles and distances

Calculate an area from previously measured points

1. >> Cogo

2. >> Area calculations

3. Select the boundary points from the list. The boundary will be created in the order that you select the points.

4. >> Calc

5. The area will be displayed on a plan view. The area and perimeter are displayed in the right hand panel. You can enter the name of area then click 'Enter'.

6. >> Store

Measure an area by selecting points in map view

1. >> Map

2. Tap the points which form the boundary of the area you want to measure. The boundary will be created in the order that you select the points so go clockwise or anticlockwise round the boundary rather than zig-zag across. Hold the stylus down anywhere on the screen and click 'Area calculations'. The area will be displayed in plan view. The area and perimeter will be displayed in the right hand panel.

3. >> Store

4. >> Esc

Export data

1. >> Jobs

2. >> Import/ Export

3. >> Export fixed format

→ 'File format' - Select file format to export (to export to Excel, select .csv)

→ 'File name' – Select file to export. Click folder to right of field to access all files

4. Select the field you want each element to appear in.

5. >> Accept

6. Select the data you want to export e.g. 'All points'

7. A dialogue box 'Transfer complete' will appear. Click 'OK'

ANSWERS TO EXERCISES

CHAPTER 4

Ex 1) 54.0m

Ex 2) a) 1:129.5

b) 37.2m

CHAPTER 8

Ex 1) a) TBM12 (BS), TBM14 (IS), C1 (IS), C2 (IS), C3 (IS), C4 (IS), TBM12 (FS)

b) 51.669

c)1mm

d)51.197, 51.201, 51.196, 51.199

e)2mm

Ex 2) a) HPC= 9.524, RLs: 8.125, 8.332, 7.823,8.001, 9.250, 9.207, 9.186, 9.316, 9.252

b) 3mm closing error

Ex 3) a) 36.517

b) 0mm closing error

Ex 4) 57.6m

CHAPTER 10

Ex 1) 3mm/55m

CHAPTER 12

Ex 1) 4"

Ex 2) 5"

Ex 3) 1mm/5mm

CHAPTER 19

Ex 1) a) 161°18'20"

     b)91.320m

Ex 2) 295°29'35"

Ex 3) a) 71°42'14"

     b) S2-S3 = 18°08'34", S3-S4 = 299°05'27", S4 – S5 = 275°04'55", S5-S1 = 202°19'07"

     c) 279.671m

     d) Misc E = 0.006m, Misc N = 0.006m

     e) S1 = (106.743Me, 87.187mN), S2 = (183.110mE, 112.436mN), S3 = (197.545mE, 156.490mN),

      S4 = (170.877mE, 171.328mN), S5 = (142.327mE, 173.866mN)

f) 1: 35,000

CHAPTER 20

Ex 1) Top (0.035) Btm -(0.037) Top (0.106) Btm (-0.105) Top (-0.322) Btm (0.322). Max error = 2mm

b) 12.535, 12.641, 12.319

c) 1mm

d) 9.411 – 9.412 = 12.499 – 12.500 = -0.001

CHAPTER 22

Ex 1) a) 37.600m

     b) 37.550m

     c) 39.550m

     d)124mm down, 161mm down, 58mm down, 279mm up

Ex 2) a) 0.327m

     b) 7mm/m run, 0.67%, 1:50

CHAPTER 23

Ex 1) a)12.3m

     b) 20.750m, 21.000m

CHAPTER 26

Ex 1) a) AI= 785.531m, IB = 359.210m

      b) 25°

      c) 44.339m

      d) 87.266m

      e) CH_{TP1} = 1016.198m, CH_{TP2} = 1103.464m

      f) 86.576m

      g) 4.741m

CHAPTER 28

Ex 1) 14.066m

Ex 2) 25.327m

Ex 3) 96°34'10"

Ex 4) 99°37'00"

Ex 5) 329.625

Ex 6) 83°16'59"

Ex 7) 0.213m

BIBLIOGRAPHY

Uren, P. and Price, B. (2010). Surveying for Engineers. Fifth Edition. Palgrave MacMillan

Smith, Jim (1997). The management of setting out in construction: ICE Design and Practice Guide. Thomas Telford Publications

Irvine, W. and Maclennan, F (2006). Surveying for Construction. Fifth Edition. McGraw-Hill Education

Sadgrove, B M and Danson, E. (2007). Setting-out procedures for the modern built environment. Second Edition. CIRIA

BS 5964-1:1990, ISO 4463-1:1989 Building setting out and measurement. Methods of measuring, planning and organization and acceptance criteria.

Index

Symbols